Measurement and Calibration Requirements for Quality Assurance to ISO 9000

WILEY SERIES IN QUALITY AND RELIABILITY ENGINEERING

Editor:
Patrick D.T. O'Connor

Electronic Component Reliability:
Fundamentals, Modelling, Evaluation and Assurance
Finn Jensen

Integrated Circuit Failure Analysis:
A Guide to Preparation Techniques
Friedrich Beck

Measurement and Calibration
Requirements for Quality
Assurance to ISO 9000
Alan S. Morris

Measurement and Calibration Requirements for Quality Assurance to ISO 9000

Alan S. Morris

University of Sheffield, UK

JOHN WILEY & SONS

Chichester • New York • Weinheim • Brisbane • Singapore • Toronto

Based on previous publication *Measurement and Calibration for Quality Assurance*, Prentice Hall, 1991, ISBN 0-13-567652-5

Baffins Lane, Chichester,
West Sussex PO19 1UD, England

National 01243 779777
International (+44) 1243 779777

e-mail (for orders and customer service enquiries): cs-books@wiley.co.uk

Visit our Home Page on http://www.wiley.co.uk
or
http://www.wiley.com

Reprinted June 1999

Other Wiley Editorial Offices

John Wiley & Sons, Inc., 605 Third Avenue,
New York, NY 10158-0012, USA

WILEY-VCH Verlag GmbH, Pappelallee 3,
D-69469 Weinheim, Germany

Jacaranda Wiley Ltd, 33 Park Road, Milton
Queensland 4064, Australia

John Wiley & Sons (Asia) Pte Ltd, 2 Clementi Loop #02-01,
Jin Xing Distripark, Singapore 129809

John Wiley & Sons (Canada) Ltd, 22 Worcester Road,
Rexdale, Ontario M9W 1L1, Canada

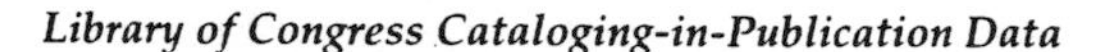

Library of Congress Cataloging-in-Publication Data

Morris, Alan S., 1948–
Measurement and calibration requirements for quality assurance to ISO 9000 / Alan S. Morris. — [Rev. ed.]
p. cm. — (Wiley series in quality and reliability engineering)
Rev. ed. of: Measurement and calibration for quality assurance. New York: Prentice Hall, 1991.
Includes bibliographical references and index.
ISBN 0-471-97685-7
1. Mensuration. 2. Quality assurance. 3. ISO 9000 Series Standards. I. Morris, Alan S., 1948– Measurement and calibration for quality assurance. II. Title. III. Series.
T50.M64 1997
670.42 — dc21 97-22234
CIP

British Library Cataloguing in Publication Data

A catalogue record for this book is available from the British Library

ISBN 0 471 97685 7

Typeset in 10/12pt Palatino by Keytec Typesetting, Bridport, Dorset
Printed and bound by Antony Rowe Ltd, Eastbourne
This book is printed on acid-free paper responsibly manufactured from sustainable forestation, for which at least two trees are planted for each one used for paper production

To the memory of my Father

Contents

Preface

Since the author's successful earlier text based on BS 5750, *Measurement and Calibration for Quality Assurance* (Prentice-Hall, 1991) which covered the general requirements of quality assurance and the measurement and calibration aspects in particular, there have been significant developments in the quality assurance movement. Prominent in these developments has been the increased emphasis on total quality management and a recognition of the necessity of having customer involvement in defining target quality levels for the quality system implemented.

The movement towards systems which ensure the quality of services provided and goods manufactured by companies is now a world-wide one. At the present time, almost all large companies around the world have ISO 9000-registered quality systems in operation. Although the quality movement is not quite so well developed in small- to medium-sized enterprises, this sector is now catching up rapidly in the design and implementation of quality systems. Even those companies who in the past have been quite lukewarm to the need for certified quality systems are now being encouraged into it by customer demands.

Quality is now the essential ingredient in the success of a company, for, without it, their market share is likely only to spiral downwards. However, achieving ISO 9000 registration is not the end of the quality story, but rather the beginning of what has to be a continual process of maintaining and further improving quality levels. Such a steady programme of quality improvement is necessary because a company's competitors are likely to be improving their quality and, in the world-wide market in which everyone now operates, such improvement is therefore necessary for survival. This movement towards a continual improvement in quality is reflected in the published quality standards, which are revised and toughened at regular intervals. This text is based on the 1994 version of ISO 9000, which supersedes earlier versions of ISO 9000, BS 5750 (the previous British equivalent) and EN 29000 (the previous European Union equivalent). This latest ISO 9000 standard gives directions on what are regarded as the minimum acceptable standards for quality system management, and in particular places greater emphasis on the involvement of customers when the requirements of quality systems are being specified.

Quality assurance has an interdisciplinary nature and, if it is to be introduced successfully, its implementation must be company-wide. Bridges must be built between all functions in a company and especially between engineering and management functions. One way to achieve this is to create quality circles. These involve regular meetings between representatives of all functions in a company and they ensure that quality control is a corporate, cohesive policy which permeates through the entire company instead of being a fragmented policy which operates effectively in some departments only. However, whether it is achieved by quality circles or by some other means, an essential component in quality assurance is full management commitment to it and the provision of adequate funding for the creation and maintenance of quality control systems. The role of management is to assign responsibility for meeting quality targets, to delegate authority for operating quality control systems and to create accountability for the operation of all aspects of a quality plan. They must also ensure that training programmes at an appropriate level in quality assurance are provided for all personnel in their company. Such training ensures that personnel fully appreciate the importance of quality assurance and understand the contribution that they are required to make towards achieving it.

Measurement and calibration procedures are an essential component within quality control systems, both for monitoring the values of quality-related process parameters at various stages of production and also for inspecting and testing the final product. Measurements at intermediate and final stages of production provide the means for assessing the degree of success of the quality control systems operated. Such measurements are the means by which the customer is assured of the quality of the product. This requires in turn that the quality and accuracy of the measurements must also be guaranteed by a properly managed system of instrument calibration.

The author has taken the opportunity to mention the relevance of the measurement and calibration procedures discussed to environmental management systems. Consideration of the environment is of growing concern to most manufacturing companies, both because of the steady development of environmental protection legislation and also because of the need to satisfy customers that the company is manufacturing products responsibly in a way that minimizes environmental damage. There is therefore brief mention of the measurement and calibration requirements of the recently published environmental management standard ISO 14000.

Quality control system management and quality measurement techniques are traditionally treated as separate subjects which are described in separate texts. Unfortunately, this mode of treatment fails to take proper account of the interdependence of these two subjects. The great value of the present text is that it covers both subjects within one book and is thus able to provide an integrated approach. Besides avoiding the need to consult separate texts, this enables measurement and calibration procedures to be considered in the proper context with respect to their importance in the operation of quality control systems and the attainment of quality assurance targets.

The method of approach in the text has been to present the subject of quality assurance first, highlighting its importance and describing the procedures involved. The general requirements of ISO 9000 are covered, with particular emphasis on the changes compared with the previous version of the standard. This is followed by a discussion of the general principles of the measurement and calibration procedures necessary for the operation of quality control systems, and a description of the mechanisms for assessing the sources of measurement error and quantifying their effect. Specific chapters are devoted to the subjects of system reliability, statistical process control and final product sampling and inspection. Software quality metrics, which is often the weak link in quality measurement in manufacturing systems which are increasingly dominated by computer systems, is also covered. Finally, a series of chapters is devoted to a number of quality-related process parameters which are commonly measured. Each of these final chapters reviews the appropriate measuring instruments available and discusses the relevant calibration procedures.

The target readership for this book includes both engineering personnel, who are concerned with the day-to-day development, implementation and maintenance of quality systems, and management personnel and company directors, who oversee quality systems and therefore must have an awareness and understanding of all activities relevant to quality assurance. In recognition of this dual role, the technical level of this text has been deliberately kept as low as possible. Even where necessary technical detail has been provided for the benefit of staff implementing quality control and measurement/calibration procedures, the aim has been to provide explanations in simple terms for the benefit of the non-technical reader.

Series Foreword

Modern engineering products, from individual components to large systems, must be designed and manufactured to be reliable in use. The manufacturing processes must be performed correctly, and with the minimum of variation. All of these aspects impact upon the costs of design, development, manufacture and use, or, as they are often called, the product's life cycle costs. The challenge of modern competitive engineering is to ensure that life cycle costs are minimised, whilst achieving requirements for performance and time to market.

If the market for the product is competitive, improved quality and reliability can generate very strong competitive advantages. We have seen the results of this in the way that many products, particularly Japanese cars, machine tools, earthmoving equipment, electronic components and consumer electronic products have won dominant positions in world markets in the last 30 to 40 years. Their domination has been largely the result of the teaching of the late W. Edwards Deming, who taught the fundamental connections between quality, productivity and competitiveness. Today this message is well understood by nearly all engineering companies that face the new competition, and those that do not understand lose position or fail.

Concurrently with the philosophy and methods that took root initially in Japan and then spread back to the West where most originated, methods were developed in the USA to address the problems of unsatisfactory quality and reliability of military equipment. These included formal systems for quality and reliability management (MIL-Q-9858 and MIL-STD-758) and methods for predicting and measuring reliability (MIL-STD-721, MIL-HDBK-217, MIL-STD-781). MIL-Q-9858 was the model for the international standard on quality systems (ISO 9000), and the methods for quantifying reliability have been similarly developed and applied to other types of product.

The methods developed in the West were driven to a large extent by the customers, particularly the military. They reacted to perceived low achievement ~~by~~ the imposition of standards and procedures, whilst their suppliers saw little motivation to improve, since they were paid for spares and repairs. By contrast, the Japanese quality movement was led by industry, who had learned how quality provided the key to greatly increased productivity and competitiveness.

These two streams of development epitomize the difference between the deductive mentality applied by the Japanese to industry in general, and to engineering in particular, in contrast to the more inductive Western approach. The deductive approach seeks to generate continuous improvements across a broad front, and new ideas are subjected to careful evaluation. The inductive approach leads to inventions and 'breakthroughs', and to greater reliance on 'systems' for control of people and processes. The deductive approach allows a clearer view, particularly in discriminating between sense and nonsense. However, it is not conducive to the development of radical new ideas. Obviously these traits are not exclusive, and most engineering work involves elements of both. However, the overall tendency of Japanese thinking shows up in their enthusiasm and success in industrial teamwork and in the way that they have adopted the philosophies of Western teachers such as Deming and Drucker, whilst their Western competitors have found it more difficult to break away from the mould of 'scientific' management, with its reliance on systems and more rigid organisations and procedures.

Unfortunately the development of quality and reliability engineering has been afflicted with more nonsense than any other branch of engineering. This has been the result of the development of methods and systems for analysis and control that contravene the deductive logic that quality and reliability are achieved by knowledge, attention to detail, and continuous improvement on the part of the people involved. Of course Western minds have also made great positive contributions: we need only recall Shewhart's invention of statistical process control and Fisher's invention of statistical experiments, and of course Deming was an American. Therefore it can be difficult for students, teachers, engineers and managers to discriminate effectively, and many have been led down wrong paths.

In this series we will attempt to provide a balanced and practical source covering all aspects of quality and reliability engineering and management, related to present and future conditions, and to the range of new scientific and engineering developments that will shape future products. I hope that the series will make a positive contribution to the teaching and practice of engineering.

Patrick D.T. O'Connor
August 1994

CHAPTER 1

Introduction

Quality in the goods and services that they provide is now the essential ingredient in the success of companies. Indeed, it is the main factor in determining whether a company succeeds or fails. Competition in the market-place is intense and international, and assurance to customers of high standards in quality control has become of vital importance. In consequence, it is essential that companies implement sound and effective quality control and quality assurance systems. The ISO 9000 series of standards has been developed in response to this need: the standards set out procedures for ensuring that an effective quality system which meets customers' needs is established and maintained.

However, high standards in the goods and services that they provide is not the only concern that companies must now have in the area of quality. Although this is currently the quality factor that determines whether or not a company survives, companies increasingly have to give attention to other aspects of quality that they have control of. Firstly, they must have regard to the quality of the working environment in terms of the job satisfaction and safety of their employees. Safety issues must be dealt with in accordance with health and safety legislation, which is constantly getting ever more rigorous around the world, and job satisfaction is obviously important in achieving the general quality goals of a company. If workers feel that the company looks after them, they are more likely to be inclined to work in the best interests of the company.

The quality of the performance of a company in terms of the return that it gives investors is a further aspect of quality which a company must address. A good performance will mean that the company attracts more investment, and this can be applied in further improving its performance in terms of the quality of the goods and services provided.

The final factor in this consideration of the wider issues of quality is that of environmental management. Interest in organizing production methods so that they minimize the damage to the environment is growing rapidly round the world, and appropriate legislation is beginning to appear in some countries. Guidance and rules for implementing environmental management systems that ensure compliance with stated environmental policies and objectives are now published in standards such as ISO 14000 [1] and BS 7750 [2]. These two standards are very similar and contain rules that follow similar principles to those defined in ISO 9001–9003 (see Section 1.2) for quality assurance in manufacturing and service functions. The management structures developed to satisfy ISO 9001–9003 are usually suitable for extension to BS 7750/ISO 14000. Also, similar rules with regard to the control of measurement and calibration procedures are specified, and all such procedures described later in this book with respect to satisfying ISO 9000 are equally relevant to BS 7750/ISO 14000.

1.1 GENERAL APPROACH TO QUALITY SYSTEMS

For a quality control system to operate successfully, several conditions have to be satisfied. Firstly, the procedures instituted must be orientated towards preventing quality problems occurring, rather than being mere fault detection systems which allow faults to be put right before someone complains about them. Secondly, when introducing quality control procedures, it is very important that employees at all levels in a company are aware of the reason for them, understand fully how to operate them, and co-operate enthusiastically in implementing them. Thirdly, whilst it is sensible to appoint a quality control manager to design and monitor quality control systems, the responsibility for quality control must never be seen as being the responsibility of this one person alone. All personnel in a company must be encouraged to share in the duty of maintaining good quality and take pride in it.

It is also important that quality control procedures should evolve and develop over a period of time. They must not be implemented and then stay the same for ever afterwards. Rather, regular review is necessary to ensure that quality control procedures continue to be efficient and remain the most appropriate as technological developments take place. Such reviews must also take changes in market forces into account and fully monitor customer satisfaction, need and expectations.

Measurement is an essential ingredient in the operation of quality control systems, both for ensuring that process variables within a manufacturing process are maintained within acceptable limits and also for checking that the final parameters of the product are within specified tolerances. If the quality control system is to operate satisfactorily, then the measurements themselves must be of high quality. Firstly, this means that only properly calibrated instruments and transducers must be used for making measurements, and it requires that the company must therefore establish and maintain appropriate calibration equipment to ensure this. Secondly, it requires that all measurement errors are identified, quantified and compensated for. The primary concern of this book is with these quality-related measurement and calibration procedures. However, before covering these matters in detail, it is sensible to cover briefly the more general principles applicable to quality systems as a whole. A fuller coverage of such general quality assurance principles can be found elsewhere [3].

1.2 ISO 9000 QUALITY STANDARDS

The term ISO 9000 is often used loosely to refer globally to a complete set of five documents which are numbered ISO 9000 to ISO 9004 [4] and which collectively are concerned with setting out procedures that are designed to achieve high standards in quality assurance. They impose a requirement on suppliers of goods and services to establish and maintain an effective, economical and demonstrable system which ensures that what they supply conforms to specified requirements.

The first document in the set, ISO 9000, is not in fact a standard itself but consists merely of guidelines for selecting and using ISO 9001 to ISO 9003. These latter three documents are the actual quality assurance standards, and their applicability is explained in Section 1.3. The set is completed by ISO 9004, which is again not a standard but a document which provides guidelines for developing and implementing quality systems.

The ISO quality standards have now been adopted by most countries around the world, but they are often published in individual countries with slightly different prefixes or codes so that they harmonize with pre-existing national coding systems for standards. The British Standards Institute for example issues the documents with the prefix BS.EN, the European Committee for Standardization (CEN) uses the prefix EN, and the American National Standards Institute (ANSI) and American Society for Quality Control (ASQC) jointly publish a set of documents with the codes ANSI/ASQC Q90–ANSI/ASQC Q94. Thus the corresponding equivalents of ISO 9001 are BS.EN.ISO 9000, EN.ISO 9000 and ANSI/ASQC Q91. A full set of National equivalents to ISO 9000 around the world can be found elsewhere [3].

It is important to note that the set of ISO 9000 quality standards defines what elements quality systems should have, but it does not prescribe how these elements should be implemented in any particular situation. This is necessarily so because every situation and application is different. Hence, any company considering adopting ISO 9000 standards is advised to seek the advice of a quality consultant who has experience of the most appropriate way of defining and implementing the necessary quality system components in the particular type of industry concerned.

Standards such as ISO 9000 are subject to constant review in the light of user experience, and it is almost inevitable therefore that the current version (1994) will be superseded during the lifetime of this book. Any reader who is contemplating implementing a quality assurance system is therefore strongly advised to seek out and read the latest versions of the standards. However, readers should not be unduly alarmed by this because, based on the past experience of revisions, any future changes made are likely to be minor.

Whilst obviously only the latest standard is of direct relevance, it is useful to look briefly at the historical development of quality standards, especially for those readers who may be more familiar with earlier, now superseded standards than with the present ISO 9000 series. The first quality assurance document was BS 5750 (parts 0–4), published by the British Standards Institute (BSI) in 1979. This had a companion document BS 5781 which provided guidance on satisfying the measurement and calibration requirements specified in BS 5750 for quality-related measurements. Then in 1987, the International Standards Organization published ISO 9000–9004 which was effectively a revised and enhanced version of BS 5750. One major part of the revision was to include a summary of BS 5781 within the new ISO 9000 series standards. The identical revised document was also published by the BSI in 1987 with the pre-existing code BS 5750, and the European Committee for Standardization (CEN) published identically worded documents as EN 29000–EN 29004.

The most significant changes in the latest version (1994) of the ISO 9000 family of documents compared with the previous version are, firstly, the increased emphasis on meeting customer requirements and, secondly, the requirement to demonstrate the achievement of continuous improvements in quality. Meeting customer needs is of course always a compromise between the quality level that a customer wants and what he is prepared to pay for, as increasing the quality level inevitably increases production costs to some extent and therefore raises the price of the product. Nevertheless, this is a significant change from the previous standard which, at least in theory, allowed the supplier to define whatever level of quality he wanted. That standard merely required that the supplier had procedures in place that guaranteed that the product met the quality level specified. Of course, no sensible supplier would have disregarded customer requirements entirely in defining the target quality level. However, it was market forces rather than the standard that was the prime motivator.

A brief summary of the contents of the current set of documents in the ISO 9000 family is given below. However, anyone intending to implement quality systems is strongly advised to read the full standards as well.

1.2.1 ISO 9000: Quality Management and Quality Assurance Standards - Guidelines

[British standard: BS.EN.ISO 9000, European standard EN.ISO 9000, USA Standard: ANSL/ASQC Q90]

This is actually published as four separate parts. Part 1 (ISO 9000-1) gives general guidance about quality management and the implementation of quality assurance procedures. It assists the reader in understanding the requirements of the set of ISO 9000 quality assurance standards (ISO 9001–ISO 9003) and advises which one is applicable in different situations. It explains that each standard can be adopted if required as a contractual quality agreement between a supplier and customer and also gives some brief guidelines about how the standards can be modified and adapted to tailor them to special contractual situations.

Part 2 (ISO 9000-2) provides more detailed guidelines on the implementation of ISO 9001–9003. Part 3 (ISO 9000-3) gives specific guidance about the application of ISO 9001 where software either is the product or is Part of the product. Finally, Part 4 (ISO 9000-4) gives guidance on developing and managing a programme which assures customers about the reliability and maintainability of products.

1.2.2 ISO 9001: Quality Systems – Quality Assurance in Design, Development, Production, Installation and Servicing Functions

[British standard: BS.EN.ISO 9001, European standard EN.ISO 9001, USA Standard: ANSL/ASQC Q91]

This quality standard describes a set of procedures for the assurance of quality in all functions associated with the manufacture of a product, including product design, development, production, inspection, test, installation and servicing. The requirements to be satisfied are summarized below, although anyone contemplating implementing the standard is strongly advised to read the full standard (ISO 9001):

1. With regard to the design process, the company must establish and maintain procedures that control the product design and ensure that specified requirements are met. Particular duties imposed are to identify clearly the person responsible for the design process, to use qualified personnel in the design team and to provide the team with adequate resources. All design procedures must be documented, and mechanisms for modifying design procedures in response to changing product requirements must also be documented.

2. Production processes must be carried out under controlled conditions which guarantee the quality of the product. Similar control must be exercised over the quality of raw materials and part-finished components bought in from subcontractors and suppliers external to the company to ensure that they conform to specified requirements.

3. Any product installation procedures that can affect quality must be rigidly controlled.

4. Handling, storage, packaging and delivery functions must be designed and controlled to avoid damage to products.

5. A system of records must be established and maintained which shows that the product has passed inspections and/or functional tests.

6. Equipment used to make quality-related measurements at any stage of production, inspection or test must be controlled so that measurement uncertainty is known. Calibration must be carried out at defined intervals of time.

7. Where a product is designed for a particular customer, all design, manufacture and testing procedures must be agreed with the customer to ensure that they meet with his expectations and needs. Alternatively, if there is no specific customer, these procedures must be carried out with due regard to the needs of the market-place.

8. The company's quality policy must be defined at executive management level and must be understood, implemented and maintained at all levels in the company.

9. The company must provide adequate resources for the operation and auditing of the quality system implemented.

10. The responsibility, authority and interrelation of all personnel who manage, perform or verify work affecting quality must be specified. One designated person must be given overall responsibility for ensuring that the requirements of ISO 9001 are met. This person must have sufficient authority to instigate remedial action if any nonconformance occurs in the quality control procedures being operated. Similarly, this designated person must have the power to limit further production, delivery or installation of a product until breaches in the quality control system have been rectified.

11. All personnel involved in procedures affecting quality control must be adequately trained.

12. The quality system established must be periodically reviewed to ensure that it continues to be effective, and records of the reviews must be kept.

13. The company must provide, control, calibrate and maintain inspection, measuring and test equipment suitable to demonstrate the performance of the product to the specified requirements. Equipment must be used in a manner that ensures that the measurement uncertainty is known.

14. Measurement and calibration for this equipment must be in accordance with controlled and documented procedures.

15. The inspection/test status of products must be identified by a documented system of markings. A proper system must be in operation which deals effectively with products that fail quality tests. A procedure for responding to such failures by reviewing the effectiveness of quality control procedures must also be documented.

16. Proper documentation must be maintained for all aspects of the quality assurance system. This documentation will typically take the form of a quality manual.

17. Copies of the quality manual, or appropriate parts of it, must be available at all locations where operations essential to the effective functioning of the quality system are carried out.

18. All documentation must be dated, have its revision number identified, and be maintained in a clean and legible condition.

19. A documented system must be established whereby the quality manual is modified or has additions made to it to keep it up to date. The responsibility for this should be clearly specified and should normally rest with the person who originally produced the quality manual.

20. The quality management system must contain documented provision for disposing of obsolete documentation.

1.2.3 ISO 9002: Quality Systems – Quality Assurance in Production, Installation and Servicing Functions

[British standard: BS.EN.ISO 9002, European standard EN.ISO 9002, USA Standard: ANSL/ASQC Q92]

The requirements of ISO 9002 are almost identical with those specified in ISO 9001 except that clauses relating to the control of design and development procedures are omitted.

1.2.4 ISO 9003: Quality Systems – Quality Assurance in Final Inspection and Test Functions

[British standard: BS.EN.ISO 9003, European standard EN.ISO 9003, USA Standard: ANSL/ASQC Q93]

The requirements of ISO 9003 are almost identical with those specified in ISO 9001 except that clauses relating to the control of design, development and production procedures are omitted.

1.2.5 ISO 9004: Quality Management and Quality System Elements

[British standard: BS.EN.ISO 9004, European standard EN.ISO 9004, USA Standard: ANSL/ASQC Q94]

ISO 9004 consists of four parts which together provide further guidance on developing and implementing quality systems. The general requirements of quality system management, documentation, audits, costing, training and the control of measurement and testing equipment are covered. Part 1 (ISO 9004-1) gives general guidelines, Part 2 (ISO 9004-2) gives more specific guidelines where a supplier is providing a service and part 3 (ISO 9004-3) gives more specific guidelines where a supplier is providing processed materials. Part 4 (ISO 9004-4) gives advice on how implement a system designed to achieve continuous improvements in quality.

1.3 APPLICABILITY OF ISO 9001 – ISO 9003

ISO 9001 is the standard of most relevance to manufacturing companies. It lays down procedures for ensuring quality assurance in all aspects of manufacturing, including product design, manufacture, installation and testing.

ISO 9002 is a subset of ISO 9001 intended for companies that only provide a manufacturing, installation and test function (i.e. no product design). It also applies to companies that only provide a service function (e.g. specialized cleaning services).

ISO 9003 is an even smaller subset of ISO 9001. It applies to companies that only provide a supply, inspection and test service (i.e. no product manufacture).

1.4 ISO 14000 STANDARDS FOR ENVIRONMENTAL MANAGEMENT SYSTEMS

Although this book was primarily written to cover the measurement and calibration requirements of ISO 9000, it is also appropriate to look briefly at the requirements of ISO 14000 standards, as these make similar measurement and calibration demands to ISO 9000, and indeed are similar in other respects also. The major difference between ISO 9000 and 14000 standards is in their application. Whereas ISO 9000 is concerned with quality in goods manufactured and services provided, ISO 14000 covers environmental concerns of air and water pollution, waste minimization/disposal and good practice in product use.

ISO 14000 [1] follows closely on from the procedures previously recommended in the British standard BS 7750 [2], but some important new requirements have been added. A sensible approach to explaining these standards for environmental management systems is to consider first the clauses in ISO 14000 that are carried over from BS 7750, and then to consider what new requirements have been introduced.

BS 7750 was based in many respects on ISO 9000 and the earlier BS 5750 standard. It required the same commitment by the board-level management of a company to the environmental management system set up, and demanded similar standards of resourcing, management control, personnel training, company-wide adoption, effectiveness measurement, effectiveness review and documentation. ISO 14000 encompasses all of these principles but gives greater emphasis in certain areas such as establishing an environmental strategic plan, refining system auditing procedures and fostering continuing improvements in environmental management standards.

1.5 SUMMARY OF MEASUREMENT AND CALIBRATION REQUIREMENTS

ISO 9001–9003 lay down procedures to be followed when selecting, using, calibrating, controlling and maintaining measurement standards and measuring equipment. A supplementary document, ISO 10012-1 [5], defines the necessary procedures in more detail and also gives some guidance about implementation. As explained in the last section, these procedures are also applicable to measurement and calibration requirements associated with environmental management systems conforming to ISO 14000.

The main measurement and calibration requirements are summarized below, but anyone implementing certified quality assurance systems is advised to read the full details in ISO 10012-1 as well.

1. Measuring equipment must be provided for the quantification of all quality-related parameters, and this must have adequate metrical characteristics of accuracy etc.

2. The list of all instruments used to make quality-related measurements must be documented.

3. An effective system for the control and calibration of measuring equipment must be established and maintained.
   Note Complete in-plant calibration is not essential if these services are obtained from sources that comply with the requirements of the standards.

4. All equipment used to make quality measurements, and all calibration instruments, must be handled carefully and used in such a way that their accuracy and fitness are protected.

5. All measurements, whether for purposes of calibration or product assessment, must take into account all the errors and uncertainties in the measurement process.

6. Objective evidence that the measurement system is effective must be available to customers.

7. Calibration must be performed by equipment traceable to national standards.

8. All personnel performing calibration functions must have adequate training.

9. Calibration procedures must be documented.

10. The calibration system must be periodically and systematically reviewed to ensure its continued effectiveness.

11. A separate calibration record must be kept for each measuring instrument. Each record must demonstrate that the instrument is capable of performing measurements within the designated limits and should contain at least the following information:

    a. a description of the instrument and a unique identifier,
    b. the calibration date,
    c. the calibration results,
    d. the calibration interval (plus date when next calibration due).

    Also, depending on the type of instrument involved, some or all of the following information is also required:

    a. the calibration procedure,
    b. the permissible error limits,
    c. a statement of the cumulative effects of uncertainties in calibration data,
    d. the environmental conditions required for calibration,
    e. the source of calibration used to establish traceability,
    f. details of any repairs or modifications which might affect the calibration status,
    g. any use limitations of the instrument.

12. Each instrument must be labelled to show its calibration status and any usage limitations (but only where it is practicable to do so).

13. Any instrument which has failed or is suspected (or known) to be out of calibration must be withdrawn from use and labelled conspicuously to prevent accidental use.

14. Adjustable devices must be sealed to prevent tampering.

1.6 INTERPRETATION OF MEASUREMENT AND CALIBRATION REQUIREMENTS

The general procedures specified by ISO 9001–9003 for meeting quality assurance targets are fairly easy to understand. In order to comply with the standards, it is necessary to institute and maintain a quality control and measurement system which ensures that the quality of manufactured goods or services does not deviate outside stated error bounds.

What error bounds are stated depends on the situation. Where a product or service is specifically designed for one customer, the appropriate quality level is that agreed with the customer. This might in some cases even be specified in a contractual agreement between the supplier and customer. In some situations, statutory standards apply which are legally enforceable. For example, weighing scales used for commercial purposes have to comply with published standards of accuracy. In other cases, consensus standards apply, often drawn up by trade associations. These represent a general agreement about what level of quality is acceptable in any individual case. If none of these situations pertain, then the supplier must evaluate what typical customers in the market-place might reasonably expect in the way of quality and set error bounds accordingly.

Once the manufacturer or service provider has set appropriate error bounds, she/he must then institute a system which measures the product at suitable intervals of time and ensures that it does not go outside those stated error bounds. In making such measurements, all instruments used must be calibrated at appropriate intervals to ensure the accuracy of the measurements made, in accordance with procedures laid down in the standards. All of these procedures for product measurement and instrument calibration must be fully documented, and this documentation must be made available to customers if required.

The major area of difficulty in conforming to these standards is in properly interpreting the requirements for providing and maintaining measurement and calibration equipment. Instrument calibration ensures that the measuring accuracy of each instrument involved in the measurement process is known over its whole measurement range, when it is used under specified environmental conditions. This knowledge is gained by comparing the output of the instrument under test against the output of an instrument of known accuracy when the same input is applied to both instruments. What this fails to establish is what the accuracy of the instrument is when it is used in environmental conditions different from those in which it is calibrated, because the characteristics of any instrument vary with ambient conditions, as described in Chapter 5.

As instruments are used in practice to make measurements which are used for quality control functions, it is clearly necessary to establish what variation in instrument characteristics occurs when they are used in environmental conditions which differ from those applying during calibration. In other words, we need to quantify how environmental condition changes affect the performance of an instrument. Knowledge of the way in which environmental conditions affect particular instruments also allows measurement procedures to be established which minimize the effect of ambient condition changes and thereby improve the quality of measurements. This subject is again covered in detail in Chapter 5.

The necessary procedure is therefore to use the calibration of the instrument under standard conditions as a baseline, and make appropriate modifications to account for the variation in characteristics due to differing environmental conditions existing in its normal place of use. Having done this, it is then possible to quantify the quality of measurements made by calculating the bounds of measurement error of each instrument as it is used in its normal operating environment.

Unfortunately, the quality of measurements does not remain at a constant level for ever because the characteristics of any instrument change over a period of time. The magnitude and the rate of this change are influenced by many factors which are difficult or impossible to predict theoretically. It is therefore necessary to apply practical experimentation to determine the rate of such changes in instrument characteristics over a period of time.

Once the maximum permissible measurement error has been defined, knowledge of the rate at which the characteristics of an instrument change allows calculation of a time interval which represents the moment in time when an instrument will have reached the bounds of its acceptable performance level. The instrument must be recalibrated either at this time or earlier. This measurement error level which an instrument reaches just before recalibration is the error bound which must be quoted in the documented specifications for the instrument.

Another requirement is to calculate the cumulative error in a measurement system when the measurement obtained is derived from the output of more than one instrument. Appropriate procedures for doing this are considered in Chapter 5.

Finally, the whole calibration procedure must be fully documented, in a manner as indicated in Chapter 3. These necessary steps can be summarized as follows:

a. To calibrate all measuring instruments under specified environmental conditions so that their measuring accuracy is known over the whole measurement range.

b. To establish the variation in instrument characteristics under environmental conditions different from the calibration conditions.

c. To establish measurement procedures which minimize the effect of environmental condition changes on the measuring instruments.

d. To calculate the bounds of measurement error of an instrument under normal operating conditions.

e. By practical experimentation to determine the rate of change of instrument characteristics over a period of time.

f. Hence to determine the frequency at which instruments should be recalibrated and the maximum possible measurement error when the instrument has drifted furthest from its specification immediately before calibration.

g. To combine all instrument measurement error levels into a figure which expresses the cumulative error level of the whole measurement process (where a measurement is composed of the outputs of more than one instrument).

h. To document all measurement and calibration procedures.

1.7 STANDARD MEASUREMENT UNITS

Before going on to discuss calibration procedures in detail, the importance of applying standard measuring units should be understood. Such understanding is aided by briefly looking at the historical development of standard units. In the early stages of human civilization, simple measurement systems were developed to regulate barter trade, using measurement units based on convenient parts of the human torso such as the foot and hand. Such units were inexact, since hands and feet vary in size. This system therefore only allowed an approximate level of equivalence to be established about the relative value of quantities of different commodities, and more exact units were eventually developed. The first of these was a unit of length (the metre) defined as 10^{-7} times the polar quadrant of the earth. A platinum bar made to this length was established as a standard of length in the early part of the nineteenth century. This was superseded by a superior quality standard bar in 1889, manufactured from a platinum–iridium alloy. Since that time, technological research has enabled further improvements to be made in the accuracy of the standard used for defining length. Firstly, in 1960, the standard metre was redefined in terms of $1.650\,763\,73 \times 10^6$ wavelengths of the radiation from Krypton 86 in vacuum. More recently, in 1983, the metre was redefined yet again as the length of path travelled by light in an interval of 1/299 792 458 seconds.

In a similar fashion, standard units for the measurement of other physical quantities have been defined, and the accuracy with which they can be measured has progressively improved over the years. The latest standards for defining the units used for measuring the fundamental physical quantities of length, mass, time, temperature, electric current, luminous intensity and matter are given in Table 1.1. Such standard definitions of these quantities act as primary standards and form the foundation on which all

Table 1.1. Definitions of standard units

Physical quantity	Standard unit	Definition
Length	metre	The length of path travelled by light in an interval of 1/299 792 458 seconds.
Mass	kilogram	The mass of a platinum–iridium cylinder kept in the International Bureau of Weights and Measures, Sèvres, Paris.
Time	second	$9.192\,631\,770 \times 10^9$ cylces of radiation from vaporized caesium 133 (an accuracy of 1 in 10^{12} or one second in 36 000 years)
Temperature	degrees	The temperature difference between absolute zero kelvin and the triple point of water is defined as 273.16 K.
Current	Ampere	One ampere is the current flowing through two infinitely long parallel conductors of negligible cross-section placed one metre apart in vacuum and producing a force of 2×10^{-7} newtons per metre length of conductor
Luminous intensity	candela	One candela is the luminous intensity in a given direction from a source emitting monochromatic radiation at a frequency of 540 terahertz (Hz $\times 10^{12}$) and with a radiant density in that direction of 1.4641 mW/steradian. (1 steradian is the solid angle which, having its vertex at the centre of a sphere, cuts off an area of the sphere surface equal to that of a square with sides of length equal to the sphere radius.)
Matter	mole	The number of atoms in a 0.012 kg mass of carbon 12.

calibration procedures depend. They represent the highest standard of accuracy achievable in the measurement of each quantity.

The early establishment of standards for the measurement of physical quantities proceeded in several countries at broadly parallel rates, and in consequence, several sets of units emerged for measuring the same physical variable. For instance, length can be measured in yards or metres or several other units. Apart from the major units of length, subdivisions of standard units exist such as feet, inches, centimetres and millimetres, with a fixed relationship between each fundamental unit and its subdivisions.

Yards, feet and inches belong to the Imperial System of units, which is characterized by having varying and cumbersome multiplication factors relating fundamental units to subdivisions such as 1760 (miles to yards), 3 (yards to feet) and 12 (feet to inches). This Imperial system is the one which was used predominantly in Britain until a few years ago.

The metric system is an alternative set of units which includes for instance the unit of the metre and its centimetre and millimetre subdivisions for measuring length. All multiples and subdivisions of basic metric units are related to the base by factors of ten and such units are therefore much easier to use than Imperial units. However, in the case of derived units such as velocity, the number of alternative ways in which these can be expressed in the metric system can lead to confusion.

As a result of this, an internationally agreed set of standard units (SI units or Systèmes Internationales d'Unités) has been defined, and strong efforts are being made to encourage the adoption of this system throughout the world. In support of this effort, the SI system of units will be used exclusively in this book.

The full range of fundamental SI measuring units and the further set of units derived from them are given in Appendix 1. Two supplementary units for measuring angles are also included within this first appendix. Following this, conversion tables relating common Imperial and metric units to their equivalent SI units are to be found in Appendix 2.

REFERENCES

1. ISO 14000: *Environmental management systems,* International Organization for Standards, Geneva (1996) (also published by the British Standards Institution as BS.EN.ISO 14000).
2. BS7750: *Environmental management systems,* British Standards Institution, London (1994).
3. Badiru, A. B., *Industry's guide to ISO 9000,* John Wiley and Sons (1995).
4. ISO 9000: *Quality Management and Quality Assurance standards,* International Organization for Standards, Geneva (1994).
   (Individual parts are published as ISO 9000, ISO 9001, ISO 9002, ISO 9003 and ISO 9004).
5. ISO 10012-1: *Quality assurance requirements for measuring equipment,* International Organization for Standards, Geneva (1992).

CHAPTER 2

Quality Systems in Manufacturing and Service Provision

High standards of quality in products and services are essential factors in determining a company's ability to compete in international markets. In order to achieve these goals, administrative and technical system networks must be developed which establish a quality control system and ensure that the required standards are met and maintained. It is important that these networks identify all activities in a manufacturing system or service provision that impact on quality, and combine to produce a cohesive and effective quality control system. After such a system has been put into operation, it is also extremely important to monitor its continued effectiveness by generating and evaluating quality-related data during the manufacturing process or provision of the service. The use of up-to-date technology is also of fundamental importance in achieving high quality products, but a good quality system remains essential to ensuring that such technology is used efficiently and effectively.

ISO 9000 standards prescribe a number of procedures which, if applied correctly, successfully achieve these quality goals. Sometimes (unjustifiably), criticisms are directed at the standards, accusing them of being too bureaucratic, too costly to implement and requiring unnecessary documentation, especially when applied to small companies. In truth, the usual reason for these problems is that the consultant responsible for advising the company concerned about ISO 9000 certification has over-specified the requirements. The ISO 9000 standards have been drafted very carefully so as to leave room for intelligent interpretation according to the needs of the industry involved and the circumstances prevailing. The only mandatory requirement is that the measures put in place are sufficient to establish a reliable quality control system that satisfies customers and assures them of its integrity, and a good consultant will avoid unnecessary bureaucracy and cost as long as this is achieved. Thus the only word of caution to companies adopting ISO 9000 is to choose their consultant with care.

It is important to note that ISO 9000 only prescribes a minimum set of standards. In some industrial sectors, market forces may dictate that higher levels of quality are achieved, and extended versions of ISO 9000 are being developed for such cases. For example, in the motor industry, Ford, Chrysler and General Motors have jointly developed a standard called QS 9000 which encompasses all of the clauses in ISO 9000 but also adds some sector-specific requirements.

2.1 TOTAL QUALITY MANAGEMENT

Total Quality Management (TQM) is the latest 'buzzword' to hit the quality assurance scene. There is no universally accepted standard definition of TQM. Different organizations and companies define and apply it in somewhat different ways. However, a global definition which encompasses most of the different interpretations is that TQM is an integrated approach to quality which operates in all parts of a company and encompasses a style of management aimed at achieving the long-term success of a company by linking quality with customer satisfaction.

TQM requires that the quality of the company's product (whether the supply of manufactured goods or the provision of a service) should be the company's number one priority, and that there should be an ongoing commitment to progressively increasing quality still further. To be successful in this, it is essential to achieve the total involvement of all personnel in a company and foster a common commitment to quality which is shared by everyone. Everyone must understand what level of quality is required of them and they must be motivated to produce it. It is extremely important that high quality is achieved in all areas of a company. Islands of quality here and there are of no use at all: if workers in a few departments are achieving quality improvements but they see other departments who are not, they will quickly lose interest in working hard to achieve quality targets themselves.

The maintenance of such motivation to achieve high quality is often assisted by the use of anonymous questionnaires from time to time to measure the morale of the workforce, remove any irritants in the workplace and also assess the attitudes and degree of co-operation between different departments in a company. Schemes whereby workers are rewarded for making suggestions that lead to quality improvements or elimination of quality problems are also generally very useful. Display boards where quality successes are reported can also be strong motivators in fostering a quality culture.

Quality circles can be a very useful method of attaining these ideals in a manufacturing company. A quality circle (see Figure 2.1) consists of a group of people who collectively represent all functions within the company that can have an effect on quality. For instance, such a group must represent

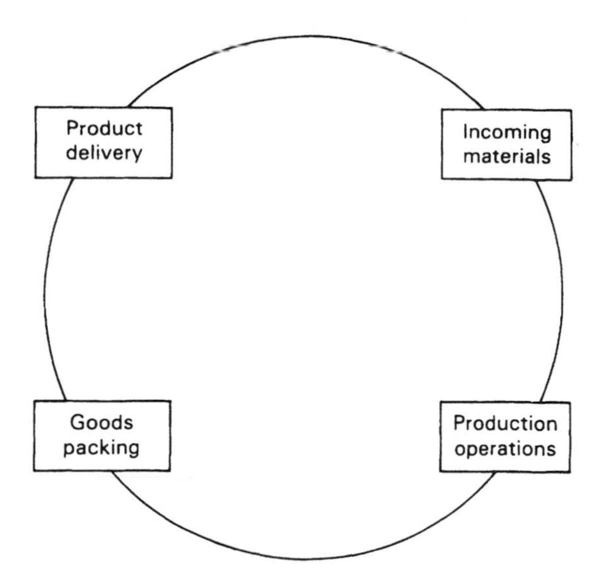

Figure 2.1. The quality circle

goods packing and delivery sections as well as shop-floor operatives from the production departments. The discussion which takes place at periodic meetings of the quality circle fulfils several functions. Firstly, it ensures that thoughts about quality maintain a high profile throughout the company. Secondly, it provides a feedback mechanism whereby breakdowns in or difficulties with a quality system can be reported and suggestions for improvement made. Thirdly, by giving everyone such a personal involvement in achieving high quality, an atmosphere is generated in which people have a pride in their work, understand the reasons for the quality control procedures implemented and are fully committed to their operation.

Some effort is usually necessary to get quality circles going in the first place but, once established, they can be very successful. Experience shows that people do not readily volunteer for things, and making general appeals to the workforce for volunteers to form a quality circle is unlikely to generate much response. However, if people are targeted and specifically asked, they usually agree to participate and take pride in being part of a quality circle.

Whether achieved by quality circles or otherwise, communication paths are extremely important for achieving quality targets, and they must exist from the shop-floor right to the top of the company. Personnel on the shop-floor involved in production operations are a valuable source of information with regard to production problems, quality lapses and the likely causes, and mechanisms must exist for this information to be rapidly transmitted to the company management. For example, the workers associated with a manufacturing cell will be the first to pick up any quality problems in the materials going into that cell, and a rapid means of communicating this information must be available. Communication in the reverse direction is also very important, with management arranging meetings to explain quality issues, describe improvement plans, discuss performance targets and report on progress made.

On top of this communication upwards and downwards in a company, sideways communication between different departments is also essential. It is not sufficient for each department to operate its own quality control system; the quality system must operate as a cohesive whole across the whole company. This can only be achieved by complete co-operation and constant dialogue between the different parts of a company. Such dialogue also needs to extend to component suppliers, because, in the philosophy of total quality management, they must be treated in exactly the same way as departments within the company.

The determination to maintain the momentum of a TQM system and ensure that it is successful must come from the managing board of a company. They must be fully committed, give full support and provide everything that is needed to achieve high quality, including adequate financial resources. This commitment of the company management to quality must be highly visible, because any apparent lack of commitment on their part will spread like a plague through the rest of the company. However, whilst they must provide the lead, they should delegate as much as possible the task of achieving high quality. Production personnel on the shop-floor should be given the opportunity to decide themselves how to tackle quality problems, for it is the people who are intimately acquainted in this way with the production environment who will often come up with the optimum solutions to quality problems. Only if they fail to do so should the management intervene.

At departmental level within a company, it can be very useful to ask every member of the department the following questions on an annual basis: 'What process are you involved in?'; 'What are the key measures of performance for the process?'; 'How do you monitor performance using these measures?'; 'What can you do to improve performance?' A further useful annual exercise is to construct a flow chart of the processes that the department is responsible for and to ask, 'Why do we carry out each operation?'; 'Is each operation necessary?' and 'Can the process route be simplified?'

One possible danger in operating a TQM system is that complacency can creep in. Once this belief that our quality is OK has been adopted throughout a factory, a false sense of security can be generated. So the quality is OK? But who has judged this to be so and against what standards? Clearly, some independent confirmation that quality is being maintained is necessary. Feedback from customers is perhaps the best arbiter, and this is therefore a very important part of TQM. Customers should be asked what their impressions of a product are both immediately after delivery and throughout its lifetime. The timing of product failures can give important clues about the fundamental cause of failure and enable appropriate remedial action to be taken. Early failures in a product soon after it has been delivered to a customer are usually due to manufacturing faults, whereas design faults are the most probable cause of faults occurring later on in the lifetime of a product.

Once implemented successfully, TQM systems operate to the very positive good of the company in generating customer goodwill and expanding the customer base and sales. However, the benefits do not end there. The increased confidence and work satisfaction that a TQM system generates in the workforce increases productivity and reduces production costs. Production costs are reduced particularly because, by preventing faults occurring rather than just detecting faults after they have occurred, expensive reworking and the losses incurred through having to scrap rejected products are avoided. A good TQM system also offers the possibility of introducing a just-in-time (JIT) manufacturing system. This greatly reduces production costs by eliminating the need for stockpiles of work in hand. With JIT, production schedules are arranged to start production so that the finished product comes out of the production system and is delivered to the customer just when he needs it. Of course, this means that there must be a very low probability of delays in the production system, because a JIT system has no ability to compensate for delays. Any delays in production automatically mean late delivery to the customer. However, JIT is achievable providing an efficient TQM system is being operated.

2.2 CUSTOMER SATISFACTION

The three activities necessary to maintain customer satisfaction and a high profile in the market-place can be described as quality control, quality management and quality assurance. Whilst these are described in this chapter with reference to manufactured products, they apply equally well to the production of computer software systems and to the provision of services. Quality defines the fitness for purpose and conformity with specification of a product. Quality control comprises the operational techniques and programmes which are implemented to achieve and sustain the required product quality. Quality management describes the managerial aspects of implementing the quality control system and the procedures involved in monitoring and maintaining the effectiveness of such systems. Quality assurance is a global term which embraces all functions and activities involved in quality control, both technical and managerial. The purpose of quality assurance is to act as a communication channel, whereby the customer is made aware of the quality control system associated with a particular product and is assured of its effectiveness in maintaining product quality. Documentation is therefore an essential part of quality assurance, wherein all procedures for quality control and management are expressed. Such documentation must be freely available to customers on request, and indeed this availability is specified as one of the conditions to be met in satisfying ISO 9001–9003.

2.3 QUALITY POLICY AND OBJECTIVES

The first step in developing a quality control system must be to establish what the general customer needs and expectations are. Caution must be exercised however in responding to such customer demands. Due regard must be taken of the cost involved in meeting the quality levels demanded by customers. The minimum acceptable quality level can normally be set as that achieved by competing manufacturers in the market-place or as defined by published quality standards established by consensus agreement for particular types of product or service. Improvements in quality beyond this level obviously have strong marketing advantages, but only if the cost of better quality levels is not reflected in a significantly higher product price. It is therefore extremely important to have knowledge of the quality-level–cost relationship and to define target quality standards appropriately. Figure 2.2 shows the typical relationship between quality level and cost. Assuming that quality control procedures are better than those of competitors in the market-place, the manufacturer has the freedom to pitch the target quality level and cost anywhere within the shaded area shown in Figure 2.2, i.e. he can sell a product of better quality at the same price or a product of the same quality at a lower price.

As well as ensuring the quality of a product immediately after delivery to a customer, the objectives of quality control procedures must also include consideration of the reliability of the product over its working life. Good design procedures and adequate quality control procedures during manufacture are essential elements in achieving high reliability in a product. It is also important to make efforts at the design stage of a product to ensure that faults which do eventually occur can be quickly and easily rectified. The

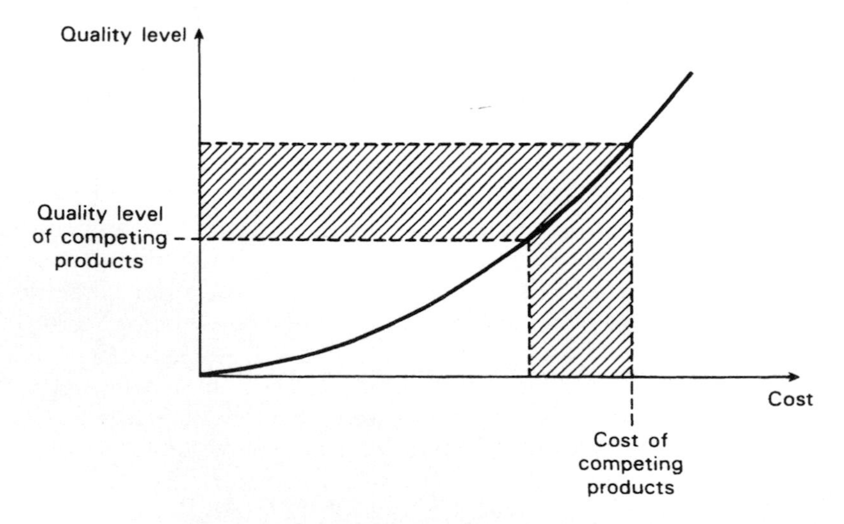

Figure 2.2. Quality level–cost relationship

time taken to repair a product is just as important as its fundamental reliability in many circumstances. The subject of reliability is considered further in Chapter 9.

Legislation governing mandatory quality control standards and product liability is obviously an additional factor which must be taken into account in shaping the quality control system developed. Such statutory requirements vary from country to country and due regard for these differences must be taken by any company involved in, or likely to become involved in, export markets. Whilst the fines imposed as a penalty for failing to comply with such legislation might be seen as only a minor irritation which does not have any significant effect on the financial well-being of a company, a very much more serious penalty is involved in respect of the inevitable loss of reputation that a company would suffer when prosecuted.

Once the quality level targets have been defined, a strategic plan can be formulated for attaining this required quality level. This will become established as a company policy document and must state clearly what the purpose of the quality control system is, i.e. to ensure the consistency of a product in meeting public interests, laws and regulations and to provide total customer satisfaction in so far as cost factors allow. The strategic plan should include mechanisms for assuring product reliability and meeting promised delivery dates as well as controlling initial product quality, as these are also important parameters in maintaining goodwill and customer satisfaction. A reputation for prompt delivery of high quality and reliable products will not only keep the current customer base of a company happy but will also rapidly attract new customers.

One very important point about the establishment of quality control systems and the development of a strategic plan to implement them is that these must never be allowed to become stagnant. Technological change brings about continual improvements in manufacturing systems and it is essential to take full account of these. As manufacturing efficiency increases, the quality level attainable at a given cost also increases. As technological developments make improved quality levels economically viable, a company's quality control system must evolve to improve product quality to this new level. Failure to apply the latest technology and take advantage of the improved product quality levels which such developments generate will rapidly decrease a company's market competitiveness, as it can be assumed that most competitors of a company in the market-place will be implementing such manufacturing system and quality control improvements.

2.4 QUALITY PLAN FOR MANUFACTURING OPERATIONS

Quality control in manufacturing must be implemented by a carefully designed and properly documented strategic plan. This will ensure that all quality-related work activities will be organised in a logical fashion and will contribute efficiently towards meeting the company's stated product quality objectives. By using such a strategic plan, the required quality levels will be established and maintained at minimum cost. The emphasis in the quality plan should be on designing procedures which prevent quality problems occurring rather than on correcting defects after they have occurred. To meet this requirement, it is necessary to design quality and reliability into a product rather than trying to 'bolt it on' afterwards. Prevention is always better than cure!

As far as the technical aspects of quality control are concerned, this means that the requirements to be satisfied must be considered at the design stage of a product. This would not be possible of course if a quality control system were being implemented for an existing product. However, it is sometimes feasible even in such cases to make design changes to a product. This makes achievement and maintenance of a particular quality level easier.

The design specifications of a product are heavily constrained by customer requirements. No degree of conformance to a particular quality level is going to satisfy a customer if the basic design of the product means that it is unable to do everything that the customer wants. If a bus operator orders a fleet of buses seating eighty passengers, he is unlikely to be satisfied if vehicles with a capacity of seventy are supplied, whatever level of reliability and fuel efficiency is claimed for them.

Whilst accepting that product specifications must therefore meet customer demands, various design modifications are often possible which enable quality levels to be met more easily during the manufacturing processes without affecting the product's ability to perform its required functions. It is these possibilities which must be investigated at the design stage. Subsequently, therefore, when the manufacturing process is being planned, it can be assumed that the product involved has been designed in such a way as to allow the required quality levels to be achieved most easily during its manufacture, subject to satisfying the constraint that the product must be able to carry out all the functions required by the customer.

The essential activities in establishing a quality plan are shown schematically in Figure 2.3. The first stage is to identify what the quality objectives are. Specifying the target quality level for a product requires customer demands to be carefully balanced against quality improvement costs, whilst paying due regard to what a company's competitors in the market-place are doing. A fuller assessment of the costs involved in quality control is given in Section 2.7 . Account must also be taken of product liability laws, which require that 'all reasonable steps be taken to ensure the quality of a product'.

Having defined the quality level required, the next step involves carefully analysing the manufacturing process to examine how this target quality level can be achieved and maintained at minimum cost. To do this properly, the whole manufacturing process must be broken down into separate elements. The potential contribution of each element towards achieving the required product quality level must be considered and a quality control plan for each element established accordingly.

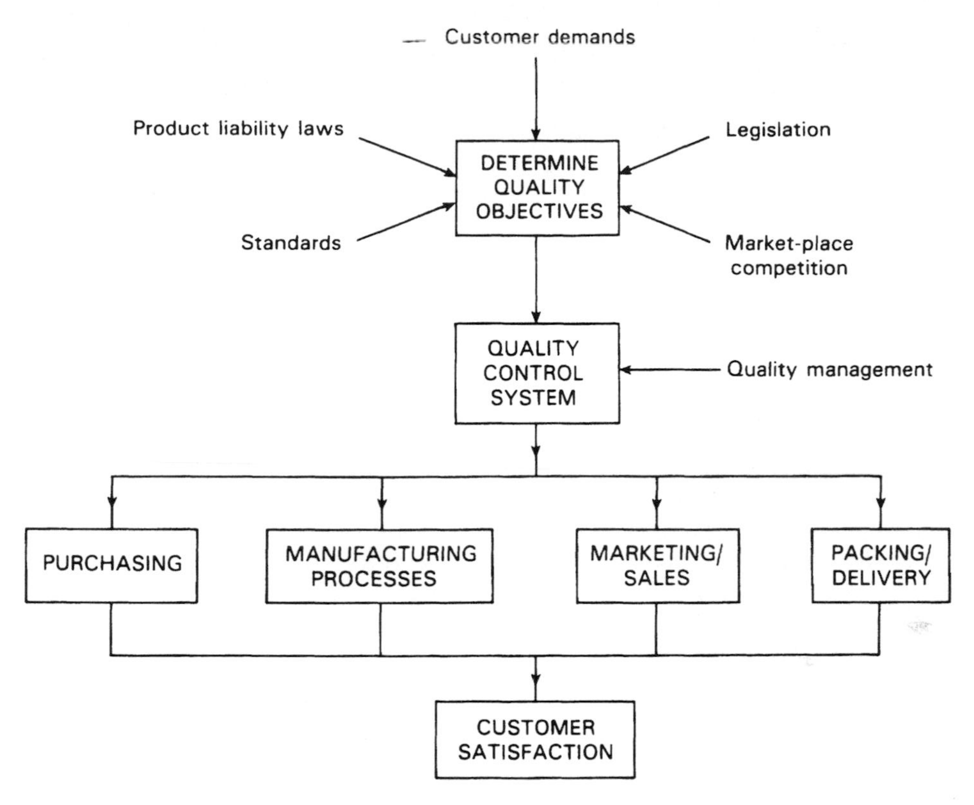

Figure 2.3. Key activities in establishing a quality plan

In this context, the 'manufacturing process' must be regarded as encompassing all the activities of a manufacturing company, including such functions as purchasing, marketing, personnel, finance, legal and secretarial, as well as the technical aspects of production. Each of these elements in a company's organisational structure must have clearly defined quality objectives and strategies which contribute towards a cohesive overall policy of quality control and customer satisfaction. Only when such a corporate strategy of quality control is instigated will the full potential benefits of quality management be realized. This is particularly important at any contact points between the company and customers. The impression given by individuals can do much good, or much harm, to the company image. If secretarial staff do not deal promptly, efficiently and courteously with customers for example, especially with respect to telephone calls, then orders can easily be lost and the benefits of good quality control in technical aspects diminished. The technical aspects of quality control cannot be considered in isolation therefore, but must be part of a cohesive overall management plan. In other words, quality management should not be considered as being fundamentally different from any other area of management, except in so far as the details of its practical implementation differ. The standard management procedures of planning, organizing, directing and controlling should be applied equally in quality management as in any other management function. This allows quality to be managed in an effective way, whereby planning provides a basis for organizing which in turn enables the quality control system established to be directed and controlled in an efficient manner.

Assurance about the quality of components and raw materials bought in from suppliers lower down the production chain is an essential extension of any quality control system operating in a factory. It is no good instituting careful quality control systems to cover the manufacturing operations taking place within the factory if the raw components coming in at the start of the process from another supplier might be defective.

This requirement is most easily met if the raw component supplier also operates a certified (e.g. to ISO 9001) quality assurance system. Indeed, in industries such as the aerospace one, it is mandatory that all components be traceable to a chain of suppliers further down the production line who are all certified as operating an approved quality assurance system. In the absence of such quality certification of suppliers, documented sampling procedures must be established to check the quality of all incoming materials.

The identification of all the separate activities in a company that can affect the attainment of a quality goal is an essential part of the quality control philosophy. However, when the quality control procedures for these separate elements are designed, care must be taken to take full account of their interaction with the rest of the manufacturing system rather than considering them in isolation. Efforts to reduce the cost of quality control in particular parts of a manufacturing system must not simply transfer the cost elsewhere. Failure to take account of this coupling between elements of a quality control system could, for example, result in efforts to reduce inspection costs causing a much greater increase in failure costs. Savings must therefore be real, not imaginary!

In designing a quality control system with a dual purpose of improving product quality levels and minimizing production costs, the size of elemental costs is very relevant. It is much more advantageous to pursue a small reduction in a large cost than a large reduction in a small cost. With this reasoning, it is clear that some costs might be regarded as insignificant and insufficient to justify spending money in trying to reduce them. What that threshold cost level ought to be is of course a subjective decision. However, a clear guide line emerges that efforts to make large cost savings should be instituted before the cost-effectiveness of smaller potential savings is investigated.

Finally, mention must be made of the need for the quality plan to include specification of the planned programme of measurements designed to ensure that the required quality is being met and maintained. Such control of product quality at each stage of the manufacturing process is an essential part of the strategic quality control plan. The details of all inspection procedures specified, their prescribed frequency and the measurement techniques required must also be fully documented.

2.5 QUALITY PLAN FOR SERVICE FUNCTIONS

Most of the general principles described above for achieving quality in manufacturing operations apply equally to service functions. However, additional attention needs to be addressed towards certain special characteristics that service functions have. Prominent among these is the fact that the interaction between the service supplier and customer is usually much greater than it is where someone is just supplying manufactured goods to a customer. Thus, when attending a customer's premises, smartness, politeness, courtesy and good timekeeping are very important. It is also essential that the person providing the service has adequate training, skill and competence for the task and performs it efficiently. Finally, communication with the customer must not be neglected: the service provider should listen to the customer, find out exactly what he/she wants and provide feedback as the service operation proceeds.

2.6 QUALITY PLAN FOR SOFTWARE PRODUCTION

Most of the general principles for planning a quality system outlined in Section 2.4 are also applicable to software development. However, the process of software production is different in some significant respects from the ordinary design and manufacturing operations for a hardware product, and these differences must be given special attention [1]. Perhaps the greatest difficulty with software is the issue of verification. Proving that software does not contain faults is much more difficult (some would say impossible) than it is for an ordinary manufactured product, and so the design and development programme has to be that much more rigorous.

The starting point for software development must be an agreement between the supplier and customer specifying an unambiguous set of functional requirements in terms of performance, reliability, security and safety for the software. Agreed mechanisms must also be documented for the verification and validation of the software to show that it meets the stated requirements. Getting things right from the start during this specification phase is crucially important, because any attempt to change requirements part way through the software production process is a recipe for disaster which is highly likely to lead to the creation of faults in the software.

Following the specification phase, the detailed design and development of the software can commence. Special effort must be put into this phase because of the great difficulty in proving the absence of faults once the software has been written. This means that rigorous software engineering principles must be applied which aim to ensure that the software product performs to specification with high standards of reliability and maintainability. A number of special software design and production methodologies have evolved which aim to achieve these ideals and prevent any possibility of faults being built into software when it is written. Many specialist software engineering texts are available which describe such methodologies, e.g. [2]. It is not appropriate in this book to go into these in detail or to try to compare them and recommend one as against another. The truth of the matter is that most of them work quite well and choice between them is therefore largely a matter of user preference. The only important criterion is that one of these methodologies is used and applied rigorously.

Following the development phase comes the verification and validation phase. Part of this consists of making a binary assessment of whether the software does or does not satisfy each specified requirement, and this is relatively easy to carry out. However, the issue of software quality measurement, quantifying the probability that the software doesn't contain faults that will emerge when unanticipated combinations of input parameters etc. occur, is a much larger and more difficult problem. In view of this, a separate chapter (Chapter 10) has been written to cover this issue of software quality metrics.

2.7 QUALITY CONTROL SYSTEM COSTING

An essential factor in designing a quality control system is the ability to make an accurate assessment of the costs involved in developing and operating the system. Whilst good quality is necessary to maintain customer satisfaction and goodwill, product price is also very important. It is no use making a very high quality product if its price is too high. The target quality level therefore has to be carefully chosen with regard to the cost of achieving that level. For this to be possible, the costs of designing and implementing all aspects of the quality control system must be accurately known. These costs can be divided into the three areas associated with (a) design and modifications aimed at fault prevention, (b) quality inspection and appraisal costs, and (c) the cost of rejected products. This classification is convenient because the three areas correspond in financial terms to investment money (design), running costs (appraisal) and negative profit (rejects). Figure 2.4 summarizes these costs.

Before going on to discuss quality control system costs in detail, two general points need to be established. Firstly, it must be recognized that quality control system costs and the savings generated are not necessarily static. Changes can occur in both over a period of time, and therefore in much of the following discussion about estimating quality control costs and savings should be regarded as being procedures which are repeated at regular intervals of time rather than being 'once only' exercises. Such regularly updated cost information is also a necessary foundation on which continuing improvements in quality control systems can be built.

Secondly, the accounting system used to assess quality control costs must be appropriate. Some accounting practices can grossly distort the sizes of both absolute and relative quality control costs by including full overheads in direct labour charges rather than just marginal costs. Such unjustified overheads often make a quality control system appear uneconomic, result-

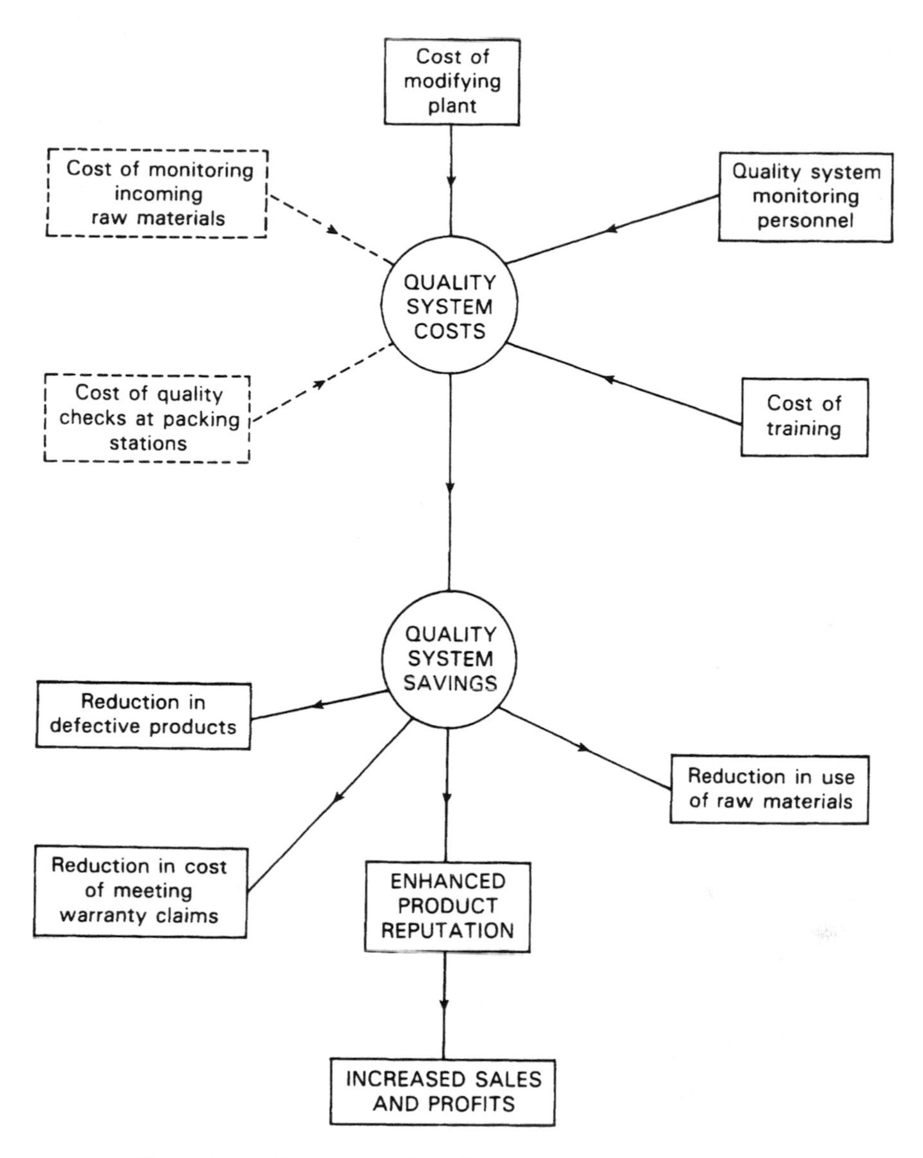

Figure 2.4. Summary of quality system costs and savings

ing in non-implementation and a net loss to the company of the financial benefits that the quality control system would bring if costed sensibly.

As well as using sensible accounting procedures, it is also important that the cost estimates used in the quantification of quality system costs are as accurate as possible. Such credibility of the cost estimates is extremely important because, without credibility, the whole argument for implementing a quality control system may be seriously weakened. Extreme care in cost estimation is vital because the exposure of a single weakness in any elemental part of the estimate can undermine confidence in the whole. For this reason costs should be produced by, or at least endorsed by, a company's accounts department, as this puts the necessary 'stamp of approval' on the estimates. In view of their crucial importance, independent corroboration of cost estimates, by perhaps consulting engineers, should also be sought.

Unfortunately, the quantification of many quality cost factors is difficult because they are generated for other reasons besides the pursuance of good quality. For example, purchasing personnel already have a natural function in a company but may expend additional time and effort to ensure that components bought in are of good quality. Similarly, personnel employed to pack components may be asked to check them for defects as they are packed. Quantifying this extra time requires apportionment of an employee's time between quality-related and other functions. This is particularly prone to inaccuracy, especially when the employee concerned is made responsible for defining this apportionment via time sheets, as the data is then reliant upon his personal, subjective judgement about what work is quality-related and what isn't. The plant design process itself can also generate 'grey-area' costs where a new plant is being designed rather than an existing one being modified. The incremental cost of such design modifications to improve quality cannot be easily quantified.

Some well-defined quality costs are the costs of training relevant personnel and the costs of employing additional persons to fill posts in quality system monitoring and management. The technical cost of redesigning and modifying existing manufacturing plant to optimize quality in a product can also be quantified via an in-depth study by experienced engineering personnel of the technical points involved.

Quality control costs can therefore be divided into those which can be accurately estimated and those which cannot. However, with the caveat that bad estimates are counter-productive, attempts should always be made to estimate all costs if at all possible, as failure to present a full picture is an irritation to those mandated to take a decision about the viability of a proposed quality control system. Even if certain costs cannot be estimated at all, it is still useful to mention them because it keeps them in view and encourages future attempts to measure them.

Assessment of the financial benefits accruing from implementing quality control procedures is even more difficult to quantify. The three major elements contributing to financial savings are the production costs of substandard products, the value of raw materials wasted in such substandard products and the increased sales generated by a reputation for good quality. It would appear that the first two of these elements can be calculated fairly easily. However, if substandard components have reached the customer, then accounting and purchasing functions in a company become involved in processing the rejects. As this cost is less easy to quantify, the savings arising from reducing reject levels are equally ill defined. Sloppy procedures in such areas as the generation and processing of rejection notes at the inspection stage can also seriously undermine the accuracy and credibility of savings estimates.

The third cost-saving element is also difficult to quantify. A simplistic way of measuring increased sales would be to monitor the sales figures over a period of time before and after the introduction of a quality control system. However, this fails to take account of other significant factors, such as whether the total market for a product expands or contracts during the periods in which sales are considered. Quality control systems can only serve to maintain or increase the share that a particular manufacturer has in a market. If the total market contracts, some decrease in sales may occur even with a quality control system in operation. The important point though is that the decrease would be even greater without the quality control system.

Any product warranties in existence impose a further complication in the quality control system cost–savings equation because of the large time-gap between implementing the quality control system and the end of the warranty period when the reduction in warranty claims can be accurately calculated. A similarly large time-delay exists between quality control implementation and its effect on a company's reputation and hence its translation into increased sales.

Further information on quality costing can be found in reference [3].

2.8 QUALITY CONTROL SYSTEM IMPLEMENTATION

Quality policy must be more than a statement of intent: it must be manifest in a specific course of action designed to achieve the quality objective defined. The quality control system designed must be implemented and maintained in an efficient and effective manner. Key components in doing this (see Figure 2.5) are hardware design and implementation, training,

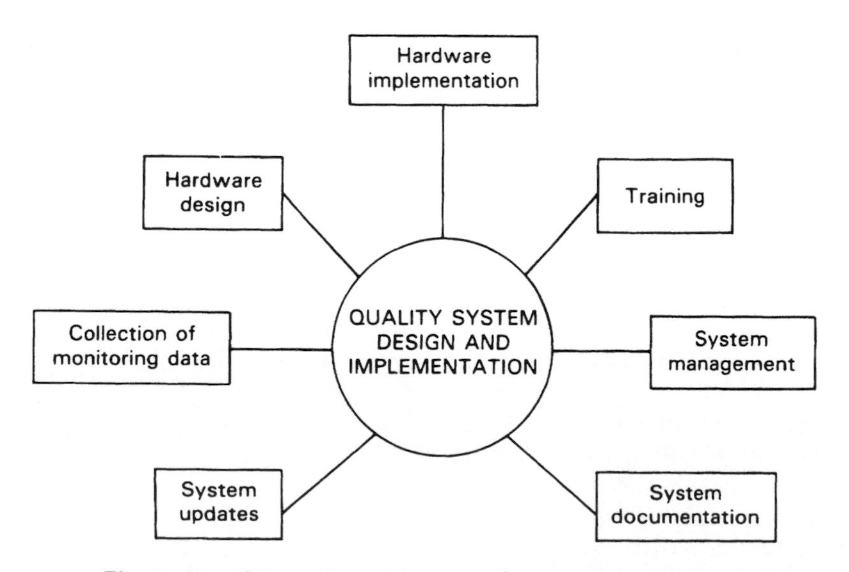

Figure 2.5. Key components in quality system implementation

system management, collection of data to monitor performance, system updates and documentation.

Hardware Design

The detailed design and installation of hardware intended to fulfil some quality control function is obviously the responsibility of appropriate engineering personnel. In executing this task, proper management is essential to ensure that the system implemented operates in an efficient manner and performs the functions required of it.

Training

Training relevant personnel is particularly important in establishing and maintaining a quality control system. 'Relevant personnel' should be taken to mean the whole workforce in a company, even if training for some of these consists only of a short one- or two-day course which acquaints them with the quality control procedures operated and their importance.

Workers' attitude to training is almost as important as the training itself. Making people go on training courses is relatively easy, but ensuring that they assimilate the necessary knowledge is considerably more difficult. If the training courses are not managed properly, there is a strong likelihood that they will be treated as a welcome break from the normal working environment but their purpose will not be taken seriously. To prevent this happening, workers must see the need for courses and positively want to go on them. One way to achieve this is to set quality targets which the workforce know can only be achieved once they have obtained necessary knowledge from a training course.

To be effective, training needs must be formulated and driven at departmental level within a company and targeted towards meeting the department's quality objectives. Figure 2.6 shows some necessary procedures in an effective departmental training plan. Training plans must be formulated individually for each member of a department. Each plan must be fully discussed with the person it is designed for and the person's full agreement obtained about the details of the training plan. It is also important that the effectiveness of the training is reviewed after it has taken place. Finally, the whole area of training needs must be periodically reviewed, because, as production methods and quality improve, the competence of the workforce may also need to improve, creating a need for further training.

It is also often beneficial to extend training to customers. By offering courses to users of equipment that companies produce, the incidence of misuse of equipment by the customer can be greatly reduced and consequential failure avoided. Although blame for the failure of equipment due to the customer misusing it cannot rightly be directed towards the supplier,

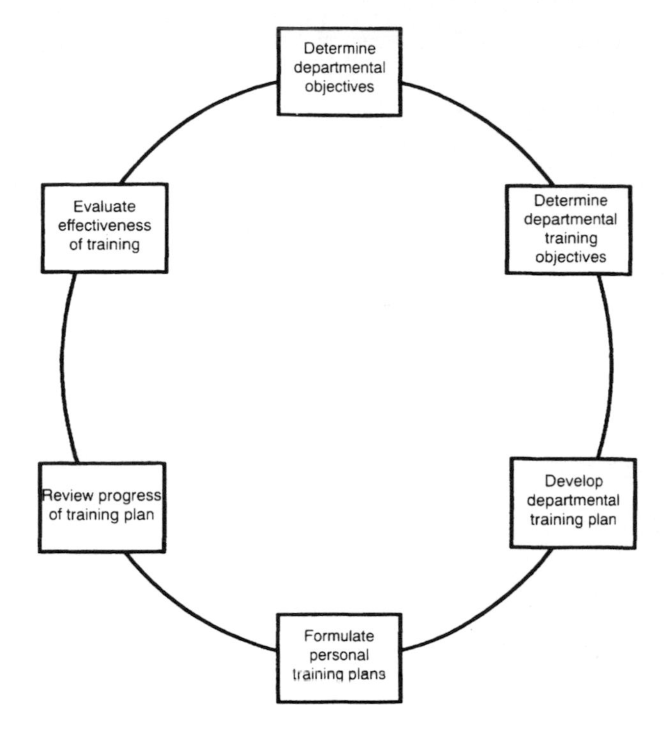

Figure 2.6. Necessary procedures for an effective training plan

the supplier is often nevertheless blamed. In such cases, protracted arguments can develop between the customer and supplier about who is at fault, and the reputation of the supplier is dented somewhat, however unfair this might be.

System Management

Management of quality control system implementation and operation has to fulfil several functions. Perhaps the most obvious is its role in ensuring that all hardware and personnel involved in quality control activities operate efficiently and within the cost estimates established at the design stage of the system. Ensuring that all staff involved have been properly trained is a necessary part of this. A second role is to make sure that the quality function interacts properly with all other general company management functions. To perform these two roles effectively requires that all the tasks involved are identified and carried out by assigning responsibility, delegating authority and creating accountability for each separate task. As many of the factors contributing to a quality control system such as accounting and purchasing functions lie outside his/her direct area of control, the quality control manager is necessarily a co-ordinator rather than a manager in many respects. This emphasizes the spirit which must be generated within a company, of quality being everyone's concern, not just that of a quality control manager, and by everyone is meant all personnel, not just departmental managers.

Whilst the achievement of quality is the concern of everyone in a factory, it is extremely important that the ultimate responsibility for setting up and maintaining a quality system falls on just one person. It is also essential that this person has sufficient authority to do the job effectively, and this is normally achieved by giving the person concerned a place on the management board of a company, usually with the title Quality Assurance Director. This level of authority is necessary so that the person concerned can control the quality system fully, and shut down production if necessary if the quality control system fails, however much the production manager may protest.

Performance Monitoring

Organizing the collection of data to monitor system performance is another duty of the quality control system manager. Necessarily included within this is a responsibility for assessing the importance of any deviation of the actual quality measured away from the target quality level. Full knowledge of the importance of such deviations from the customer's viewpoint is necessary in order to arrive at the proper decision about whether to reject the product in question. A proper framework setting out the rules for making this decision must be established, whether the inspection function is performed manually or via an automatic inspection system. This usually involves identifying the various quality characteristics associated with a product, quantifying the various deviations from standard characteristics and giving appropriate weights to indicate the relative importance of each deviation.

Suitable programmes of sampling, inspection and testing need to be established for measuring process parameters and the quality level achieved at all stages in the manufacturing process from the incoming raw materials to the outgoing finished components. This procedure can become difficult where process parameters have allowable variations within a given range due to random effects. In this case, statistical process control (see Chapter 11) can be very useful for distinguishing between parameter variations that are due to random effects and those which indicate the onset of some real production problem which will affect product quality if left uncorrected.

The cost of each elemental part of the quality control system must also be monitored continually, and cost-data collection must be extended to beyond the manufacturing cycle to assess the costs involved in products subject to warranty claims. All of this information must be presented in a suitable form for analysis and appraisal, such as by bar charts. As far as possible, this information should be combined with production information collected by the production manager to provide an overall management information system. This provides the company management with a basis for evaluating the effectiveness of the quality control system and determining the level of compliance with its procedural requirements.

Final Product Inspection

Before the widespread adoption of quality control practices during manufacturing operations, inspection of products before they were passed to customers was the standard means of ensuring that products delivered were of the required quality. However, the advent of quality assurance systems which control quality at all stages in a production system has now greatly reduced the need for final product inspection and testing, because, if the quality system is operating as it should, there ought to be no problems with the final product. Unfortunately, it is not an ideal world, and quality systems are not quite as foolproof as this. It is a fact of life that quality systems can and do fail from time to time, often when unforeseen circumstances arise. Thus, there remains a need for some product inspection and testing to provide the final feedback loop in a quality control and assurance system. It remains necessary because the costs in terms of loss of customer goodwill if quality control procedures break down and defective products are shipped out is so enormous that no manufacturer can risk this situation arising.

Whilst they therefore remain necessary, product inspection and test procedures are expensive to operate, and cost factors usually dictate that such product-testing be based on a statistically representative sample of the total number of products in a production run rather than on testing every item. Sampling allows the overall quality level of a complete production run to be predicted by testing just a few samples from the output, although, because it is a statistical procedure, there is a finite probability that defective products will not be detected. Fortunately, provided product sampling algorithms are well designed, the incidence of failure is very low. Appropriate sampling procedures are discussed in Chapter 12.

Measurement and Calibration

The measurement processes involved in quality control inspired data collection are of fundamental importance, and discussion of them is the main purpose of this book. All measurements are derived from instruments of one form or another, and a full knowledge of the possible error level in the data collected is essential. The first requirement for achieving this is that all instruments involved in quality control procedures be calibrated at prescribed intervals of time, as described in Chapter 3. Knowledge of instrument characteristics (Chapter 4), sources of error (Chapter 5), signal processing (Chapter 6) and reliability studies (Chapter 9) are further requirements. The quality control manager must ensure that data is analysed and errors assessed in a correct manner, appropriate to the measurements involved.

System Updates

One factor complicating the design and operation of quality control systems is that the quality levels demanded change continually under the influence of technological developments and market forces. This requires the quality control system in operation to be updated at various points in time to meet the new demands. Modifications to the system also become necessary if the monitoring exercise on costs and performance shows a deviation from the target cost and performance goals.

2.9 QUALITY SYSTEM DOCUMENTATION

Full documentation of all aspects of a quality control system is essential. This should take the form of a quality manual in which every procedure in the quality control system is carefully set out and the whole philosophy and purpose of the system is described. The quality level target and its justification in terms of balancing the quality costs–savings equation is a very necessary part of this. Whilst there is no particular format defined for this manual, the information must be presented in a clear, systematic and orderly way and it must include all the major elements shown in Figure 2.7. In the case of a small company, all quality system documentation would typically be bound within one manual. For larger companies, however, it is

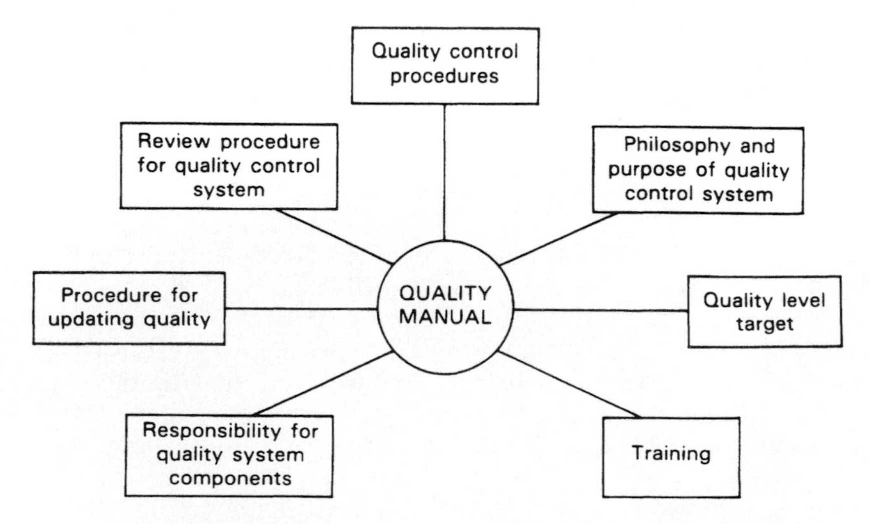

Figure 2.7. Essential components in quality system documentation

often more appropriate to maintain separate corporate and divisional quality manuals.

Each identified element in the quality control system should be documented so that the details of hardware designs (with full drawings), operational instructions, purpose, running cost and interaction with other elements are all expressed clearly. All documentation should be dated, and this is especially important in the case of engineering drawings of components which are made intermittently in batches. Records must be kept about which version of a drawing was used for any particular batch and what the results of quality control tests on it were. In addition, along with detailed information about the steps taken in the design of a product, all changes made to improve quality must also be documented.

The documentation must include details of any elements of the system that are assigned to subcontractors. The list of approved contractors must be written down and the conditions regarding the quality of materials supplied, as specified on purchase orders, must also be documented. The exact details of the agreement between purchaser and supplier must be clear so that there is no ambiguity about what quality level is expected.

Review procedures for the quality control system should also be defined in the document and the required frequency for measuring its performance given. Such inspection, testing, auditing, costing and calibration of quality control procedures must be included within the manual as quality records.

Documentation should identify clearly what the organizational structure and mode of operation of the quality control system is, including a description of the associated training procedures. Assignment of responsibility for each elemental part of the system is particularly important and care should be taken to see that activities on the boundaries between functions are adequately managed. The system must be watertight without any loopholes!

Standard procedures for modifying or making additions to the quality manual must also be established and documented within the manual. These must provide for disposing of any parts of the documentation which become redundant or outdated. To ensure that this procedure functions without mistakes occurring, it is essential that all pages within the quality manual are marked with a date and revision number.

2.10 QUALITY ASSURANCE

The importance of quality assurance in maintaining and expanding the customer base for a product cannot be overemphasized. The whole effort involved in quality control procedures is expended in striving to meet this customer satisfaction goal. The benefits coming from the achievement of good quality products, however, are greatly diminished if customers are not made aware of this achievement.

Communication to the customer about the quality control systems associated with a product is therefore an essential part of quality assurance. This is primarily the responsibility of the marketing department in a company but other people have important roles to play in this also, such as in the provision of the documentation which demonstrates that the quality control implemented is efficient and maintained satisfactorily. This documentation should show clearly that the quality system conforms to ISO 9000 and that associated measurement and calibration procedures follow the guidelines specified. Customers should be able at all times to inspect this documentation and, if necessary, visit the manufacturing plant to see the procedures described in operation.

2.11 QUALITY SYSTEM AUDITS

If a quality control system is to be operated with the intention of assuring customers about the quality of a product, then those customers must have full confidence in the effectiveness of the quality procedures in operation. Mention has already been made about the importance of allowing customers to have full access to quality system documentation and also of allowing them to make plant visits to see the systems in operation. However, whilst such access to documentation and the manufacturing plant is necessary, it is not sufficient. In addition, audits of the quality control system being operated must be carried out on a regular basis and recorded in the quality manual to ensure that the system continues to be effective and that no departures have been made from the procedures laid down. Most of these reviews can be carried out by appropriately qualified internal company personnel. However, periodically, an audit must be carried out by an independent third party if the quality system is to have full credibility. Within the UK, the main provider of such an auditing service is the British Standards Institute: similar bodies exist elsewhere around the world.

External Audits

The requirements for external audits will be presented first because these also give essential guidance to what a company should be doing when conducting an internal audit. Audits by a third party must necessarily be thorough, although if the same audit procedures have been followed previously by internal reviews (as they should be) then there will be no difficulty in 'passing' the audit and gaining recertification of the quality control system operated. Such certification of quality control procedures is only required for those parts of a factory that a company wishes to claim accredited quality assurance for. The nature of some products may be such that it is not cost-effective to operate a rigorous quality control system and, in such cases, it is perfectly permissible for a company to register and claim accreditation for quality control procedures that just apply to part of its range of products.

A major item of concern to external auditors will be satisfying themselves that there is support at the highest level of the company concerned for the quality assurance system operated. Therefore, they will require to have brief discussions with the managing director of the company to confirm this before proceeding with the audit. Thereafter, the starting point of a quality audit is generally to look at the measures taken to ensure the quality of incoming raw materials and part-finished components. Either certification of the quality control procedures operated by the supplier or else an effective sampling method for incoming material will be required. Following this, evidence will be required of the successful operation of the quality control procedures instituted to cover all aspects of the company's activities that have an effect upon product quality. Checks will be made to see that there have been no departures from the documented quality control system (other than agreed ones notified to the accrediting body) and records of the internal audits undertaken to monitor the effectiveness of quality control procedures will be inspected. Where work is being done to a particular published standard (e.g. British Standard Institution Standard), evidence will also be required that the most up-to-date version of that standard is being used. Finally, proof must be shown of the existence of, firstly, an efficient and documented measurement system which monitors quality-related parameters and, secondly, a calibration system traceable to reference standards which checks the accuracy of all measuring instruments used.

Quality system auditors will be especially keen to ensure that all personnel whose actions can have an effect on quality are working in the correct manner. They will test this out by selecting people at random to ask them what job they do, what training they have had, where the instructions are that tell them how to do the job and what specific steps they take to maintain product quality. Such questions may be particularly targeted at personnel outside the production departments in areas involving activities such as part storage, packing and delivery. Such personnel need to know things like how many components can be stacked on top of each other without causing damage or, in the case of say delivery of chemicals by tanker, what cleaning procedures are required between loads of different materials.

The auditors will have a check-list on which they mark down the conformance or otherwise of the quality system to each of the clauses in the standard (ISO 9001, 9002 or 9003 as appropriate). Nonconformance will be graded in three categories: (a) those which are a major diversion from the standard and are likely to lead to quality problems; (b) those which are less serious but might lead to failure of the quality system in some circumstances; (c) minor faults which need correcting. Any faults in category (a) will certainly mean that the auditors do not certify the quality system. Whether faults in category (b) allow certification depends on their number and severity, with the most favourable outcome being that provisional registration is given but re-auditing will be required sooner than usual. Category (c) faults will usually allow certification, with a recommendation that the faults be corrected and an expectation that this will be done well before the next audit.

The outcome of the audit will therefore either be the issue of an accreditation certificate for the quality assurance operated, or, alternatively, a statement of what corrective action needs to be taken before the system can be so certified. Once an accreditation certificate is given, arrangements will be made to re-audit the system, perhaps on an annual basis. In between such full audits, unannounced surveillance visits might be made to monitor the continued effectiveness of the quality control procedures agreed.

Internal Audits

Internal audits are necessary for two reasons. Firstly, as mentioned above, the external auditors will expect to see documentary evidence that internal audits of the quality system have been carried out at regular intervals. Secondly, they are a necessary step to ensuring that the quality system 'passes' when subjected to the external audit.

The internal auditing procedure consists of systematically going through all the questions that the external auditors will ask. This seems simple enough, but a problem can arise in that the person responsible for quality system operation in a company may be too close to it and may fail to spot deficiencies. It is much better if two or more companies can get together and swap their quality control managers for the purpose of carrying out internal quality audits. Such audits by an outside but 'friendly' auditor can often highlight problems that the company's own quality control manager has failed to see through overfamiliarity with the system.

2.12 ISO 9000 REGISTRATION

Specialized guidance can be obtained elsewhere [4] but it is appropriate here to give a brief summary of the main steps necessary to achieve ISO 9000 registration of a quality system. The first step must always be to design and document the quality system to be implemented, and the high desirability of this in obtaining the advice of consultants who have specialist knowledge of the industry in question has been emphasized earlier.

Once the quality system has been implemented, its operation must be carefully monitored by a series of internal audits until everything seems to be working satisfactorily. Once this stage is reached, the company can apply to their National Standards Organization† for an external audit, which is a necessary prerequisite of registration. The accreditation body will appoint a registrar to audit the system, and this be done at a mutually agreed time. The audit will examine whether the design, documentation and implementation of the quality system are satisfactory and whether the target quality

† In the USA, the appropriate body is the American Registrar Accreditation Board, and in the UK it is the British Standards Institute.

level is being met. Once any deficiencies that the registrar has identified have been put right, a certificate of registration will be issued and the company will be placed on a public register of companies who are operating to the appropriate ISO 9000 standard (either ISO 9001, ISO 9002 or ISO 9003).

REFERENCES

1. ISO 9000-3, International Organization for Standards, Geneva (1991).
2. Ince, D., Sharp, H. and Woodman, M., *Introduction to software project management and quality assurance*, McGraw-Hill, New York (1993).
3. BS 6143: *Guide to the economics of quality*, British Standards Institution, London (1990).
4. Lamprecht, J. L., *ISO 9000: Preparing for registration*, Marcel Dekker, New York (1992).

CHAPTER 3

Calibration Procedures

Measurements related to product quality are an essential part of quality control systems. Such measurements may be directly related to product quality where they take the form of dimensional measurements etc., or they may indirectly affect product quality where they take the form of processing temperatures etc. In either case, accuracy in such measurements is mandatory, and to achieve this accuracy, calibration of the instruments used to obtain the measurements must be carried out at a predetermined frequency. Such periodic recalibration is necessary because the characteristics of any measuring instrument change over a period of time and affect the relationship between the input and output. Changes in instrument characteristics are brought about by such factors as mechanical wear, and the effects of dirt, dust, fumes and chemicals in the operating environment. To a great extent, the magnitude of the drift in characteristics depends on the amount of use an instrument receives and hence on the amount of wear and the length of time that it is subjected to the operating environment. However, some drift also occurs even in storage, as a result of ageing effects in components within the instrument.

Further consideration is given to the reasons for this drift in characteristics in Chapter 5. It is sufficient here to accept that such drift does occur and that the rate at which characteristics change with time varies according to the type of instrument used, the frequency of use and the prevailing environmental conditions. A full knowledge of the mechanisms involved in these two factors is therefore necessary before the required frequency of instrument recalibrations can be prescribed.

When calibration is carried out for quality assurance purposes, it is important that all elements in the measurement chain right up to the final element which produces the quantity used for quality measurement are calibrated. Thus the signal recorder as well as all the components within the measuring instrument must be included in calibration procedures.

3.1 PROCESS INSTRUMENT CALIBRATION

Calibration consists of comparing the output of the process instrument being calibrated against the output of a standard instrument of known accuracy, when the same input (measured quantity) is applied to both instruments. During this calibration process, the instrument is tested over its whole range by repeating the comparison procedure for a range of inputs.

The instrument used as a standard for this procedure must be one that is kept solely for calibration duties. It must never be used for other purposes. Most particularly, it must not be regarded as a spare instrument which can be used for process measurements if the instrument normally used for that purpose breaks down. Proper provision for process instrument failures must be made by keeping a spare set of process instruments. Standard calibration instruments must be totally separate.

To ensure that these conditions are met, the calibration function must be managed and executed in a professional manner. This will normally mean setting aside a particular place within the Instrumentation Department of a Company where all calibration operations take place and where all instruments used for calibration are kept. As far as possible this should take the form of a separate room, rather than a sectioned-off area in a room used for other purposes as well. This will enable better environmental control to be applied in the calibration area and will also offer better protection against unauthorized handling or use of the calibration instruments.

All instrument characteristics are affected to some extent by environmental conditions, and any parameters given in data sheets only apply for specified conditions. Therefore, as far as practicable, these same environmental conditions should be reproduced during calibration procedures. However, specification of the level of environmental control required should be considered carefully with due regard to what level of accuracy is needed in the calibration procedure, as overspecification will lead to unnecessary expense. In practice, full air-conditioning is not normally required for calibration at this level, as it is very expensive, but sensible precautions should be taken to guard the area from extremes of heat or cold, and good standards of cleanliness should also be maintained. Useful guidance on the operation of standards facilities can be found in reference [1].

For various reasons, it is not always possible to perform calibration operations in a controlled environment. For example, it may not be convenient or possible to remove instruments from process plant, and in such cases it is standard practice to calibrate them *in situ*. In these circumstances, appropriate corrections must be made for the deviation in the calibration environmental conditions from those specified. This practice does not obviate the need to protect calibration instruments and maintain them under constant conditions in a calibration laboratory at all times other than when they are involved in calibration duties on plant.

Calibration instruments usually have a greater inherent accuracy than the process instruments that they are used to calibrate. Where instruments are only used for calibration purposes, this greater accuracy can often be achieved by specifying a type of instrument that would be unsuitable for normal process measurements. Ruggedness for instance is not a requirement, and freedom from this constraint opens up a much wider range of possible instruments. In practice, high accuracy, null-type instruments are very commonly used for calibration duties, because their requirement for a human operator is not a problem in these circumstances.

As far as management of calibration procedures is concerned, it is important that the performance of all calibration operations is assigned as the clear responsibility of just one person. That person should have total control over the calibration function, and be able to limit access to the calibration laboratory to designated approved personnel only. Only by giving this appointed person total control over the calibration function can the function be expected to operate efficiently and effectively. Lack of such rigid management will inevitably lead to unintentional neglect of the calibration system, and result in the use of equipment in an out-of-date state of calibration. Traceability to reference standards will thus be lost. Professional management is essential, so that the customer can be assured that an efficient calibration system is in operation and that the accuracy of measurements is guaranteed.

One of the clauses in ISO 9000 requires that all persons using calibration equipment be adequately trained. The manager in charge of the calibration function is clearly responsible for ensuring that this condition is met. Training must be adequate and targeted at the particular needs of the calibration systems involved. People must understand what they need to know and especially why they must have this information. Successful completion of training courses should be marked by the award of qualification certificates. These attest to the proficiency of personnel involved in calibration duties and are a convenient way of demonstrating that the ISO 9000 training requirement has been satisfied.

The calibration facilities provided within the instrumentation department of a company provide the first link in the calibration chain. Instruments used for calibration at this level are known as working standards. A fundamental responsibility in the supervision of this calibration function is to establish the frequency at which the various shop-floor instruments should be calibrated and ensure that calibration is carried out at the appropriate times.

Determination of the calibration frequency required is dependent upon several factors which require specialist knowledge. If an instrument is required to measure some quantity to an accuracy of ∓ 2 percent, then a certain amount of performance degradation can be allowed if its accuracy immediately after recalibration is ∓ 1 percent. What is important is that the pattern of performance degradation be quantified, so that the instrument can be recalibrated before its accuracy has reduced to the limit defined by the application.

The quantities which cause the deterioration in the performance of instruments over a period of time are mechanical wear, dust, dirt, ambient temperature and frequency of use. Susceptibility to these factors varies according to the type of instrument involved. The effect of these quantities on the accuracy and other characteristics of an instrument can only be quantified by someone who has an in-depth knowledge of the mechanical construction and other features of instrument. Some form of practical experimentation is normally required to determine the calibration frequency necessary in the typical operating conditions the instrument works in. Further discussion on the means of quantifying the rate of change of instrument characteristics is given in Chapter 5.

A proper course of action must be defined which describes the procedures to be followed when an instrument is found to be out of calibration, i.e. when its output is different to that of the calibration instrument when the same input is applied. The action required depends very much upon the nature of the discrepancy and the type of instrument involved. In many cases, deviations in the form of a simple output bias can be corrected by a small adjustment to the instrument (following which the adjustment screws must be sealed to prevent tampering). In other cases, the output scale of the instrument may have to be redrawn, or scaling factors altered where the instrument output is part of some automatic control or inspection system. In extreme cases, where the calibration procedure shows up signs of instrument damage, it may be necessary to send the instrument for repair or even scrap it.

Whatever system and frequency of calibration are established, it is important to review them from time to time to ensure that the system remains effective and efficient. It may happen that a cheaper (but equally effective) method of calibration becomes available with the passage of time, and such an alternative system must clearly be adopted in the interests of cost-efficiency. However, the main item under scrutiny in this review is normally whether the calibration frequency is still appropriate. Records of the calibration history of the instrument will be the primary basis on which this review is made. It may happen that an instrument starts to go out of calibration more quickly after a period of time, either because of ageing of the instrument or because of changes in the operating environment. The conditions or mode of use of the instrument may also be subject to change. As the environmental and usage conditions of an instrument may change beneficially as well as adversely, there is the possibility that the recommended calibration interval may increase as well as decrease.

Maintaining proper records is an important part of calibration procedures. A separate record, similar to that shown in Table 3.1, should be kept for every instrument in the factory, whether it is in use or kept as a spare. This record should start by giving a description of the instrument and follow this by stating what the required calibration frequency is. Each occasion when the instrument is calibrated should be recorded, and every such calibration log should show the status of the instrument in terms of the deviation from its required specification and the action taken to correct it. The calibration record is also very useful in providing feedback which shows whether the calibration frequency has been chosen correctly.

Table 3.1. Typical format for instrument record sheets

<table>
<tr><td colspan="2">Type of
Instrument:</td><td colspan="2">Company
serial number:</td></tr>
<tr><td colspan="2">Manufacturer's
part number:</td><td colspan="2">Manufacturer's
serial number:</td></tr>
<tr><td colspan="2">Measurement
limit:</td><td colspan="2">Date
introduced:</td></tr>
<tr><td colspan="4">Location:</td></tr>
<tr><td colspan="4">Instructions
for use:</td></tr>
<tr><td colspan="2">Calibration
frequency:</td><td colspan="2">Signature of
person responsible
for calibration:</td></tr>
<tr><td colspan="4">CALIBRATION RECORD</td></tr>
<tr><td>Calibration date</td><td colspan="2">Calibration results</td><td>Calibrated by</td></tr>
<tr><td></td><td colspan="2"></td><td></td></tr>
</table>

3.2 STANDARDS LABORATORIES

We have established so far that process instruments which are used to make quality-related measurements must be calibrated from time to time against a working standard instrument. As this working standard instrument is one which is kept by the Instrumentation Department of a company for calibration duties, and for no other purpose, then it can be assumed that it will maintain its accuracy over a reasonable period of time because use-related deterioration in accuracy is largely eliminated. However, over the longer term, the characteristics of even such a standard instrument will drift, mainly owing to ageing effects in components within it. Over this longer term, therefore, a programme must be instituted for calibrating this working standard instrument against one of yet higher accuracy at appropriate intervals of time. The instrument used for calibrating working standard instruments is known as a secondary reference standard. This must obviously be a well-engineered instrument which gives high accuracy and is stabilized against drift in its performance with time. This implies that it will be an expensive instrument to buy. It also requires that the environmental conditions in which it is used be carefully controlled in respect of ambient temperature, humidity, etc.

The secondary reference instrument and the controlled environment for it to operate in are clearly expensive to provide, and the establishment of a Company Standards Laboratory to provide such a calibration facility is economically viable only in the case of very large companies, where large numbers of instruments need to be calibrated across several factories. In the case of small- to medium-sized companies, the cost of buying and maintaining such equipment is not justified. Instead, they would normally use the services of one of the specialist companies who have developed a suitable Standards Laboratory for providing calibration at this level.

When the working standard instrument has been calibrated by an authorized standards laboratory, a calibration certificate will be issued [2]. This will contain at least the following information:

- the identification of the equipment calibrated
- the calibration results obtained
- the measurement uncertainty
- any use limitations on the equipment calibrated
- the date of calibration
- the authority under which the certificate is issued.

3.3 VALIDATION OF STANDARDS LABORATORIES

In the United Kingdom, the appropriate National Standards Organization for validating Standards Laboratories is the National Physical Laboratory (in the United States of America, the equivalent body is the National Bureau of Standards). This has established a National Measurement Accreditation Service (NAMAS) which monitors both instrument calibration and mechanical testing laboratories. The formal structure for accrediting instrument calibration in Standards Laboratories is known as the British Calibration Service (BCS), and that for accrediting testing facilities is known as the National Testing Laboratory Accreditation Scheme (NATLAS).

Although each different country has its own structure for the maintenance of standards, each of these different frameworks tends to be equivalent in its effect. To achieve confidence in the goods and services that move across national boundaries, international agreements have established the equivalence of the different accreditation schemes in existence.

A Standards Laboratory has to meet strict conditions [3, 4] before it is approved. These conditions control laboratory management, environment, equipment and documentation. The person appointed as head of the laboratory must be suitably qualified, and independence of operation of the laboratory must be guaranteed. The management structure must be such that any pressure to rush or skip calibration procedures for production reasons can be resisted. As far as the laboratory environment is concerned, proper temperature and humidity control must be provided, and high standards of cleanliness and housekeeping must be maintained. All equipment used for calibration purposes must be maintained to reference standards, and supported by calibration certificates which establish this traceability. Finally, full documentation must be maintained. This should describe all calibration procedures, maintain an index system for recalibration of equipment, and include a full inventory of apparatus and traceability schedules. Having met these conditions, a Standards Laboratory becomes an accredited laboratory for providing calibration services and issuing calibration certificates. This accreditation is reviewed at approximately twelve-monthly intervals to ensure that the laboratory is continuing to satisfy the conditions for approval laid down.

3.4 PRIMARY REFERENCE STANDARDS

Primary reference standards, as listed in Table 1.1, describe the highest level of accuracy that is achievable in the measurement of any particular physical quantity. All items of equipment used in Standards Laboratories as secondary reference standards have to be calibrated themselves against primary reference standards at appropriate intervals of time. This procedure is acknowledged by the issue of a calibration certificate in the standard way. National Standards Organizations maintain suitable facilities for this calibration, although, in certain cases, such primary reference standards can be located outside National Standards Organizations. For example, the primary reference standard for dimension measurement is defined by the wavelength of the orange-red line of Krypton light, and this can be realized in any laboratory equipped with an interferometer.

In certain cases (e.g. the measurement of viscosity), such primary reference standards are not available and reference standards for calibration are achieved by collaboration between several National Standards Organizations who perform measurements on identical samples under controlled conditions [5].

3.5 TRACEABILITY

What has emerged from the foregoing discussion is that calibration has a chain-like structure in which every instrument in the chain is calibrated against a more accurate instrument immediately above it in the chain, as shown in Figure 3.1. All of the elements in the calibration chain must be known so that the calibration of process instruments at the bottom of the chain is traceable to the fundamental measurement standards.

This knowledge of the full chain of instruments involved in the calibration procedure is known as traceability, and is specified as a mandatory requirement in satisfying standards such as ISO 9001. Documentation must exist which shows that process instruments are calibrated by standard instruments which are linked by a chain of increasing accuracy back to national reference standards. There must be clear evidence to show that there is no break in this chain.

To illustrate a typical calibration chain, consider the calibration of micrometers (Figure 3.2). A typical shop-floor micrometer has an uncertainty (inaccuracy) of less than 1 in 10^4. It would normally be calibrated in the Instrumentation Department Laboratory of a Company against laboratory

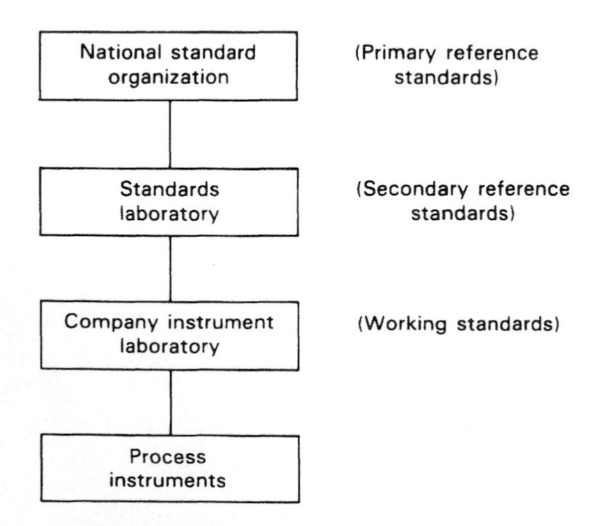

Figure 3.1. Instrument calibration chain

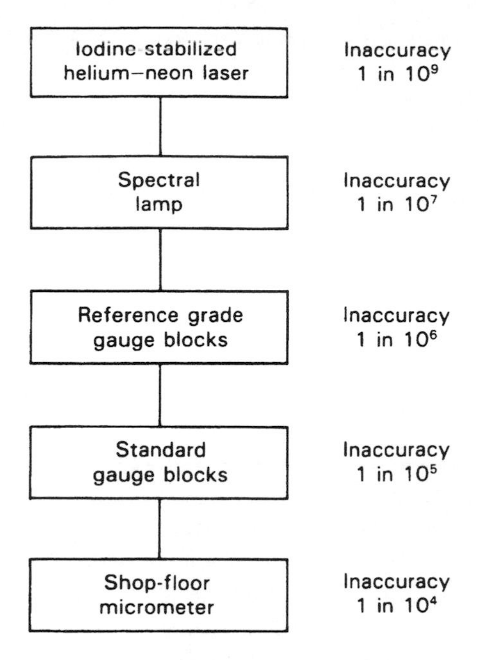

Figure 3.2. Typical calibration chain for micrometers

standard gauge blocks with a typical uncertainty of less than 1 in 10^5. A Specialist Calibration Service Company would provide facilities for calibrating these laboratory standard gauge blocks against reference grade gauge blocks with a typical uncertainty of less than 1 in 10^6. More accurate calibration equipment still is provided by National Standards Organizations. The National Physical Laboratory (UK) maintains two sets of standards for this type of calibration, a working standard and a primary standard. Spectral lamps are used to provide a working reference standard with an uncertainty of less than 1 in 10^7. The primary standard is provided by an iodine-stabilized Helium–Neon laser which has a specified uncertainty of less than 1 in 10^9. All of the links in this calibration chain must be shown in any documentation which describes the use of micrometers in making quality-related measurements.

3.6 DOCUMENTATION IN THE WORKPLACE

An essential element in the maintenance of measurement systems and the operation of calibration procedures is the provision of full documentation. This must give a full description of the measurement requirements throughout the workplace, the instruments used, and the calibration system and procedures operated. Individual calibration records for each instrument must be included within this. This documentation is a necessary part of the Quality Manual, although it may physically exist as a separate volume if this is more convenient. An overriding constraint on the style in which the documentation is presented is that it should be simple and easy to read. This is often greatly facilitated by a copious use of appendices.

The starting point in the documentation must be a statement of what measurement limits have been defined for each measurement system documented. Such limits are established by balancing the costs of improved accuracy against customer requirements, and also with regard to what overall quality level has been specified in the Quality Manual. The technical procedures required for this, which involve assessing the type and magnitude of relevant measurement errors, are described in Chapters 4 to 7. It is customary to express the final measurement limit calculated as ± 2 standard deviations, i.e. within 95 percent confidence limits (see Chapter 5 for explanation of these terms).

The instruments specified for each measurement situation must be listed next. This list must be accompanied by full instructions about the proper use of the instruments concerned. These instructions will include details about any environmental control or other special precautions which must be taken to ensure that the instruments provide measurements of sufficient accuracy to meet the measurement limits defined. The proper training courses appropriate to plant personnel who will use the instruments must also be specified.

Having disposed of the question about what instruments are used, the documentation must go on to cover the subject of calibration. Full calibration is not applied to every measuring instrument used in a workplace because ISO 9000 standards acknowledge that formal calibration procedures for some equipment are not required where it is uneconomic or technically unnecessary because the accuracy of the measurement involved has an insignificant effect on the overall quality target for a product. However, any equipment which is excluded from calibration procedures in this manner must be specified as such in the documentation. Identification of equipment in this category is a matter of informed judgement.

For instruments which are the subject of formal calibration, the documentation must specify what standard instruments are to be used for the purpose and define a formal procedure of calibration. This procedure must include instructions for the storage and handling of standard calibration instruments and specify the required environmental conditions under which calibration is to be performed. Where a calibration procedure for a particular instrument uses published standard practices, it is sufficient to include reference to that standard procedure in the documentation rather than to reproduce the whole procedure. Whatever calibration system is established, a formal review procedure must be defined in the documentation which ensures its continued effectiveness at regular intervals. The results of each review must also be documented in a formal way.

A standard format for the recording of calibration results in record sheets should be defined in the documentation and, where appropriate, the documentation must also define the manner in which calibration results are to be recorded on the instruments themselves. A separate record must be kept for every instrument present in the workplace and should include details of the instrument's description, the required calibration frequency, the date of each calibration and the calibration results on each occasion. Table 3.1 shows a typical format for such records.

The documentation must specify procedures which are to be followed if an instrument is found to be outside the calibration limits. This may involve adjustment, redrawing its scale or withdrawing it, depending upon the nature of the discrepancy and the type of instrument involved. Withdrawn instruments will either be repaired or scrapped, but, until faults have been rectified, their status must be clearly marked on them to prevent them being accidentally put back into use.

Two other items must also be covered by the calibration document. The traceability of the calibration system back to national reference standards must be defined and supported by calibration certificates (see Section 3.2). Training procedures must also be documented, specifying the particular training courses to be attended by various personnel and what, if any, refresher courses are required.

All aspects of these documented calibration procedures will be given consideration as part of the periodic audit of the quality control system which calibration procedures are instigated to support. Whilst the basic responsibility for choosing a suitable interval between calibration checks rests with the engineers responsible for the instruments concerned, the quality system auditor will require to see the results of tests which show that the calibration interval has been chosen correctly and that instruments are not going outside allowable measurement uncertainty limits between calibrations. Audits will check in particular for the existence of procedures which are instigated in response to instruments found to be out of calibration. Evidence that such procedures are effective in avoiding degradation in the quality assurance function will also be required.

REFERENCES

1. *The operation of a company standards department*, British Standards Society, London (1979).
2. NAMAS Document B 5103: *Certificates of Calibration*, NAMAS Executive, National Physical Laboratory, Middlesex, UK (1985).
3. ISO/IEC GUIDE 25: *General requirements for the calibration and competence of testing laboratories*, International Organization for Standards, Geneva (1990).
4. BS 6460: *Accreditation of testing laboratories*, British Standards Institution, London (1983).
5. ISO 5725: *Accuracy of measurement test methods and results*, International Organization for Standards, Geneva (1986) (also published by British Standards Institution as BS ISO 5725).

CHAPTER 4

Instrument Classification, Characteristics and Choice

One important aspect of quality control and assurance is the correct choice of instruments for monitoring process variables and measuring product dimensions and other quality-related variables. Knowledge of the possible error levels in measurements is essential and a necessary prerequisite for this is a proper understanding of the operational characteristics of instruments and an examination of the way in which instrument performance is specified. A convenient way to gain this knowledge is to classify instruments into different types and then to study the characteristics of each of these various instrument subgroups.

Instruments consist of one or more separate components which together serve to give an output reading which is some function of a measured physical quantity. The primary component in an instrument is a transducer which translates the measured physical quantity into another form. Further possible components within the instrument are an amplifier, an amplifier–analyser and an output display system. The term 'instrument' is used somewhat loosely throughout this text, as is fairly common practice, to describe any or all of these components.

4.1 INSTRUMENT CLASSIFICATION

Instruments can be subdivided into separate classes according to several criteria. These subclassifications are useful in broadly establishing several attributes of particular instruments such as accuracy, cost, and general applicability to different applications.

Active/Passive Instruments

Instruments are divided into active or passive ones according to whether the instrument output is entirely produced by the quantity being measured or whether the quantity being measured simply modulates the magnitude of some external power source. This might be more easily understood if it were illustrated by an example.

An example of a passive instrument is the pressure measuring device shown in Figure 4.1. The pressure of the fluid is translated into a movement of a pointer against a scale. The energy expended in moving the pointer is derived entirely from the change in pressure measured: there are no other energy inputs to the system.

An example of an active instrument is a petrol tank level indicator as sketched in Figure 4.2. Here, the change in petrol level moves a potentiometer arm, and the output signal consists of a proportion of the external voltage source applied across the two ends of the potentiometer. The energy in the output signal comes from the external power source: the primary

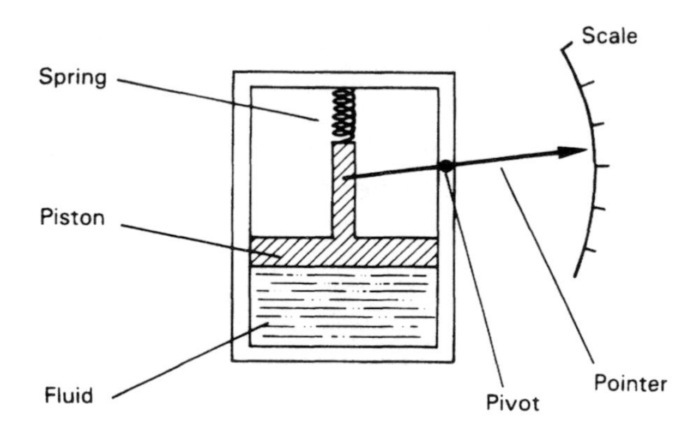

Figure 4.1. Passive pressure gauge

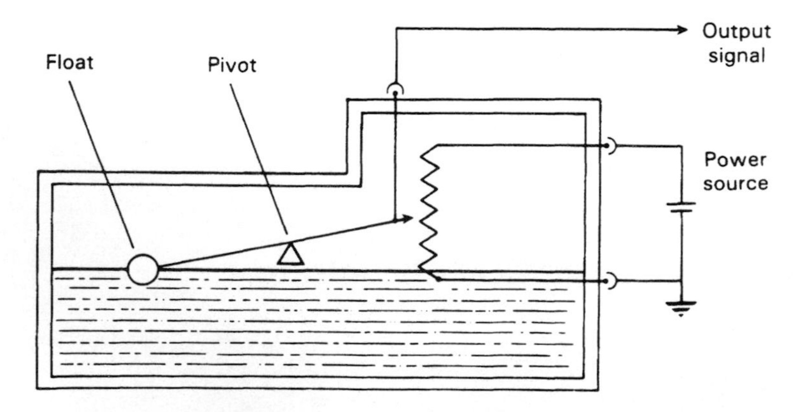

Figure 4.2. Petrol tank level indicator

transducer float system is merely modulating the value of the voltage from this external power source.

In active instruments, the external power source is usually in electrical form, but in some cases it can be in other forms of energy such as pneumatic or hydraulic.

One very important difference between active and passive instruments is the level of measurement resolution that can be obtained. With the simple pressure gauge shown, the amount of movement made by the pointer for a particular pressure change is very much determined by the nature of the instrument. Whilst it is possible to increase measurement resolution by making the pointer longer, so that the pointer tip moves through a longer arc, the scope for such improvement is clearly bounded by the practical limit on how long the pointer can conveniently be. In an active instrument however, adjustment of the magnitude of the external energy input allows much greater control over measurement resolution. Whilst the scope for improving measurement resolution is much greater incidentally, it is not infinite because of limitations placed on the magnitude of the external energy input, in consideration of heating effects and for safety reasons.

In terms of cost, passive instruments are normally of a more simple construction than active ones and are therefore cheaper to manufacture. Choice between active and passive instruments for a particular application therefore involves carefully balancing the measurement-resolution requirements against cost.

Null/Deflection-Type Instruments

The last pressure gauge is a good example of a deflection type of instrument, where the value of the quantity being measured is displayed in terms of the amount of movement of a pointer.

An alternative type of pressure gauge is the dead-weight gauge shown in Figure 4.3 which is a null-type instrument. Here, weights are put on top of

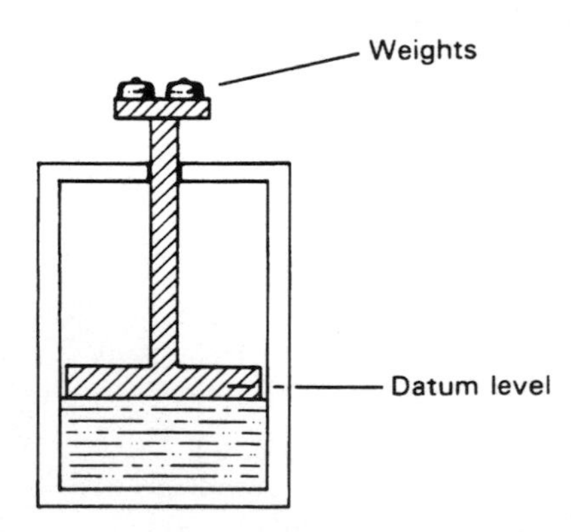

Figure 4.3. Dead-weight pressure gauge

the piston until the downwards force balances the fluid pressure. Weights are added until the piston reaches a datum level, known as the null point. Pressure measurement is made in terms of the value of the weights needed to reach this null position.

The accuracy of these two instruments depends on different things. For the first one it depends on the linearity and calibration of the spring, whilst for the second it relies on the calibration of the weights. As calibration of weights is much easier than careful choice and calibration of a linear-characteristic spring, this means that the second type of instrument will normally be the more accurate. This is in accordance with the general rule that null-type instruments are more accurate than deflection types.

In terms of usage, the deflection-type instrument is clearly more convenient. It is far simpler to read off the position of a pointer against a scale than to add and subtract weights until a null point is reached. A deflection-type instrument is therefore the one that would normally be used in the workplace. For calibration duties, however, the null-type instrument is preferable because of its superior accuracy. The extra effort required to use such an instrument is perfectly acceptable in this case because of the infrequent nature of calibration operations.

Monitoring/Control Instruments

An important distinction between different instruments is made according to whether they are suitable only for monitoring functions or whether their output is in a form that can be directly included as part of an automatic control system. Instruments which only give an audio or visual indication of the magnitude of the physical quantity measured, such as a liquid-in-glass thermometer, are only suitable for monitoring purposes. This class normally includes all null-type instruments and most passive transducers.

For an instrument to be suitable for inclusion in an automatic control system, its output must be in a suitable form for direct input into the controller. Usually, this means that an instrument with an electrical output is required, although other forms of output such as optical or pneumatic signals are used in some systems.

Analogue/Digital Instruments

An analogue instrument gives an output which varies continuously as the quantity being measured changes. The output can have an infinite number of values within the range that the instrument is designed to measure. The deflection type of pressure gauge described earlier in this chapter is a good example of an analogue instrument. As the input value changes, the pointer moves with a smooth continuous motion. Whilst the pointer can therefore be in an infinite number of positions within its range of movement, the number of different positions that the eye can discriminate between is strictly limited, this discrimination being dependent upon how large the scale is and how finely divided it is.

A digital instrument has an output which varies in discrete steps and so can only have a finite number of values. The rev counter sketched in Figure 4.4 is an example of a digital instrument. In this, a cam is attached to the revolving body whose motion is being measured, and on each revolution the cam opens and closes a switch. The switching operations are counted by an electronic counter. This system can only count whole revolutions and therefore cannot discriminate any motion which is less than a full revolution.

The distinction between analogue and digital instruments has become particularly important with the rapid growth in the application of microcomputers to automatic control systems. Any digital computer system, of which the microcomputer is but one example, performs its computations in digital form. An instrument whose output is in digital form is therefore particularly advantageous in such applications, as it can be interfaced directly with the control computer. Analogue instruments must be interfaced with the microcomputer by an analogue-to-digital (A/D) converter, which converts the analogue output signal from the instrument into an equivalent digital quantity which can be read into the computer. This conversion has several disadvantages. Firstly, the A/D converter adds a significant cost to the system. Secondly, a finite time is involved in the process of converting an analogue signal to a digital quantity, and this time can be critical in the control of fast processes where the accuracy of control depends on the speed of the controlling computer. Degrading the speed of operation of the control computer by imposing a requirement for A/D conversion thus degrades the accuracy with which the process is controlled.

4.2 STATIC INSTRUMENT CHARACTERISTICS

If we have a thermometer in a room and its reading shows a temperature of 20 °C, then it does not really matter whether the true temperature of the room is 19.5 °C or 20.5 °C. Such small variations around 20 °C are too small

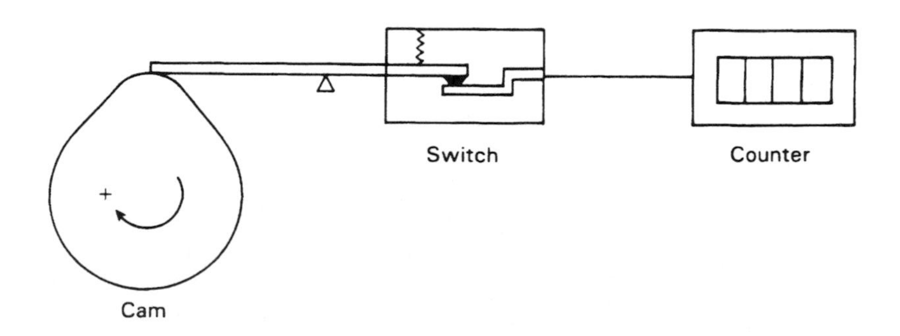

Figure 4.4. Rev counter

to affect whether we feel warm enough or not. Our bodies cannot discriminate between such close levels of temperature and therefore a thermometer with an accuracy of ±0.5 °C is perfectly adequate. If we had to measure the temperature of certain chemical processes however, a variation of 0.5 °C might have a significant effect on the rate of reaction or even the products of a process. A measurement accuracy much better than ±0.5 °C is therefore clearly required in this case.

Accuracy of measurement is thus one consideration in the choice of instrument for a particular application. Other quantities such as sensitivity, linearity and the reaction to ambient temperature changes are further considerations. These attributes are collectively known as the static characteristics of instruments, and are given in the data sheet for a particular instrument. It is important to note that the values quoted for instrument characteristics in such a data sheet only apply when the instrument is used under specified standard calibration conditions. Due allowance must be made for variations in the characteristics when the instrument is used in other conditions.

The various static characteristics are defined in the following paragraphs. More formal definitions can be found in references [1] and [2].

Accuracy

Accuracy is the extent to which a reading might be wrong, and is often quoted as a percentage of the full-scale reading of an instrument. If, for example, a pressure gauge of range 0–10 bar has a quoted inaccuracy of ±1.0 per cent f.s. (±1 per cent of full-scale reading), then the maximum error to be expected in any reading is 0.1 bar. This means that when the instrument is reading 1.0 bar, the possible error is 10 percent of this value. For this reason, it is an important system design rule that instruments are chosen such that their range is appropriate to the spread of values being measured, in order that the best possible accuracy be maintained in instrument readings. Thus, if we were measuring pressures with expected values between 0 and 1 bar, we would not use an instrument with a range of 0–10 bar.

Tolerance

Tolerance is a term which is closely related to accuracy and defines the maximum error which is to be expected in some value. Whilst it is not, strictly speaking, a static characteristic of measuring instruments, it is mentioned here because the accuracy of some instruments is sometimes quoted as a tolerance figure.

Tolerance, when used correctly, describes the maximum deviation of a manufactured component from some specified value. Crankshafts for instance are machined with a diameter tolerance quoted as so many microns, and electric circuit components such as resistors have tolerances of perhaps 5 percent. One resistor chosen at random from a batch having a nominal value 1000 Ω and tolerance 5 percent might have an actual value anywhere between 950 Ω and 1050 Ω.

Precision/Repeatability/Reproducibility

Precision is a term which describes an instrument's degree of freedom from random errors. If a large number of readings are taken of the same quantity by a high precision instrument, then the spread of readings will be very small.

High precision does not imply anything about measurement accuracy. A high precision instrument may have a low accuracy. Low accuracy measurements from a high precision instrument are normally caused by a bias in the measurements, which is removable by recalibration.

The terms repeatability and reproducibility mean approximately the same but are applied in different contexts as given below. Repeatability describes the closeness of output readings when the same input is applied repetitively over a short period of time, with the same measurement conditions, same instrument and observer, same location and same conditions of use maintained throughout. Reproducibility describes the closeness of output readings for the same input when there are changes in the method of measurement, observer, measuring instrument, location, conditions of use and time of measurement. Both terms thus describe the spread of output readings for the same input. This spread is referred to as repeatability if the measurement conditions are constant and as reproducibility if the measurement conditions vary.

The degree of repeatability or reproducibility in measurements from an instrument is an alternative way of expressing its precision. Figure 4.5 illustrates this more clearly. The figure shows the results of tests on three industrial robots which were programmed to place components at a particular point on a table. The target point was at the centre of the concentric circles shown and the black dots represent the points where each robot actually deposited components at each attempt. Both the accuracy and precision of Robot 1 is shown to be low in this trial. Robot 2 consistently puts the component down at approximately the same place but this is the wrong point. Therefore, it has high precision but low accuracy. Finally, Robot 3 has both high precision and high accuracy, because it consistently places the component at the correct target position.

Range or Span

The range or span of an instrument defines the minimum and maximum values of a quantity that the instrument is designed to measure.

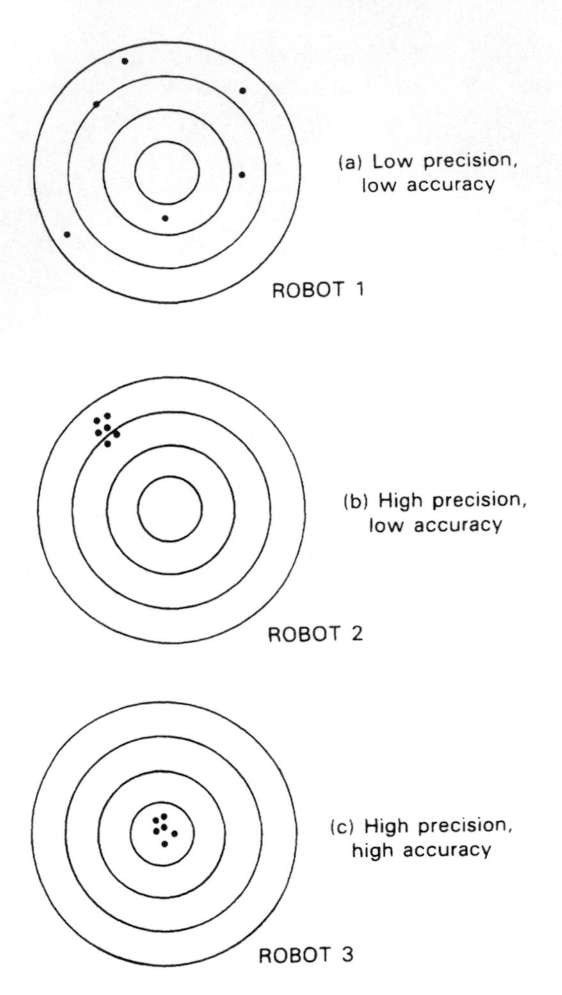

Figure 4.5. Comparison of accuracy and precision

Bias

Bias describes a constant error which exists over the full range of measurement of an instrument. This error is normally removable by calibration. Bathroom scales are a common example of instruments which are prone to bias. It is quite usual to find that there is a reading of perhaps 1 kg with no one standing on the scales. If someone of known weight 70 kg were to get on the scales, the reading would be 71 kg, and if someone of known weight 100 kg were to get on the scales, the reading would be 101 kg. This constant bias of 1 kg can be removed by calibration: in the case of bathroom scales this normally means turning a thumbwheel with the scales unloaded until the reading is zero.

Linearity

It is normally desirable that the output reading of an instrument is linearly proportional to the quantity being measured. The X's marked on Figure 4.6 show a plot of the typical output readings of an instrument when a sequence of input quantities are applied to it. Normal procedure is to draw a good-fit straight line through the X's, as shown in this figure. (Whilst this can often be done with reasonable accuracy by eye, it is always preferable to apply a mathematical least-squares line-fitting technique, as described in Chapter 7). The non-linearity is then defined as the maximum deviation of any of the output readings marked X from this straight line. Non-linearity is usually expressed as a percentage of the full-scale reading.

Sensitivity of Measurement

The sensitivity of measurement is a measure of the change in instrument output which occurs when the quantity being measured changes by a given amount. Sensitivity is thus the ratio:

$$\frac{\text{scale deflection}}{\text{value of measured quantity causing deflection}}$$

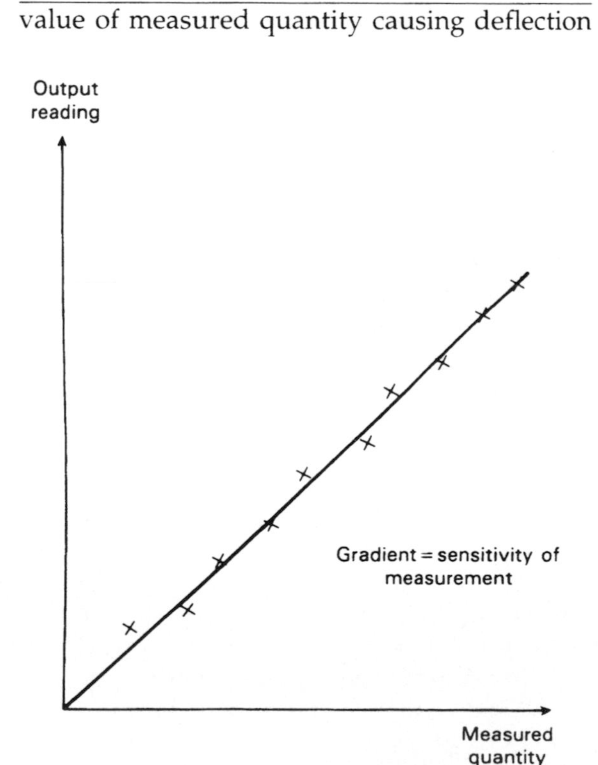

Figure 4.6. Instrument output characteristic

The sensitivity of measurement is therefore the slope of the straight line drawn on Figure 4.6. If, for example, a pressure of 2 bar produces a deflection of 10 degrees in a pressure transducer, the sensitivity of the instrument is 5 degrees/bar. (Assuming that the relationship between pressure and the instrument reading is a straight-line one)

Sensitivity to Disturbance

All calibrations and specifications of an instrument are only valid under controlled conditions of temperature, pressure etc. These standard ambient conditions are usually defined in the instrument specification. As variations occur in the ambient temperature etc., certain static instrument characteristics change, and the sensitivity to disturbance is a measure of the magnitude of this change.

Such environmental changes affect instruments in two main ways, known as zero drift and sensitivity drift.

Zero Drift

Zero drift describes the effect where the zero reading of an instrument is modified by a change in ambient conditions. Typical units by which zero drift is measured are volts/°C, in the case of a voltmeter affected by ambient temperature changes. The effect of zero drift is to impose a bias in the instrument output readings: this is normally removable by recalibration in the usual way. A typical change in the output characteristic of a pressure gauge subject to zero drift is shown in Figure 4.7(a).

Sensitivity Drift (Scale factor drift)

Sensitivity drift defines the amount by which an instrument's sensitivity of measurement varies as ambient conditions change. An alternative name for this phenomenon is scale factor drift. Many components within an instrument are affected by environmental fluctuations, such as temperature changes: for instance, the modulus of elasticity of a spring is temperature dependent. Figure 4.7(b) shows what effect sensitivity drift can have on the output characteristic of an instrument. Sensitivity drift is measured in units of the form (angular degree/bar)/°C.

If an instrument suffers both zero drift and sensitivity drift at the same time, then the typical modification of the output characteristic is as shown in Figure 4.7(c).

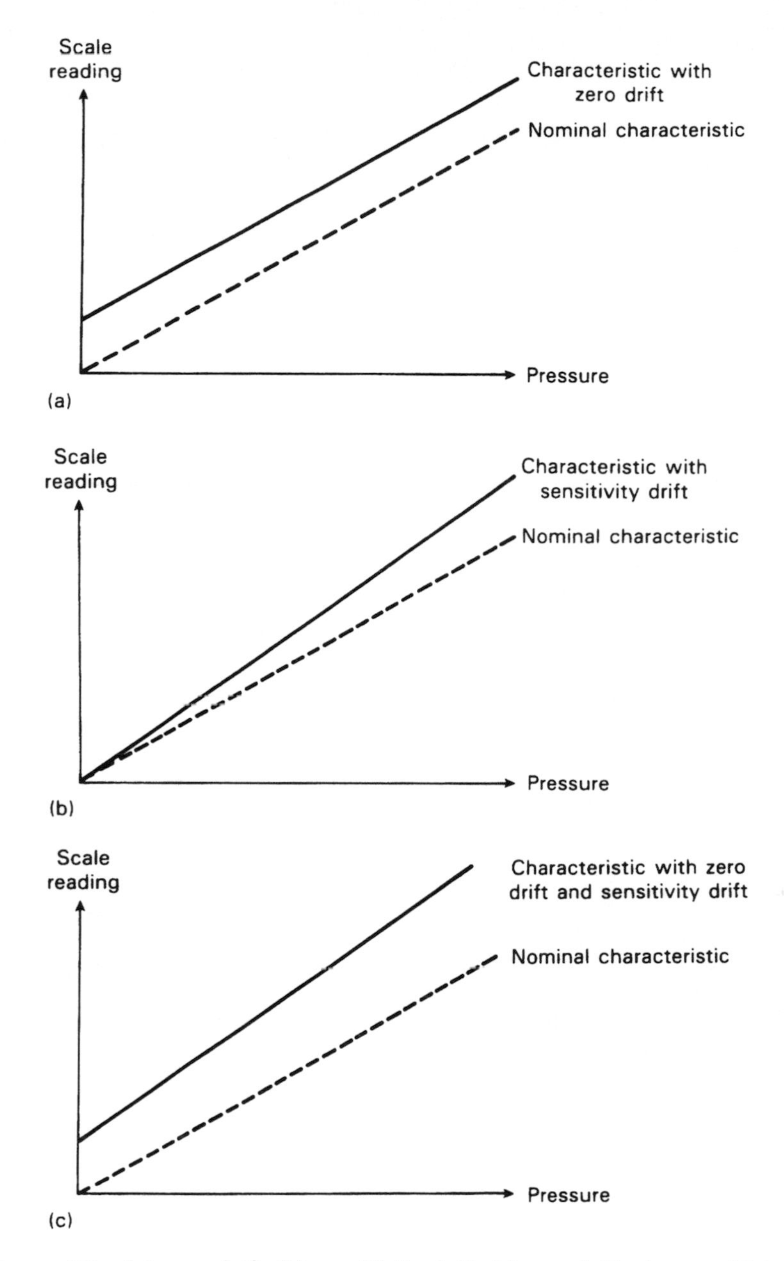

Figure 4.7. (a) zero drift; (b) sensitivity drift; (c) zero drift plus sensitivity drift

Hysteresis

Figure 4.8 illustrates the output characteristic of an instrument which exhibits hysteresis. If the measured quantity input to the instrument is steadily increased from a negative value, the output reading varies in the manner shown in curve (a). If the input variable is then steadily decreased,

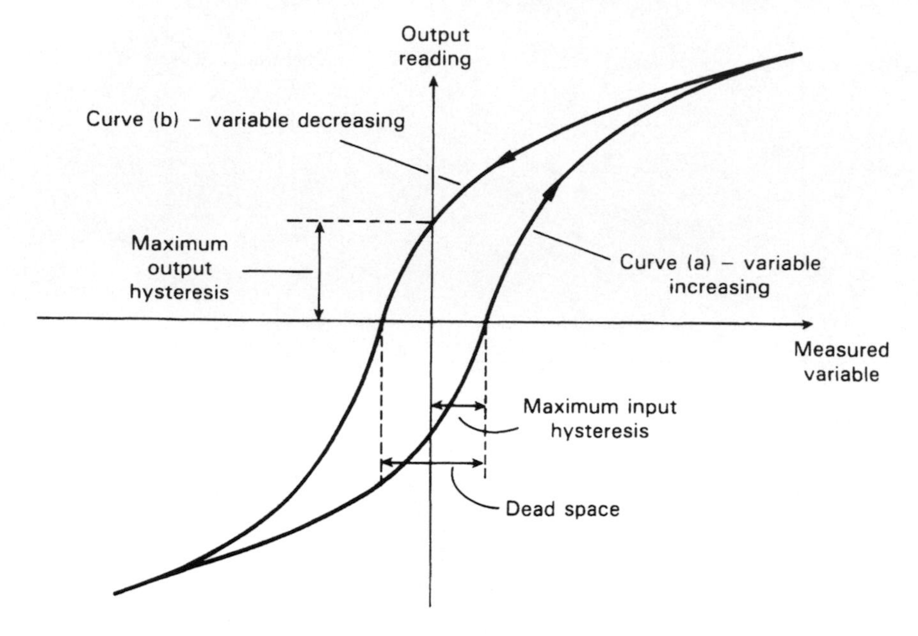

Figure 4.8. Instrument characteristic with hysteresis

the output varies in the manner shown in curve (b). The non-coincidence between these loading and unloading curves is known as hysteresis.

Two quantities, maximum input hysteresis and maximum output hysteresis, are defined as shown in the figure. These are normally expressed as a percentage of the full-scale input or output reading respectively.

Dead Space

Dead space is defined as the range of different input values over which there is no change in output value. Any instrument which exhibits hysteresis also displays dead space, as marked on Figure 4.8. Some instruments which do not suffer from any significant hysteresis can still exhibit a dead space in their output characteristics however. Backlash in gears is a typical cause of dead space, and results in the sort of instrument output characteristic shown in Figure 4.9.

Threshold

If the input to an instrument is gradually increased from zero, the input will have to reach a certain minimum level before the change in the instrument output reading is of a large enough magnitude to be detectable. This minimum level of input is known as the threshold of the instrument. Manufacturers vary in the way that they specify threshold for instruments.

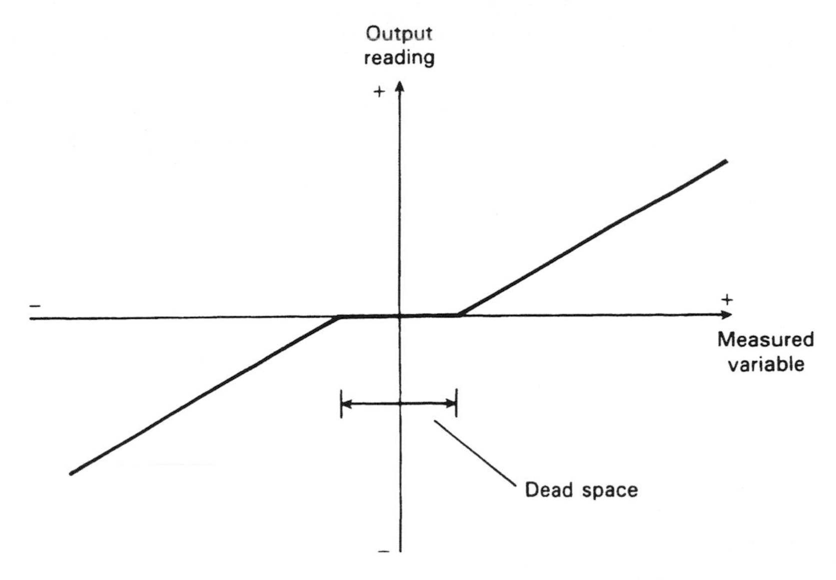

Figure 4.9. Instrument characteristic with dead space

Some quote absolute values, whereas others quote threshold as a percentage of full-scale readings.

As an illustration, a car speedometer typically has a threshold of about 15 km/h. This means that if the vehicle starts from rest and accelerates, no output reading is observed on the speedometer until the speed reaches 15 km/h.

Resolution

When an instrument is showing a particular output reading, there is a lower limit on the magnitude of the change in the input measured quantity which produces an observable change in the instrument output. Like threshold, resolution is sometimes specified as an absolute value and sometimes as a percentage of full-scale deflection.

One of the major factors influencing the resolution of an instrument is how finely its output scale is divided into subdivisions. Using a car speedometer as an example again, this has subdivisions of typically 20 km/h. This means that when the needle is between the scale markings, we cannot estimate speed more accurately than to the nearest 5 km/h. This figure of 5 km/h thus represents the resolution of the instrument.

4.3 DYNAMIC INSTRUMENT CHARACTERISTICS

The static characteristics of a measuring instrument, such as the accuracy of the reading etc., are concerned only with the steady-state reading that the instrument settles down to. The dynamic characteristics describe the behaviour between the time a measured quantity changes value and the time when the instrument output attains a steady value in response. As with static characteristics, any values for dynamic characteristics quoted in instrument data sheets only apply when the instrument is used under specified environmental conditions. Outside these calibration conditions, some variation in the dynamic parameters can be expected.

Various types of dynamic characteristic can be classified, known as zero-order, first-order and second-order characteristics. The practical effects of dynamic characteristics in the output of an instrument can be understood without resorting to formal mathematical analysis. However, this analysis is presented in Appendix 3 for interested readers.

In a *zero-order instrument*, the dynamic characteristics are negligible and the instrument output responds and reaches its final reading almost instantaneously following a step change in the measured quantity applied at its input. A potentiometer, which measures motion, is a good example of such an instrument, where the output voltage changes approximately instantaneously as the slider is displaced along the potentiometer track.

In a *first-order instrument*, the output quantity q_o in response to a step change in the measured quantity q_i varies with time in the manner shown in Figure 4.10. The time constant τ of the step response is the time taken for the output quantity q_o to reach 63 percent of its final value.

The liquid-in-glass thermometer is a good example of a first-order instrument. It is well known that, if a mercury thermometer at room temperature is plunged into boiling water, the mercury does not rise instantaneously to a level indicating 100 °C but instead approaches a reading of 100 °C in the manner indicated in Figure 4.10.

A large number of other instruments also belong to this first-order class: this is of particular importance in control systems where it is necessary to take account of the time lag that occurs between a measured quantity changing in value and the measuring instrument indicating the change. Fortunately, the time constant of many first-order instruments is small relative to the dynamics of the process being measured, and so no serious problems are created.

It is convenient to describe the characteristics of second-order instruments in terms of three parameters K (static sensitivity), ω (undamped natural frequency) and ϵ (damping ratio). These are formally defined in Appendix 3.

The manner in which the output reading of a *second-order instrument* changes following a change in the measured quantity applied to its input depends on the value of these three parameters. The damping ratio parameter, ϵ, controls the shape of the output response and responses of a second-order instrument for various values of ϵ are shown in Figure 4.11. For case (a) where $\epsilon = 0$, there is no damping and the instrument output exhibits constant amplitude oscillations when disturbed by any change in the physical quantity measured. For light damping of $\epsilon = 0.2$, represented by case (b), the response to a step change in input is still oscillatory but the

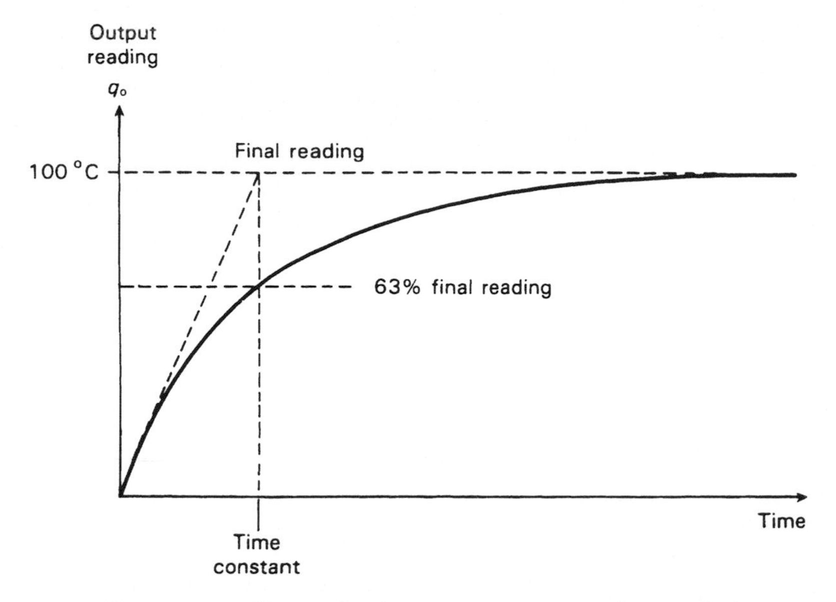

Figure 4.10. First-order instrument response characteristic

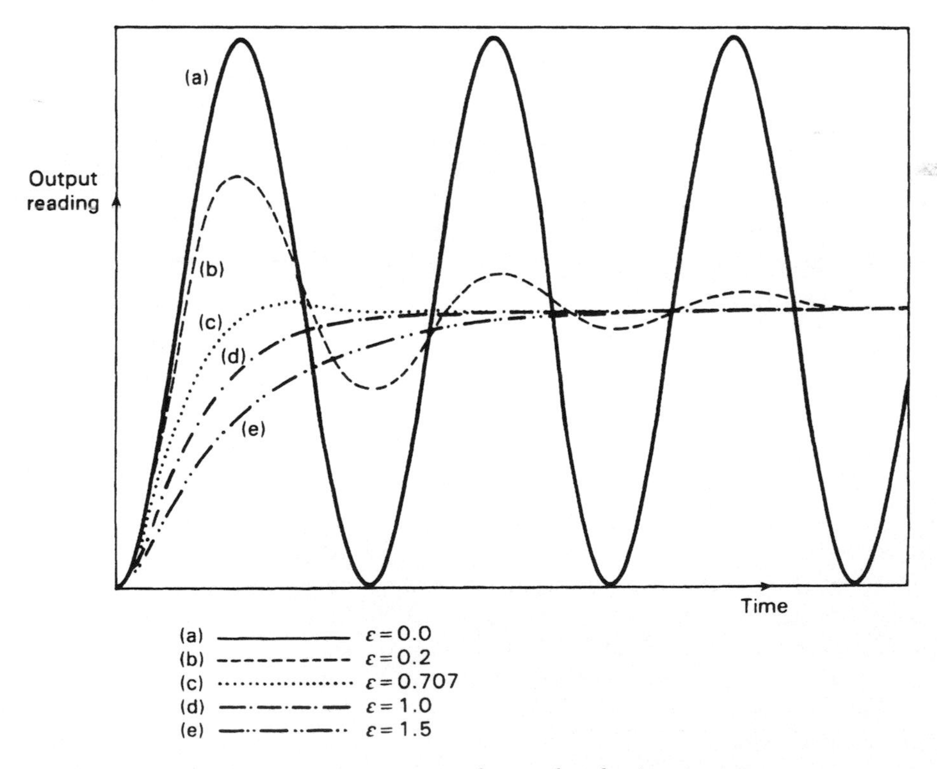

Figure 4.11. Responses of second-order instruments

oscillations gradually die down. Further increase in the value of ϵ reduces oscillations and overshoot still more, as shown by curves (c) and (d), and finally the response becomes very overdamped as shown by curve (e) where the output reading creeps up slowly towards the correct reading. Clearly, the extreme response curves (a) and (e) are grossly unsuitable for any measuring instrument. If an instrument were to be only ever subjected to step inputs, then the design strategy would be to aim towards a damping ratio of 0.707, which gives the critically damped response (c). Unfortunately, most of the physical quantities that instruments are required to measure do not change in the mathematically convenient form of steps, but rather in the form of ramps of varying slopes. As the form of the input variable changes, so the best value for ϵ varies, and choice of ϵ becomes one of compromise between those values that are best for each type of input variable behaviour anticipated. Commercial second-order instruments, of which the accelerometer is a common example, are generally designed to have a damping ratio (ϵ) somewhere in the range of 0.6–0.8.

The dynamics characteristics of instruments are of particular importance in control systems where it is necessary to take account of any time lag that occurs between a measured quantity changing in value and the measuring instrument indicating the change. This may limit the frequency at which the instrument output can be read and cause consequential difficulties in controlling rapidly changing variables.

4.4 COST, DURABILITY AND MAINTENANCE

The static and dynamic characteristics discussed so far are those features which form the technical basis for a comparison between the relative merits of different instruments. In assessing the relative suitability of different instruments for a particular measurement situation however, considerations of cost, durability and maintenance are also of great importance.

Cost is very strongly correlated with the performance of an instrument, as measured by its characteristics. Increasing the accuracy or resolution of an instrument, for example, can only be done at a penalty of increasing its manufacturing cost. Instrument choice therefore proceeds by specifying the minimum characteristics required by a measurement situation and then searching manufacturers' catalogues to find an instrument whose characteristics match those required. As far as accuracy is concerned, it is usual to specify maximum measurement uncertainty levels which are 10 percent of the tolerance levels of the parameter to be measured. To select an instrument whose accuracy and other characteristics are superior to the minimum levels required would only mean paying more than necessary for a level of performance which is greater than that needed.

As well as purchase cost, other important factors in the assessment exercise are the maintenance requirements and the instrument's durability. Maintenance requirements must be taken into account, as they also have cost implications. With regard to durability, it would not be sensible to spend £400 on a new instrument whose projected life was five years if an instrument of equivalent specification with a projected life of ten years was available for £500. However, this consideration is not necessarily simple, as the projected life of instruments often depends on the conditions in which the instrument will have to operate.

As a general rule, a good assessment criterion is obtained if the total purchase cost and estimated maintenance costs of an instrument over its life are divided by the period of its expected life. The figure obtained is thus a cost per year. However, this rule becomes modified where instruments are being installed in a process whose life is expected to be limited, perhaps in the manufacture of a particular model of car. Then, the total costs can only be divided by the period of time an instrument is expected to be used, unless an alternative use for the instrument is envisaged at the end of this period.

To summarize therefore, instrument choice is a compromise between performance characteristics, ruggedness and durability, maintenance requirements and purchase cost. To carry out such an evaluation properly, the instrument engineer must have a wide knowledge of the range of instruments available for measuring particular physical quantities, and he/she must also have a deep understanding of how instrument characteristics are affected by particular measurement situations and operating conditions.

REFERENCES

1. BS 5233: *Glossary of terms used in metrology* (incorporating BS 2643), British Standards Institution, London (1986).
2. ISO 3534: *Statistics – Vocabulary and Symbols*, International Organization for Standards, Geneva (1985) (also published by British Standards Institution as BS.ISO 3534).

CHAPTER 5

Sources of Measurement Error

5.1 INTRODUCTION

This chapter is concerned with identifying the various errors which exist in a measurement system, and suggesting mechanisms for reducing their magnitude and effect. A discussion is also included about the way in which the separate error components are combined in order to calculate the overall measurement system error level.

It is extremely important in any measurement system to reduce errors in instrument output readings to the minimum possible level and to quantify the maximum error which may exist in any output reading. A prerequisite in this is a detailed analysis of the sources of error which exist. Such errors in measurement data can be divided into two groups, known as random errors and systematic errors.

Random errors are perturbations of the measurement either side of the true value caused by random and unpredictable effects, such that positive errors and negative errors occur in approximately equal numbers for a series of measurements made of the same quantity. Such perturbations are mainly small, but large perturbations occur from time to time, again unpredictably. Random errors often arise when measurements are taken by human observation of an analogue meter, especially where this involves interpolation between scale points. Electrical noise can also be a source of random errors. To a large extent, random errors can be overcome by taking the same measurement a number of times and extracting a value by averaging or other statistical techniques, as discussed in Section 5.2.1. However, any quantification of the measurement value and statement of error bounds remains a statistical quantity. Because of the nature of random errors and the fact that large perturbations in the measured quantity occur from time to time, the best that we can do is to express measurements in probabilistic terms: we may be able to assign a 95 percent or even 99 percent confidence level that the measurement is a certain value within error bounds of say ±1 percent, but we can never attach a 100 percent probability to measurement values which are subject to random errors.

Systematic errors describe errors in the output readings of a measurement system that are consistently on one side of the correct reading, i.e. either all the errors are positive or they are all negative. Two major sources of systematic errors are system disturbance during measurement and the effect of environmental inputs, as discussed in Sections 5.3.1 and 5.3.2. Other sources of systematic error include bent meter needles, the use of uncalibrated instruments, poor cabling practices and the thermal generation of electromotive forces (commonly known as thermal e.m.f.'s). The last two sources are considered in Section 5.3.3. Even when systematic errors due to the above factors have been reduced or eliminated, some errors remain which are inherent in the manufacture of an instrument. These are quantified by the accuracy figure quoted in the published specifications contained in the instrument data sheet.

Finally, a word must be said about the distinction between random and systematic errors. Error sources in the measurement system must be examined carefully to determine what type of error is present, random or systematic, and to apply the appropriate treatment. In the case of manual data measurements, a human observer may make a different observation at each attempt, but it is often reasonable to assume that the errors are random and that the mean of these readings is likely to be close to the correct value. However, this is only true as long as the human observer is not introducing a systematic parallax-induced error as well by persistently reading the position of a needle against the scale of an analogue meter from the same side rather than from directly above. In that case, correction would have to be made for this systematic error, or bias, in the measurements before statistical techniques were applied to reduce the effect of random errors.

5.2 RANDOM ERRORS

Random errors in measurements are caused by random, unpredictable variations in the measurement system, and they can only be quantified in probabilistic terms. An introduction to the concepts of probability is given in Appendix 4 for the benefit of any reader who is not familiar with them. In practice, because of their nature, random errors can largely be eliminated by calculating the mean or median of the measurements. The degree of confidence in the calculated mean/median values can be quantified by calculating the standard deviation or variance of the data, these being parameters which describe how the measurements are distributed about the mean/median value. All of these terms are explained more fully in Section 5.2.1.

Because of the unpredictability of random errors, any error bounds placed on measurements can only be quantified in probabilistic terms. Thus, if we say that the possible error in a measurement subject to random errors is ±2 percent of the measured value, we are only implying that this is probably true, i.e. there is say a 95 percent probability that the error level does not exceed ± 2 percent.

The distribution of measurement data about the mean value can be displayed graphically by frequency distribution curves, as discussed in Section 5.2.2. Calculation of the area under the frequency distribution curve gives the probability that the error will lie between any two chosen error levels.

5.2.1 Statistical Analysis of Data

In the analysis of measurements subject to random errors, various parameters can be extracted. Formal definitions of these and their means of calculation are given in the following sections. It should be noted that 'hand calculation' of these parameters is rarely necessary nowadays, as many standard computer packages are available for their calculation, and this facility is also provided by many personal calculators and also intelligent instruments.

Mean and Median Values

In any measurement situation subject to random errors, the normal technique is to take the same reading a number of times, ideally using different observers, and extract the most likely value from the measurement data set. For a set of n measurements $x_1, x_2 \ldots x_n$, the most likely true value is the *mean* given by:

$$x_{\text{mean}} = \frac{x_1 + x_2 + \cdots x_n}{n} \tag{5.1}$$

This is valid for all data sets where the measurement errors are distributed equally about the line of zero error, i.e. where the positive errors are balanced in quantity and magnitude by the negative errors.

When the number of values in the data set is large however, calculation of the mean value is tedious, and it is more convenient to use the median value, this being a close approximation to the mean value. The *median* is given by the middle value when the measurements in the data set are written down in ascending order of magnitude.

For a set of n measurements $x_1, x_2, \ldots x_n$ written down in ascending order of magnitude, the median value is given by:

$$x_{\text{median}} = x_{(n+1)/2}$$

Thus, for a set of 9 measurements $x_1, x_2, \ldots x_9$ the median value is x_5. For an even number of data values, the median value is midway between the centre two values, i.e. for ten measurements $x_1 \ldots x_{10}$, the median value is given by $(x_5 + x_6)/2$.

Table 5.1. Set of mass measurements subject to random errors

81.4	80.9	81.1	81.6	81.1	81.3	80.6	81.2	81.7	81.1
80.8	81.3	81.0	81.5	81.1	81.3	80.8	81.6	81.0	

Suppose that, in a particular measurement situation, a mass is measured by a beam balance, and the set of readings in grams shown in Table 5.1 is obtained at a particular time by different observers. The mean value of this set of data is 81.18 g, calculated according to equation 5.1. The median value is 81.1 g, which is the middle value if the data values are written down in ascending order, starting at 80.6 g and ending at 81.7 g.

Standard Deviation and Variance

The probability that the mean or median value of a data set represents the true measurement value depends on how widely scattered the data values are. If the values of mass measurements in Table 5.1 had ranged from 79 g up to 83 g, our confidence in the mean value would be much less. The spread of values about the mean is analysed by first calculating the deviation of each value from the mean. For any general value x_i, the deviation d_i is given by:

$$d_i = x_i - x_{\text{mean}}$$

The extent to which n measurement values are spread about the mean can now be expressed by the *standard deviation* σ, where σ is given by:

$$\sigma = \sqrt{\left[\frac{d_1^2 + d_2^2 + \cdots + d_n^2}{n-1}\right]} \tag{5.2}$$

This spread can alternatively be expressed by the *variance* V, which is the square of the standard deviation, i.e.

$$V = \sigma^2$$

Readers should note that some books give an alternative definition for σ, which is also used by some calculators for evaluating the standard deviation function. This alternative expression has (n) instead of $(n-1)$ in the denominator and is properly called the *root-mean-squared deviation*. It is given by:

$$\text{deviation}_{\text{rms}} = \sqrt{\left[\frac{d_1^2 + d_2^2 + \cdots + d_n^2}{n}\right]} \tag{5.3}$$

The difference between expressions (5.2) and (5.3) arises because (5.3) is based on an infinite data set, whereas in the case of measurements we are always concerned with only finite data sets [1]. For a finite set of measurements (d_i), $i = 1, n$, the mean x_m will differ from the true mean μ of the infinite data set that the finite set d_i is part of. If somehow we knew the true mean μ of a set of measurements, then the deviations d_i could be calculated as the deviation of each data value from the true mean and it would be correct to calculate σ using (n) as in equation (5.3) instead of $(n - 1)$ as in equation (5.2). However, in normal situations, using $(n - 1)$ in the denominator produces a value of the standard deviation which is statistically closer to the correct value.

Example 5.1
The following measurements were taken of the current in a circuit. (The circuit was in steady state and therefore, although the measurements varied owing to random errors, the current was actually constant.)

21.5 mA, 22.1 mA, 21.3 mA, 21.7 mA, 22.0 mA, 22.2 mA,

21.8 mA, 21.4 mA, 21.9 mA, 22.1 mA,

Calculate the mean value, the deviations from the mean and the standard deviation.

Solution
Mean value $= \sum(\text{data values})/10 = 218/10 = 21.8$ mA
Now draw table of measurements and deviations:

Measurement	21.5	22.1	21.3	21.7	22.0	22.2	21.8	21.4	21.9	22.1
Deviation from mean	−0.3	+0.3	−0.5	−0.1	+0.2	+0.4	0.0	−0.4	+0.1	+0.3
$(\text{deviations})^2$	0.09	0.09	0.25	0.01	0.04	0.16	0.0	0.16	0.01	0.09

$\sum(\text{deviations})^2 = 0.90$
n = number of measurements = 10

$$\sum(\text{deviations})^2/(n-1) = \sum(\text{deviations})^2/9 = 0.10$$

$$\sqrt{\left[\sum(\text{deviations})^2/9\right]} = 0.316$$

Thus, standard deviation = 0.32 mA (the nature of the measurements does not justify expressing the standard deviation to any accuracy greater than two decimal places).

5.2.2 Frequency Distributions

A further and very powerful way of analysing the way in which measurements deviate from the mean value is to use graphical techniques. The simplest way of doing this is by means of a *histogram*, where bands of equal width across the range of measurement values are defined and the number of measurements within each band is counted. Figure 5.1 shows a histogram of measurements drawn from the set of mass data in Table 5.1 by choosing bands 0.3 g wide. There are for instance nine measurements in the range between 81.05 g and 81.35 g, and so the height of the histogram at this point is nine units (NB The scaling of the bands was deliberately chosen so that no measurements fell on the boundary between different bands and caused ambiguity about which band to put them in). Such a histogram has the characteristic shape shown by truly random data, with symmetry about the mean value of the measurements.

As the number of measurements increases, smaller bands can be defined for the histogram, which retains its basic shape but then consists of a larger number of smaller steps on each side of the peak. In the limit, as the number of measurements approaches infinity, the histogram becomes a smooth curve known as the *frequency distribution curve* of the measurements, as shown in Figure 5.2. The ordinate of this curve is the frequency of occurrence of each measurement value, $F(x)$, the abscissa is the magnitude,

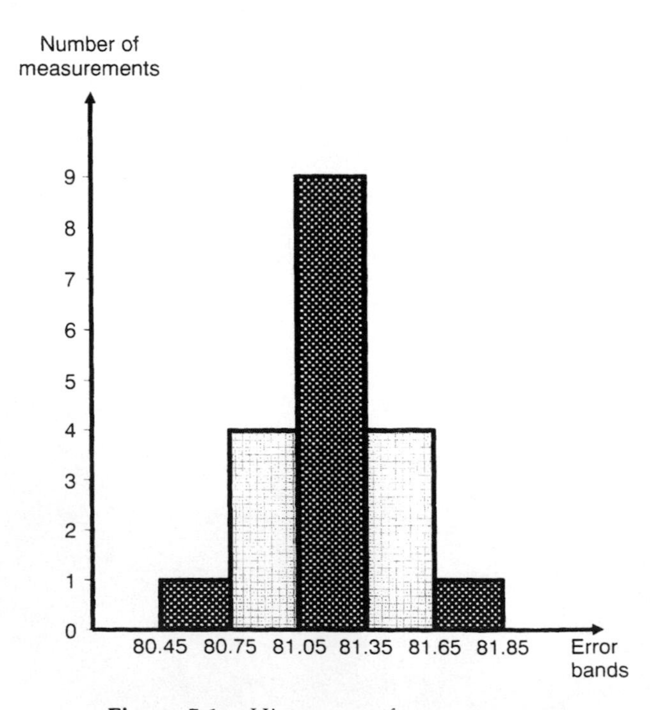

Figure 5.1. Histogram of measurements

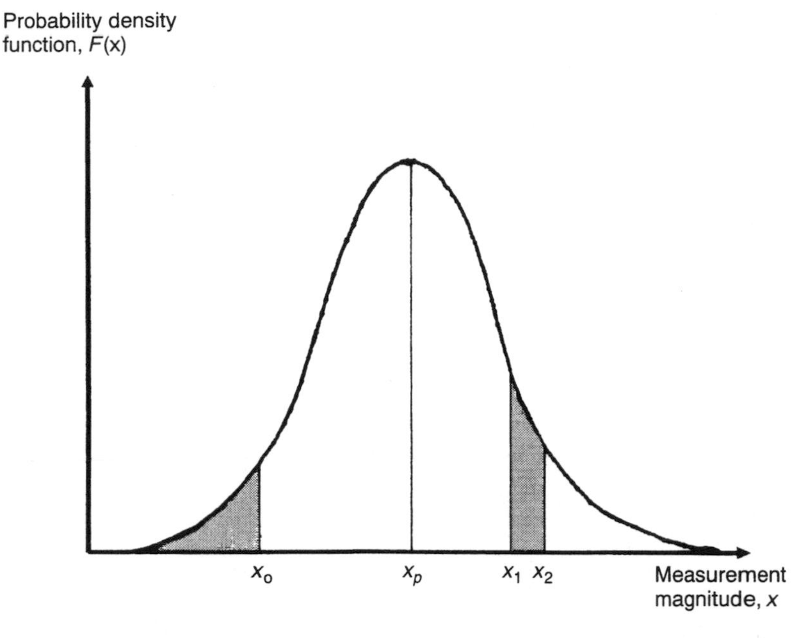

Figure 5.2. Frequency distribution curve of measurements

x, and x_p is the most probable data value. If the errors are truly random, x_p is the mean value of the measurements.

The symmetry of the measurements about the mean data value in Figures 5.1 and 5.2 is very useful for showing graphically that the measurement data has only random errors. However, in order to quantify the magnitude and distribution of the measurement errors, it is more useful to draw histograms and frequency distribution curves of the errors. To do this, the mean of the measurement data values is calculated first, then the error in each measurement in terms of its deviation from this mean value is calculated. Error bands of equal width are then defined and a histogram of errors drawn, as shown in Figure 5.3, according to the number of error values falling within each band. Provided that the errors are only random, this histogram has symmetry about the line of zero error. As the number of measurements increases, smaller error bands can be defined for the error histogram and in the limit, as the number of measurements approaches infinity, the histogram becomes a smooth curve as before. In this case, the curve is known as a *frequency distribution curve* of the errors, as shown in Figure 5.4. The ordinate of this curve is the frequency of occurrence of each error level, $F(E)$, and the abscissa is the error magnitude, E. The error magnitude E_p corresponding with the peak of the frequency distribution curve is the value of error that has the greatest probability. If the errors are entirely random in nature, then the value of E_p will equal zero. Any non-zero value of E_p indicates systematic errors in the data, in the form of a bias which is often removable by recalibration.

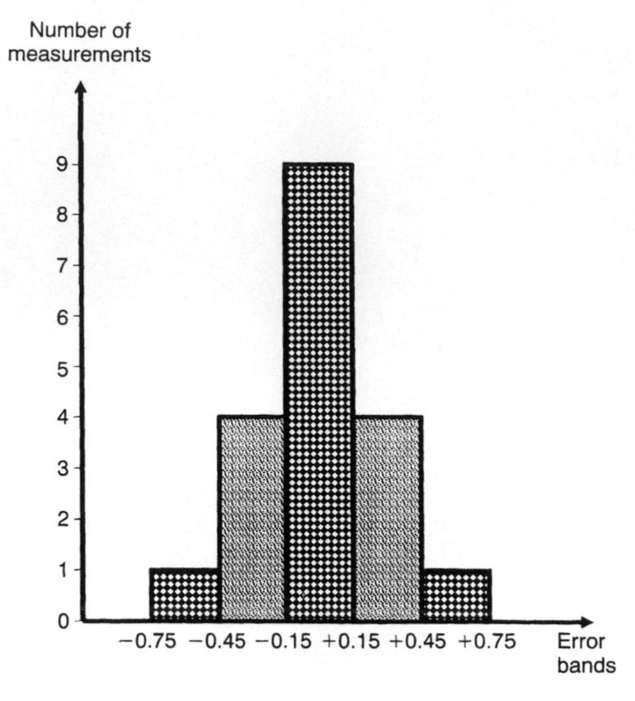

Figure 5.3. Histogram of errors

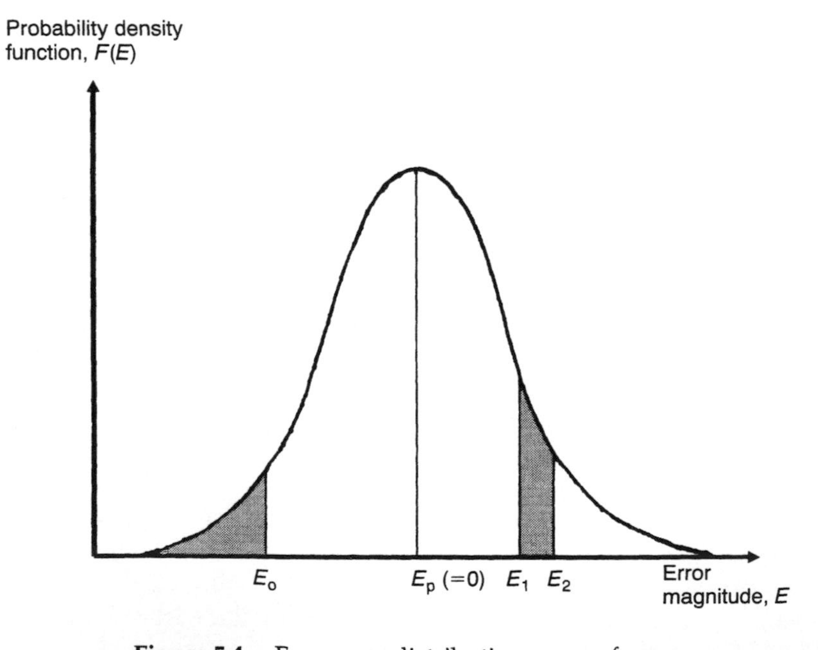

Figure 5.4. Frequency distribution curve of errors

If the height of the frequency distribution of errors curve is normalized such that the area under it is unity, then the curve in this form is known as a *probability curve*, and the height $F(E)$ at any particular error magnitude E is known as the *probability density function* (p.d.f.). The condition that the area under the curve is unity can be expressed mathematically as:

$$\int_{-\infty}^{\infty} F(E)\,\mathrm{d}E = 1$$

The probability that the error in any one particular measurement lies between two levels E_1 and E_2 can be calculated by measuring the area under the curve contained between two vertical lines drawn through E_1 and E_2, as shown by the right-hand shaded area in Figure 5.4. This can be expressed mathematically as:

$$P(E_1 \leqslant E \leqslant E_2) = \int_{E_1}^{E_2} F(E)\,\mathrm{d}E \tag{5.4}$$

Expression (5.4) is often known as the *error function*.

Of particular importance for assessing the maximum error likely in any one measurement is the *cumulative distribution function* (c.d.f.). This is defined as the probability of observing a value less than or equal to E_o, and is expressed mathematically as:

$$P(E \leqslant E_o) = \int_{-\infty}^{E_o} F(E)\,\mathrm{d}E \tag{5.5}$$

Thus the c.d.f. is the area under the curve to the left of a vertical line drawn through E_o, as shown by the left-hand shaded area in Figure 5.4.

Three special types of frequency distribution known as the Gaussian, Binomial and Poisson distributions exist, and these are very important because most data sets approach closely to one or other of them. The distribution of relevance to data sets containing random measurement errors is the Gaussian one.

Gaussian Distribution

A Gaussian curve is defined as a normalized frequency distribution where the frequency and magnitude of quantities are related by the expression:

$$F(x) = \frac{1}{\sqrt{\sigma(2\pi)}} \cdot \exp\left[-(x-m)^2/2\sigma^2\right] \tag{5.6}$$

where m is the mean value of the measurement set x and the other quantities are as defined before. It is only applicable to data that has only random errors, i.e. where no systematic errors exist. Equation (5.6) is particularly useful for analysing a Gaussian set of measurements and predicting how many measurements lie within some particular defined range.

If the measurement errors E are calculated for all measurements such that $E = x - m$, then the curve of error frequency $F(E)$ plotted against error magnitude E is a Gaussian curve known as the error frequency distribution curve. The mathematical relationship between $F(E)$ and E can then be derived by modifying equation (5.6) to give:

$$F(E) = \frac{1}{\sqrt{\sigma(2\pi)}} \cdot \exp[-E^2/2\sigma^2] \tag{5.7}$$

Most measurement data sets such as the values of mass in Table 5.1 fit to a Gaussian distribution curve because, if errors are truly random, small deviations from the mean value occur much more often than large deviations, i.e. the number of small errors is much larger than the number of large ones. Alternative names for the Gaussian distribution curve are the *normal distribution* or the *bell-shaped distribution*.

The Gaussian distribution curve is symmetrical about the line through the mean of the measurement values, which means that positive errors away from the mean value occur in equal quantities to negative errors in any data set containing measurements subject to random error. If the standard deviation is used as a unit of error, the curve can be used to determine what probability there is that the error in any particular measurement in a data set is greater than a certain value. By substituting the expression for $F(E)$ (5.7) into the probability equation (5.4), the probability that the error lies in a band between error levels E_1 and E_2 can be expressed as:

$$P(E_1 \leqslant E \leqslant E_2) = \int_{E_1}^{E_2} \frac{1}{\sqrt{\sigma(2\pi)}} \cdot \exp[-E^2/2\sigma^2]\,\mathrm{d}E \tag{5.8}$$

Equation (5.8) can be simplified by making the substitution:

$$z = E/\sigma. \tag{5.9}$$

(The quantity z is sometimes known as the *standard normal deviate*). Then:

$$P(E_1 \leqslant E \leqslant E_2) = \int_{z_1}^{z_2} \frac{1}{\sqrt{\sigma(2\pi)}} \cdot \exp[-z^2/2]\,\mathrm{d}z = F(z) \tag{5.10}$$

Even after carrying out this simplification, equation (5.10) still cannot be evaluated by the use of standard integrals, and numerical integration has to be used instead. To simplify the burden involved in this, values of $F(z)$, which is known as the *error function*, are normally read from pre-computed tables of the integral $F(z)$ for various values of z. Table 5.2 shows such a tabulation of values and this is usually known as an *error function table*.

Error Function Tables

In Table 5.2, $F(z)$ represents the proportion of data values that are less than or equal to z and is equal to the area under the normalized probability curve to the left of z. Study of the table will show that $F(z) = 0.5$ for $z = 0$. This shows that, as expected, the number of data values $\leqslant 0$ is 50 percent of the total. This must be so if the data only has random errors.

Table 5.2. Computed error function values

z	0.00	0.01	0.02	0.03	0.04	0.05	0.06	0.07	0.08	0.09
					$F(z)$					
0.0	0.5000	0.5040	0.5080	0.5120	0.5160	0.5199	0.5239	0.5279	0.5319	0.5359
0.1	0.5398	0.5438	0.5478	0.5517	0.5557	0.5596	0.5636	0.5675	0.5714	0.5753
0.2	0.5793	0.5832	0.5871	0.5910	0.5948	0.5987	0.6026	0.6064	0.6103	0.6141
0.3	0.6179	0.6217	0.6255	0.6293	0.6331	0.6368	0.6406	0.6443	0.6480	0.6517
0.4	0.6554	0.6591	0.6628	0.6664	0.6700	0.6736	0.6772	0.6808	0.6844	0.6879
0.5	0.6915	0.6950	0.6985	0.7019	0.7054	0.7088	0.7123	0.7157	0.7190	0.7224
0.6	0.7257	0.7291	0.7324	0.7357	0.7389	0.7422	0.7454	0.7486	0.7517	0.7549
0.7	0.7580	0.7611	0.7642	0.7673	0.7703	0.7734	0.7764	0.7793	0.7823	0.7852
0.8	0.7881	0.7910	0.7939	0.7967	0.7995	0.8023	0.8051	0.8078	0.8106	0.8133
0.9	0.8159	0.8186	0.8212	0.8238	0.8264	0.8289	0.8315	0.8340	0.8365	0.8389
1.0	0.8413	0.8438	0.8461	0.8485	0.8508	0.8531	0.8554	0.8577	0.8599	0.8621
1.1	0.8643	0.8665	0.8686	0.8708	0.8729	0.8749	0.8770	0.8790	0.8810	0.8830
12.	0.8849	0.8869	0.8888	0.8906	0.8925	0.8943	0.8962	0.8980	0.8997	0.9015
1.3	0.9032	0.9049	0.9066	0.9082	0.9099	0.9115	0.9131	0.9147	0.9162	0.9177
1.4	0.9192	0.9207	0.9222	0.9236	0.9251	0.9265	0.9279	0.9292	0.9306	0.9319
1.5	0.9332	0.9345	0.9357	0.9370	0.9382	0.9394	0.9406	0.9418	0.9429	0.9441
1.6	0.9452	0.9463	0.9474	0.9484	0.9495	0.9505	0.9515	0.9525	0.9535	0.9545
1.7	0.9554	0.9564	0.9573	0.9582	0.9591	0.9599	0.9608	0.9616	0.9625	0.9633
1.8	0.9641	0.9648	0.9656	0.9664	0.9671	0.9678	0.9686	0.9693	0.9699	0.9706
1.9	0.9713	0.9719	0.9726	0.9732	0.9738	0.9744	0.9750	0.9756	0.9761	0.9767
2.0	0.9772	0.9778	0.9783	0.9788	0.9793	0.9798	0.9803	0.9808	0.9812	0.9817
2.1	0.9821	0.9826	0.9830	0.9834	0.9838	0.9842	0.9846	0.9850	0.9854	0.9857
2.2	0.9861	0.9864	0.9868	0.9871	0.9875	0.9878	0.9881	0.9884	0.9887	0.9890
2.3	0.9893	0.9896	0.9898	0.9901	0.9904	0.9906	0.9909	0.9911	0.9913	0.9916
2.4	0.9918	0.9920	0.9922	0.9924	0.9926	0.9928	0.9930	0.9932	0.9934	0.9936
2.5	0.9938	0.9940	0.9941	0.9943	0.9945	0.9946	0.9948	0.9949	0.9951	0.9952
2.6	0.9953	0.9955	0.9956	0.9957	0.9959	0.9960	0.9961	0.9962	0.9963	0.9964
2.7	0.9965	0.9966	0.9967	0.9968	0.9969	0.9970	0.9971	0.9972	0.9973	0.9974
2.8	0.9974	0.9975	0.9976	0.9977	0.9977	0.9978	0.9979	0.9979	0.9980	0.9981
2.9	0.9981	0.9982	0.9982	0.9983	0.9984	0.9984	0.9985	0.9985	0.9986	0.9986
3.0	0.9986	0.9987	0.9987	0.9988	0.9988	0.9989	0.9989	0.9989	0.9990	0.9990
3.1	0.9990	0.9991	0.9991	0.9991	0.9992	0.9992	0.9992	0.9992	0.9993	0.9993
3.2	0.9993	0.9993	0.9994	0.9994	0.9994	0.9994	0.9994	0.9995	0.9995	0.9995
3.3	0.9995	0.9995	0.9995	0.9996	0.9996	0.9996	0.9996	0.9996	0.9996	0.9996
3.4	0.9997	0.9997	0.9997	0.9997	0.9997	0.9997	0.9997	0.9997	0.9997	0.9998
3.5	0.9998	0.9998	0.9998	0.9998	0.9998	0.9998	0.9998	0.9998	0.9998	0.9998
3.6	0.9998	0.9998	0.9998	0.9999	0.9999	0.9999	0.9999	0.9999	0.9999	0.9999

Use of Error Function Tables

It will be observed that Table 5.2, in common with most published error function tables, only gives $F(z)$ for positive values of z. For negative values of z, we can make use of the following relationship because the frequency distribution curve is normalized:

$$F(-z) = 1 - F(z) \tag{5.11}$$

$F(-z)$ is the area under the curve to the left of $(-z)$, i.e. it represents the proportion of data vales $\leqslant -z$.

Example 5.2
How many measurements in a data set subject to random errors lie outside boundaries of $+\sigma$ and $-\sigma$, i.e. how many measurements have an error $<|\sigma|$?

Solution
The required number is represented by the sum of the two shaded areas in Figure 5.5. This can be expressed mathematically as

$$P[E < -\sigma \text{ or } E > +\sigma] = P[E < -\sigma] + P[E > +\sigma].$$

For $E = -\sigma$, $z = -1.0$ (from equation 5.9), using an error function table

$$P[E < -\sigma] = F(-1) = 1 - F(1) = 1 - 0.8413 = 0.1587$$

Similarly, for $E = +\sigma$, $z = +1.0$, the error function table gives

$$P[E > +\sigma] = 1 - P[E < +\sigma] = 1 - F(1) = 1 - 0.8413 = 0.1587$$

(This last step is valid because the frequency distribution curve is normalized so that the total area under it is unity.) Thus,

$$P[E < -\sigma] + P[E > +\sigma] = 0.1587 + 0.1587 = 0.3174 \approx 32 \text{ percent}$$

Therefore, 32 percent of the measurements lie outside the $\pm\sigma$ boundaries, i.e. 32 percent of the measurements have an error greater than $|\sigma|$. It follows that 68 percent of the measurements lie inside the boundaries of $\pm\sigma$.

The above analysis shows that, for Gaussian-distributed data values, 68 percent of the measurements have errors which lie within the bounds of $\pm\sigma$. Similar analysis shows that boundaries of $\pm 2\sigma$ contain 95.4 percent of data points, and extending the boundaries to $\pm 3\sigma$ encompasses 99.7 percent

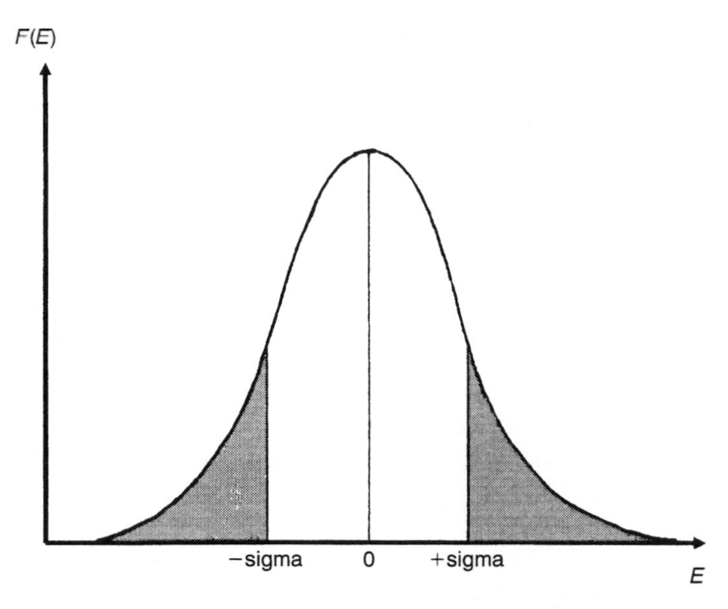

Figure 5.5. $\pm\sigma$ Boundaries on measurement errors

of data points. The probability of any data point lying outside particular error boundaries can therefore be expressed by the following table.

Error boundaries	% of data points within boundary	Probability of any particular data point being outside boundary
$\pm\sigma$	68.0	34.0%
$\pm 2\sigma$	95.4	4.6%
$\pm 3\sigma$	99.7	0.3%

Standard Error of the Mean

The foregoing analysis is only strictly true for measurement sets containing infinite populations. It is not of course possible to obtain an infinite number of data values, and some error must therefore be expected in the calculated mean value of the practical, finite data set available. If several subsets are taken from an infinite data population, then, by the central limit theorem, the means of the subsets will form a Gaussian distribution about the mean of the infinite data set. The error in the mean of a finite data set is usually expressed as the standard error of the mean, α, which is calculated as:

$$\alpha = \sigma / \sqrt{n}$$

This tends towards zero as the number of measurements in the data set is expanded towards infinity. The value obtained from a set of n measurements, $x_1, x_2 \ldots x_n$ is then expressed as:

$$x = x_{\text{mean}} \pm \alpha$$

For the data set of mass measurements in Table 5.1, $n = 19$, $\sigma = 0.318$ and $\alpha = 0.073$. The mass can therefore be expressed as 81.18 ± 0.07 kg (68 percent confidence limit). However, it is more usual to express measurements with 95 percent confidence limits ($\pm 2\sigma$ boundaries). In this case, $2\sigma = 0.636$, $2\alpha = 0.146$ and the value of the mass can therefore be expressed as 81.18 ± 0.15 kg (95 percent confidence limits).

5.3 SYSTEMATIC ERRORS

Systematic errors in the output of many instruments are due to factors inherent in the manufacture of the instrument arising from tolerances in the components of the instrument. They can also arise owing to wear in instrument components over a period of time. In other cases, systematic errors are introduced by either the effect of environmental disturbances or through the disturbance of the measured system by the act of measurement. These various sources of systematic error, and ways in which the magnitude of the errors can be reduced, are discussed below.

5.3.1 System Disturbance Due to Measurement

Disturbance of the measured system by the act of measurement is one source of systematic error. If we were to start with a beaker of hot water and wished to measure its temperature with a mercury-in-glass thermometer, then we should take the thermometer, which would be initially at room temperature, and plunge it into the water. In so doing, we would be introducing the relatively cold mass of the thermometer into the hot water and a heat transfer would take place between the water and the thermometer. This heat transfer would lower the temperature of the water. Whilst in this case the reduction in temperature would be so small as to be undetectable by the limited measurement resolution of such a thermometer, the effect is finite and clearly establishes the principle that, in nearly all measurement situations, the process of measurement disturbs the system and alters the values of the physical quantities being measured.

Another example is that of measuring car tyre pressures with the type of pressure gauge commonly obtainable from car accessory shops. Measurement is made by pushing one end of the pressure gauge onto the valve of the tyre and reading the displacement of the other end of the gauge against a scale. As the gauge is used, a quantity of air flows from the tyre into the gauge. This air does not subsequently flow back into the tyre after measurement, and so the tyre has been disturbed and the air pressure inside it has been permanently reduced.

Thus, as a general rule, the process of measurement always disturbs the system being measured. The magnitude of the disturbance varies from one measurement system to the next and is affected particularly by the type of instrument used for measurement. Ways of minimizing disturbance of measured systems is an important consideration in instrument design. A prerequisite for this, however, is a full understanding of the mechanisms of system disturbance.

Measurements in electric circuits are particularly prone to errors induced through the loading effect on the circuit when instruments are applied to make voltage and current measurements. For most electrical networks, circuit analysis methods such as Thevenin's theorem are needed to analyse such loading effects. However, for the simple circuit shown in Figure 5.6, the analysis is fairly easy.

In this circuit, the voltage across resistor R_2 is to be measured by a voltmeter with resistance R_m. Here, R_m acts as a shunt resistance across R_2, decreasing the resistance between points A and B and so disturbing the circuit. The voltage E_m measured by the meter is therefore not the value of the voltage E_o that existed prior to measurement. The extent of the disturbance can be assessed by calculating the open-circuit voltage E_o and comparing it with E_m.

Starting with the unloaded circuit in Figure 5.6, the current I is given by Ohm's law as:

$$I = \frac{V}{R_1 + R_2}$$

Again, using Ohm's law, the voltage across AB is then given by:

$$E_o = IR_2 = \frac{VR_2}{R_1 + R_2} \tag{5.12}$$

With the voltmeter added to the circuit, there are now two resistances in parallel across AB, R_2 and R_m, and the expression for the resistance across AB can be written as:

$$R_{AB} = \frac{R_2 R_m}{R_2 + R_m}$$

Then, replacing R_2 by R_{AB} in equation (5.12) above, the voltage E_m measured by the meter is given by:

$$E_m = \frac{VR_{AB}}{R_1 + R_{AB}} = \frac{VR_2 R_m}{(R_2 + R_m)} \cdot \frac{1}{R_1 + \dfrac{R_2 R_m}{R_2 + R_m}} = \frac{VR_2 R_m}{R_1 R_2 + R_m(R_1 + R_2)} \tag{5.13}$$

Thus, from equations (5.12) and (5.13):

$$\frac{E_m}{E_o} = \frac{VR_2 R_m}{[R_1 R_2 + R_m(R_1 + R_2)]} \cdot \frac{(R_1 + R_2)}{VR_2} = \frac{R_m(R_1 + R_2)}{R_1 R_2 + R_m(R_1 + R_2)} \tag{5.14}$$

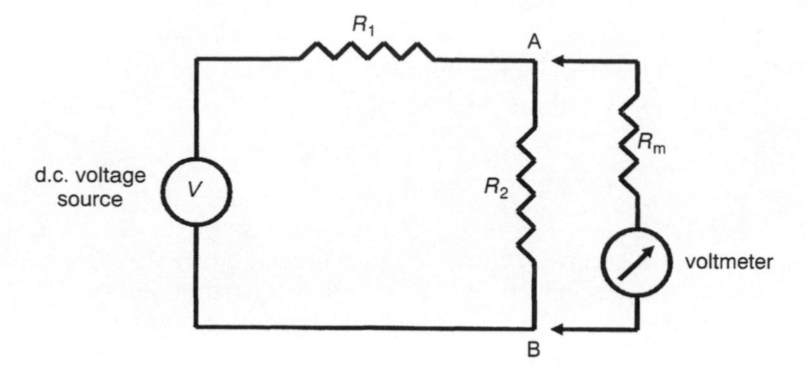

Figure 5.6. Disturbance of electrical circuit by loading

If R_m is very large compared with R_1 and R_2, then $R_m(R_1 + R_2) \gg R_1 R_2$, the denominator of equation (5.14) approaches $R_m(R_1 + R_2)$, E_m/E_o approaches unity, and thus E_m approaches E_o.

It is thus obvious that as R_m gets larger, the ratio E_m/E_o gets closer to unity, showing that the design strategy should be to make R_m as large as possible in order to minimize the disturbance of the measured system. (Note that we did not calculate the value of E_o, since this was not required in quantifying the effect of R_m.)

Example 5.4
Suppose that the components of the circuit shown in Figure 5.6 have the following values: $R_1 = 500\ \Omega$; $R_2 = 500\ \Omega$. The voltage across AB is measured by a voltmeter whose internal resistance is $4750\ \Omega$. What is the measurement error caused by the resistance of the measuring instrument?

Solution
Proceeding by substituting the given component values into equation (5.14), we obtain

$$\frac{E_m}{E_o} = \frac{4750 \times 1000}{[25 \times 10^4 + 4750 \times 1000]} = 0.95$$

Thus the error in the measured value is 5 percent.

5.3.2 Environmental Inputs to Measurement Systems

Environmental inputs are systematic measurement errors induced by changes in the ambient conditions of the environment surrounding the measurement system. In some texts, environmental errors are given the alternative name *modifying inputs* because they modify the output of a measurement system even if the real input (the measured quantity of interest) remains constant.

Environmental inputs can be caused by changes in various environmental parameters such as the ambient temperature, pressure and humidity. However, these do not affect every instrument to the same extent, and indeed some instruments are affected very little. One of the aims in instrument design is to make the instrument as insensitive as possible to the sort of ambient condition changes it is likely to meet in normal operation. Unfortunately, this design aim conflicts with other design criteria in many cases, and consequently some instruments remain significantly affected by environmental inputs.

The magnitude of environmental inputs is quantified by the sensitivity drift and zero drift parameters defined previously in Chapter 4, both of which are generally included in the published specifications for an instrument. Without proper analysis, it is impossible to establish how much of an instrument's output is due to the real input and how much is due to one or more environmental inputs. This is illustrated by the following example. Suppose that we have a small closed box weighing 0.1 kg when empty, which we think contains either a rat or a mouse. If we put the box onto bathroom scales and observe a reading of 1.0 kg, this does not immediately tell us what is in the box because the reading may be due to one of three things:

a. a 0.9 kg rat in the box (real input)

b. an empty box with a 0.9 kg bias on the scales due to a temperature change (environmental input)

c. a 0.4 kg mouse in the box together with a 0.5 kg bias (real + environmental inputs)

Thus, the magnitude of any environmental input must be measured before the value of the measured quantity, which is the real input, can be determined from the output reading of an instrument.

In any general measurement situation, it is very difficult to avoid environmental inputs, because it is either impractical or impossible to control the environmental conditions surrounding the measurement system. System designers are therefore charged with the task of either reducing the susceptibility of measuring instruments to environmental inputs or alternatively quantifying the effect of environmental inputs and correcting for them in the instrument output reading.

The techniques used to deal with environmental inputs and minimize their effect on the final output measurement follow a number of routes, as discussed below.

Careful Instrument Design

Careful instrument design is the most useful weapon in the battle against environmental inputs, by reducing the sensitivity of an instrument to environmental inputs to as low a level as possible. In the design of strain gauges for instance, the element should be constructed from a material whose resistance has a very low temperature coefficient (i.e. the variation of the resistance with temperature is very small). For many instruments, however, it is not possible to reduce their sensitivity to environmental inputs to a satisfactory level by simple design adjustments, and other measures have to be taken.

Method of Opposing Inputs

The method of opposing inputs compensates for the effect of a environmental input in a measurement system by introducing an equal and opposite environmental input which cancels it out. One example of how this technique is applied is in the type of millivoltmeter shown in Figure 5.7. This consists of a coil suspended in a fixed magnetic field produced by a permanent magnet. When an unknown voltage is applied to the coil, the magnetic field due to the current interacts with the fixed field and causes the coil (and a pointer attached to the coil) to turn. If the coil resistance is sensitive to temperature, then any environmental input to the system in the form of a temperature change will alter the value of the coil current for a given applied voltage and so alter the pointer output reading. Compensation for this is made by introducing a compensating resistance R_{comp} into the circuit, where R_{comp} has a temperature coefficient which is equal in magnitude but opposite in sign to that of the coil.

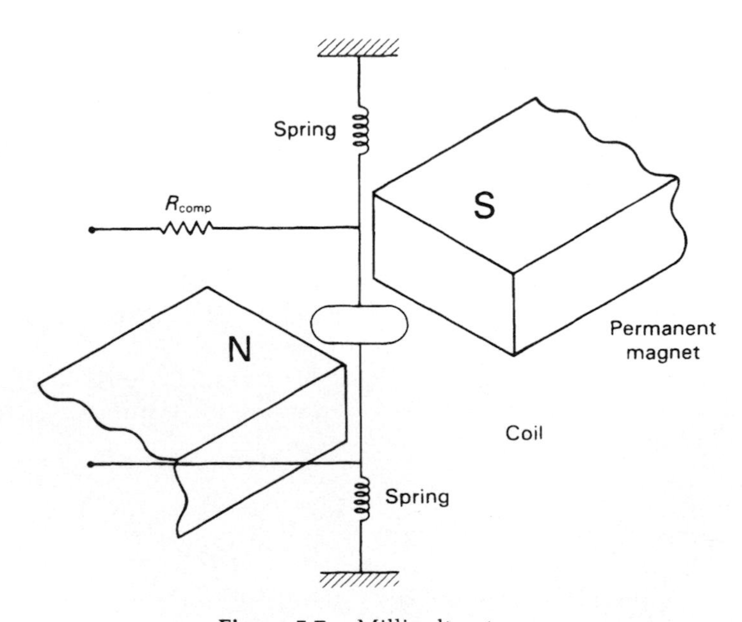

Figure 5.7. Millivoltmeter

High Gain Feedback

The benefit of adding high-gain feedback to many measurement systems is illustrated by considering the case of the voltage-measuring instrument whose block diagram is shown in Figure 5.8. In this system, the unknown voltage E_i is applied to a coil of torque constant K_c, and the torque induced turns a pointer against the restraining action of a spring with spring constant K_s. The effect of environmental inputs on the torque and spring constants is represented by variables D_c and D_s.

In the absence of environmental inputs, the displacement of the pointer X_o is given by:

$$X_o = K_c \cdot K_s \cdot E_i$$

However, in the presence of environmental inputs, both K_c and K_s change and the relationship between X_o and E_i can be affected greatly. It therefore becomes difficult or impossible to calculate E_i from the measured value of X_o.

Consider now what happens if the system is converted into a high-gain, closed-loop one, as shown in Figure 5.9, by adding an amplifier of gain constant K_a and a feedback device with gain constant K_f. Assume also that the effect of environmental inputs on the values of K_a and K_f is represented by D_a and D_f. The feedback device feeds back a voltage E_o proportional to

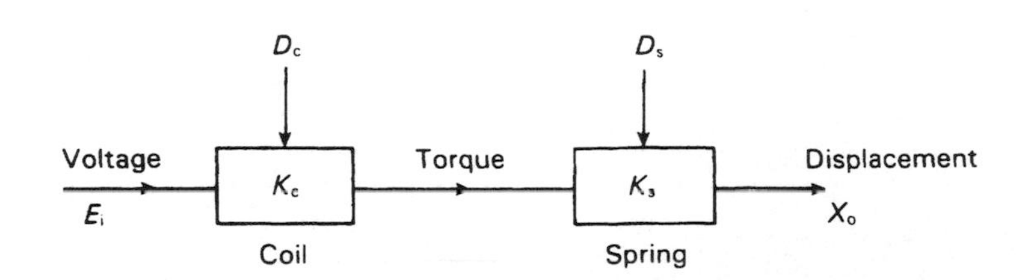

Figure 5.8. Block diagram for voltage-measuring instrument

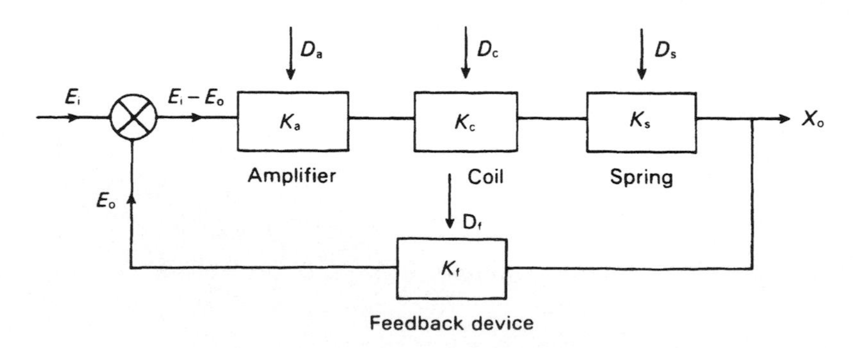

Figure 5.9. Block diagram of voltage-measuring instrument with high-gain feedback

the pointer displacement X_o. This is compared with the unknown voltage E_i by a comparator and the error is amplified.

Writing down the equations of the system, we have

$$E_o = K_f \cdot X_o$$

$$X_o = (E_i - E_o) \cdot K_a \cdot K_c \cdot K_s = (E_i - K_f \cdot X_o) \cdot K_a \cdot K_c \cdot K_s$$

thus:

$$E_i \cdot K_a \cdot K_c \cdot K_s = (1 + K_f \cdot K_a \cdot K_c \cdot K_s) \, X_o$$

i.e.

$$X_o = \frac{K_a \cdot K_c \cdot K_s}{1 + K_f \cdot K_a \cdot K_c \cdot K_s} \cdot E_i \tag{5.15}$$

Because K_a is the gain constant of a high-gain amplifier, it is very large, and therefore the product $K_f \cdot K_a \cdot K_c \cdot K_s$ is very much greater than one, and equation (5.15) can be reduced to

$$X_o \approx \frac{E_i}{K_f}$$

This is a highly important result because we have reduced the relationship between X_o and E_i to one which involves only K_f. The sensitivity of the gain constants K_a, K_c and K_s to the environmental inputs D_a, D_c and D_s has thereby been rendered irrelevant and we only have to be concerned with one environmental input D_f. Conveniently, it is usually an easy matter to design a feedback device which is insensitive to environmental inputs: this is much easier than trying to make a coil or spring insensitive. Thus high-gain feedback techniques are often a very effective way of reducing a measurement system's sensitivity to environmental inputs. One potential problem that must be mentioned however is that there is a possibility that high-gain feedback will cause instability in the system. Any application of this method must therefore include careful stability analysis of the system.

Signal Filtering

One frequent problem in measurement systems is corruption of the output reading by periodic noise, often at a frequency of 50 Hz caused by pick-up through the close proximity of the measurement system to apparatus or current-carrying cables operating on a mains supply. Periodic noise corruption at higher frequencies is also often introduced by mechanical oscillation or vibration within some component of a measurement system. The amplitude of all such noise components can be substantially attenuated by the inclusion of filtering of an appropriate form in the system, as discussed at greater length in Chapter 6. Band-stop filters can be especially useful where corruption is of one particular known frequency, or, more generally, low-pass filters are employed to attenuate all noise in the frequency range of 50 Hz and above.

Measurement systems with a low-level output, such as a bridge circuit measuring a strain-gauge resistance, are particularly prone to noise, and Figure 5.10(a) shows the typical corruption of a bridge output by 50 Hz pick-up. The beneficial effect of putting a simple passive RC low-pass filter across the output is shown in Figure 5.10(b).

5.3.3 Other Sources of Systematic Error

Wear in Instrument Components

Systematic errors can frequently develop over a period of time because of wear in instrument components. Recalibration often provides a full solution to this problem.

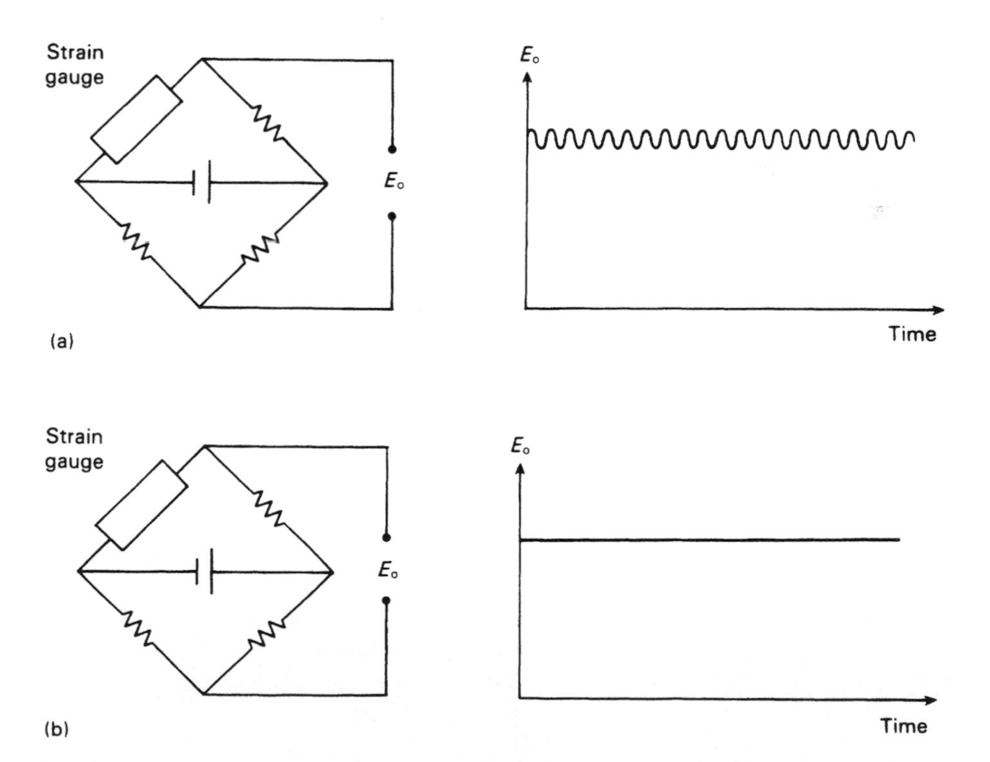

Figure 5.10. (a) Noise-corrupted output of bridge circuit measuring resistance of strain gauge; (b) effect of adding low-pass filter

Connecting Leads

In connecting together the components of a measurement system, a common source of error is the failure to take proper account of the resistance of connecting leads, or pipes in the case of pneumatically or hydraulically actuated measurement systems. In typical applications of a resistance thermometer for instance, it is common to find the thermometer separated from other parts of the measurement system by perhaps 30 metres. The resistance of such a length of 7/0.0076 copper wire is 2.5 Ω and there is a further complication that such wire has a temperature coefficient of 1 mΩ/°C.

Careful consideration therefore needs to be given to the choice of connecting leads. Not only should they be of adequate cross-section so that their resistance is minimized, but they should be adequately screened if they are thought likely to be subject to electrical or magnetic fields which could otherwise cause induced noise. Where screening is thought essential, then the routeing of cables also needs careful planning. In one application in the author's personal experience involving instrumentation of an electric-arc steelmaking furnace, screened signal-carrying cables between transducers on the arc furnace and a control room at the side of the furnace were initially corrupted by high amplitude 50 Hz noise. However, by changing the route of the cables between the transducers and the control room, the magnitude of this induced noise was reduced by a factor of about 10.

Thermal e.m.f.'s (Thermally-generated Electromotive Forces)

Whenever metals of two different types are connected together, a thermal e.m.f. is generated, which varies according to the temperature of the joint. This is known as the *thermoelectric effect* and is the physical principle on which temperature-measuring thermocouples operate (see Chapter 13). Such thermal e.m.f.'s are only a few millivolts in magnitude and so the effect is only significant when typical voltage output signals of a measurement system are of a similarly low magnitude.

One such situation is where one e.m.f.-measuring instrument is used to monitor the output of several thermocouples measuring the temperature at different points in a process control system. This requires a means of automatically switching the output of each thermocouple to the measuring instrument in turn. Nickel–iron reed-relays with copper connecting leads are commonly used to provide this switching function. This introduces a thermocouple effect of magnitude 40 μV/°C between the reed-relay and the copper connecting leads. There is no problem if both ends of the reed relay are at the same temperature because then the thermal e.m.f.'s will be equal and opposite and so cancel out. However, there are several recorded instances where, because of lack of awareness of the problem, poor design has resulted in the two ends of a reed-relay being at different temperatures and causing a net thermal e.m.f. The serious error that this introduces is clear. For a temperature difference between the two ends of only 2 °C, the thermal e.m.f. is 80 μV, which is very large compared with a typical thermocouple output level of 400 μV.

Another example of the difficulties that thermal e.m.f.'s can create becomes apparent in considering the following problem which was reported in a current-measuring system. This system had been designed so that the current in a particular part of a circuit was calculated by applying it to an accurately calibrated wire-wound resistance of value 100 Ω and measuring the voltage drop across the resistance. In calibration of the system, a known current of 20 μA was applied to the resistance and a voltage of 2.20 mV was measured by an accurate high impedance instrument. Simple application of Ohm's law reveals that such a voltage reading indicates a current value of 22 μA. What then is the explanation for this discrepancy? The answer once again is a thermal e.m.f. Because the designer was not aware of thermal e.m.f.'s, the circuit had been constructed so that one side of the standard resistance was close to a power transistor, creating a difference in temperature between the two ends of the resistor of 2 °C. The thermal e.m.f. associated with this was sufficient to account for the ten per-cent measurement error found.

5.4 ERROR REDUCTION USING INTELLIGENT INSTRUMENTS

An intelligent instrument is distinguished from a dumb (non-intelligent) instrument by the inclusion of a microcomputer and by the addition of one or more extra transducers at its input. These additions inevitably add significantly to the instrument's cost, but large reductions in measurement errors can often be achieved by using them.

The inputs to an intelligent instrument are data from a *primary transducer* and additional data from one or more *secondary transducers*. The primary transducer measures the magnitude of the main quantity of interest whilst the secondary transducers measure the magnitude of environmental parameters. For instance, in an intelligent mass-measuring instrument, the primary transducer is usually a load cell and additional secondary transducers are provided to measure environmental inputs such as the ambient temperature and atmospheric pressure.

The microcomputer performs preprogrammed signal processing functions and data manipulation algorithms on the data from the primary transducer (the measured quantity of interest), using data read from the secondary transducer(s), and outputs the processed measurement from the primary transducer for presentation at the instrument output. The effect of this computerization of the signal processing function is an improvement in the quality of the instrument output measurements and a general simplification of the signal processing task. Some examples of the signal processing that a microprocessor within an intelligent instrument can readily perform include correction of the instrument output for bias caused by environmental variations (e.g. temperature changes), and conversion to produce a linear output from a transducer whose characteristic is fundamentally non-linear. A fuller discussion about the techniques of digital signal processing can be found in Chapter 6.

An intelligent instrument behaves as a black box as far as the user is concerned, and no knowledge of its internal mode of operation is required in normal measurement situations. However, the following summary of its operating principles may be of interest to some readers.

Like computer systems in general, the computer hardware within an intelligent instrument contains the essential components of a central processing unit, some memory and an input–output interface. Both ROM (read only memory) and RAM (random access memory) types of memory are included, ROM to store the signal processing program and RAM to provide space to store input–output data. As the external data signals are usually analogue in form, it is normally necessary to include analogue-to-digital and digital-to-analogue conversion elements within the input–output interface.

The signal processing operation requires data values to be input, processed and output according to a sequence of operations defined by the computer program. It is not usual for the user to be expected to write this program. Indeed, there is rarely any provision for the user to create or modify operating programs even if he/she wished to do so. There are several reasons for this. Firstly, the signal processing needed within an intelligent instrument is usually well defined, and therefore it is more efficient for a manufacturer to produce this rather than to have each individual user produce near-identical programs separately. Secondly, better program integrity and instrument operation are achieved if a standard program produced by the instrument manufacturer is used. Finally, use of a standard program allows it to be burnt into ROM, thereby protecting it from any failure of the instrument power supply. This also facilitates software maintenance and updates, by the mechanism of the manufacturer providing a new ROM which simply plugs into the slot previously occupied by the old ROM.

Intelligent instruments offer many advantages over their non-intelligent counterparts, principally because of the improvement in accuracy achieved by processing the output of transducers to correct for errors inherent in the measurement process. The solutions which intelligent instruments offer to many problems occurring in measurement systems will be discussed at various points in later chapters.

One example of the benefit that intelligence can bring to instruments is in volume flow rate measurement, where the flow rate is inferred by measuring the differential pressure across an orifice plate placed in a fluid-carrying pipe (see Chapter 17 for more details). The flow rate is proportional to the square root of the difference in pressure across the orifice plate. For a given flow rate, this relationship is affected both by the temperature and by the mean pressure in the pipe, and changes in the ambient value of either of these cause measurement errors. A typical intelligent flow rate measuring instrument contains three transducers, a primary one measuring the pressure difference across an orifice plate and secondary ones measuring absolute pressure and temperature. The instrument is programmed to correct the output of the primary differential pressure transducer according to the values measured by the secondary transducers, using appropriate physical laws which quantify the effect of ambient temperature and pressure changes on the fundamental relationship between flow and differential pressure. The instrument is also normally programmed to convert the square root relationship between flow and signal output into a direct one, making the output much easier to interpret. Typical inaccuracy levels of such intelligent flow measuring instruments are ± 0.1 percent, compared with ± 0.5 percent for their non-intelligent equivalents, showing an improvement by a factor of five.

Intelligent instruments usually provide many other facilities in addition to those mentioned above, such as:

1. signal damping with selectable time constants;
2. switchable ranges (using several primary transducers within the instrument which each measure over a different range);
3. switchable output units (e.g. display in imperial or SI units);
4. diagnostic facilities;
5. remote adjustment and control of instrument options from up to 1500 metres away via 4-way, 20 mA signal lines.

Suitable care must always be taken when introducing a microcomputer into a measurement system to avoid creating new sources of measurement noise. This is particularly so where one microcomputer is used to process the output of several transducers and is connected to them by signal wires. In such circumstances, the connections and connecting wires can create noise through electrochemical potentials, thermoelectric potentials, offset voltages introduced by common mode impedances, and a.c. noise at power, audio and radio frequencies. Recognition of all these possible noise sources allows them to be eliminated in most cases by employing good measurement system construction practices. All remaining noise sources are usually eliminated by the provision of a set of four earthing circuits within the interface which fulfil the following functions.

- Power earth: provides a path for fault currents due to power faults.
- Logic earth: provides a common line for all logic circuit potentials.

- Analogue earth (ground): provides a common reference for all analogue signals.
- Safety earth: connected to all metal parts of equipment to protect personnel should power lines come into contact with metal enclosures.

5.4.1 Reduction of Random Errors

If a measurement system is subject to random errors, intelligent instruments can be programmed to take a succession of measurements of a quantity within a short space of time and perform simple averaging or other statistical techniques on the readings before displaying an output measurement. This is valid for reducing any form of random error, including those due to human observation deficiencies, electrical noise or other random fluctuations.

As well as displaying an average value obtained from a number of measurements, intelligent instruments are often able to display other statistical parameters of the measurements taken, such as the standard deviation, variance and standard error of the mean. All of these quantities could of course be calculated manually, but the great advantage of using intelligent instruments is their much higher processing speed and the avoidance of the arithmetic errors that are liable to occur if humans perform these functions.

5.4.2 Reduction of Systematic Errors

The inclusion of intelligence in instruments can bring about a large reduction in the magnitude of systematic errors. For instance, in the case of electrical circuits that are disturbed by the loading effect of the measuring instrument, an intelligent instrument can readily correct for measurement errors by applying equations such as (5.14) with the resistance of the measuring instrument inserted.

Intelligent instruments are particularly effective in improving the accuracy of measurements subject to environmental inputs, by the mechanism of the computer within the instrument correcting the measurement obtained from the primary transducer according to the values read by the secondary transducers. However, their ability to achieve this requires that the following preconditions be satisfied.

(a) The physical mechanism by which a measurement transducer is affected by ambient condition changes must be fully understood and all physical quantities which affect the transducer output must be identified.

(b) The effect of each ambient variable on the output characteristic of the measurement transducer must be quantified.

(c) Suitable secondary transducers for monitoring the value of all relevant ambient variables must be available for input to the intelligent instrument.

Condition (a) above means that the thermal expansion/contraction of all elements within a transducer must be considered in order to evaluate how it will respond to ambient temperature changes. Similarly, the transducer response, if any, to changes in ambient pressure, humidity, and gravitational force must be examined.

Quantification of the effect of each ambient variable on the characteristics of the measurement transducer is then necessary, as stated in condition (b). Analytic quantification of ambient condition changes from purely theoretical consideration of the construction of a transducer is usually extremely complex and so is normally avoided. Instead, the effect is quantified empirically in laboratory tests where the output characteristic of the transducer is observed as the ambient environmental conditions are changed in a controlled manner.

Once the ambient variables affecting a measurement transducer have been identified and their effect quantified, an intelligent instrument can be designed which includes secondary transducers to monitor the value of the ambient variables. Suitable transducers which will operate satisfactorily within the environmental conditions prevailing for the measurement situation must of course exist, as stated in condition (c).

5.5 TOTAL MEASUREMENT SYSTEM ERRORS

A measurement system often consists of several separate components, each of which is subject to systematic and/or random errors. Mechanisms have now been presented for quantifying the errors arising from each of these sources and therefore the total error at the output of each measurement system component can be calculated. What remains to be investigated is how the errors associated with each measurement system component combine together, so that a total error calculation can be made for the complete measurement system.

All four mathematical operations of addition, subtraction, multiplication and division may be performed on measurements derived from different instruments/transducers in a measurement system. Appropriate techniques for the various situations which arise are covered below.

5.5.1 Error in a Product

If the outputs y and z of two measurement system components are multiplied together, the product can be written as:

$$P = yz$$

If the possible error in y is $\pm ay$ and in z is $\pm bz$, then the maximum and minimum values possible in P can be written as:

$$P_{max} = (y + ay)(z + bz) \qquad P_{min} = (y - ay)(z - bz)$$

$$= yz + ayz + byz + aybz \qquad = yz - ayz - byz + aybz$$

For typical measurement system components with output errors of up to 1 or 2 percent in magnitude, both a and b are very much less than one in magnitude and thus terms in $aybz$ are negligible compared with other terms. Therefore we have:

$$P_{max} = yz(1 + a + b) \quad P_{min} = yz(1 - a - b)$$

Thus the possible error in the product P lies within the range of $\pm(a + b)$.

Example 5.5
If the power in a circuit is calculated from measurements of voltage and current in which the calculated maximum errors are respectively ± 1 percent and ± 2 percent, then the possible error in the calculated power value is ± 3 percent.

5.5.2 Error in a Quotient

If the output measurement y of one system component with possible error $\pm ay$ is divided by the output measurement z of another system component with possible error $\pm bz$, then the maximum and minimum possible values for the quotient can be written as:

$$Q_{max} = \frac{y + ay}{z - bz} \qquad Q_{min} = \frac{y - ay}{z + bz}$$

$$= \frac{(y + ay)(z + bz)}{(z - bz)(z + bz)} \qquad = \frac{(y - ay)(z - bz)}{(z + bz)(z - bz)}$$

$$= \frac{yz + ayz + byz + abyz}{z^2 - b^2z^2} \qquad = \frac{yz - ayz - byz + abyz}{z^2 - b^2z^2}$$

For $a \ll 1$ and $b \ll 1$, terms in ab and b^2 are negligible compared with the other terms. Hence:

$$Q_{max} \approx \frac{yz(1 + \mathrm{a} + \mathrm{b})}{z^2} \qquad Q_{min} \approx \frac{yz(1 - a - b)}{z^2}$$

i.e.

$$Q = \frac{y}{z} \pm \frac{y}{z}(a + b)$$

Thus the possible error in the quotient lies within the range $\pm(a + b)$.

Example 5.6
If the resistance in a circuit is calculated from measurements of voltage and current where the respective errors are ±1 percent and ±0.3 percent, the likely error in the resistance value is ±1.3 percent.

5.5.3 Error in a Sum

If the two outputs y and z of separate measurement system components are to be added together, we can write the sum as:

$$S = y + z$$

If the maximum errors in y and z are $\pm ay$ and $\pm bz$ respectively, we can express the maximum and minimum possible values of S as:

$$S_{max} = y + ay + z + bz \quad S_{min} = y - ay + z - bz$$

or

$$S = y + z \pm (ay + bz)$$

This relationship for S is not convenient because in this form the error term cannot be expressed as a fraction or percentage of the calculated value for S. Fortunately, statistical analysis can be applied which expresses S in an alternative form so that the most probable maximum error in S is represented by a quantity e, where e is given by:

$$e = \sqrt{[(ay)^2 + (bz)^2]} \tag{5.16}$$

Thus

$$S = (y + z) \pm e.$$

This can be expressed in the alternative form:

$$S = (y + z)(1 \pm f)$$

where

$$f = e/(y + z). \tag{5.17}$$

Example 5.7
A circuit requirement for a resistance of 550 Ω is satisfied by connecting together two resistors of nominal values 220 Ω and 330 Ω in series. If each resistor has a tolerance of ±2 percent, the error in the sum calculated according to equations (5.16) and (5.17) is given by:

$$e = \sqrt{[(0.02 \times 220)^2 + (0.02 \times 330)^2]} = 7.93$$

$$f = 7.93/550 = 0.0144$$

Thus the total resistance S can be expressed as:

$$S = 550\ \Omega \pm 7.93\ \Omega$$

or

$$S = 550(1 \pm 0.0144)\ \Omega \qquad \text{i.e. } S = 550\ \Omega \pm 1.4\%$$

5.5.4 Error in a Difference

If the two outputs y and z of separate measurement systems are to be subtracted from one another, and the possible errors are $\pm ay$ and $\pm bz$, then the difference S can be expressed as:

$$S = (y - z) \pm e \qquad \text{or} \qquad S = (y - z)(1 \pm f)$$

where e is calculated as above (equation 5.15), and $f = e/(y - z)$

Example 5.8
A fluid flow rate is calculated from the difference in pressure measured on both sides of an orifice plate. If the pressure measurements are 10.0 bar and 9.5 bar and the error in the pressure measuring instruments is specified as ±0.1 percent, then values for e and f can be calculated as:

$$e = \sqrt{[(0.001 \times 10)^2 + (0.001 \times 9.5)^2]} = 0.0138$$

$$f = 0.0138/0.5 = 0.0276$$

Thus the pressure difference can be expressed as 0.5 bar ± 2.8 percent.

This example illustrates very poignantly the relatively large error which can arise when calculations are made based on the difference between two measurements.

5.5.5 Total Error when Combining Multiple Measurements

The final case to be covered is where the final measurement is calculated from several measurements which are combined in a way which involves more than one type of arithmetic operation. For example, the density of a rectangular-sided solid block of material can be calculated from measurements of its mass divided by the product of measurements of its length, height and width. The errors involved in each stage of arithmetic are cumulative, and so the total measurement error can be calculated by adding together the two error values associated with the two multiplication stages involved in calculating the volume and then calculating the error in the final arithmetic operation when the mass is divided by the volume.

Example 5.9
A rectangular-sided block has edges of lengths a, b and c, and its mass is m. If the values and possible errors in quantities a, b, c and m are as shown below, calculate the value of density and the possible error in this value.

$$a = 100 \text{ mm} \pm 1\%, \quad b = 200 \text{ mm} \pm 1\%,$$

$$c = 300 \text{ mm} \pm 1\%, \quad m = 20 \text{ kg} \pm 0.5\%.$$

Solution
Value of $ab = 0.02 \text{ m}^2 \pm 2\%$ [possible error = 1% + 1% = 2%]
Value of $(ab)c = 0.006 \text{ m}^3 \pm 3\%$ [possible error = 2% + 1% = 3%]
Value of $\dfrac{m}{(abc)} = \dfrac{20}{0.006} = 3330 \text{ kg/m}^3 \pm 3.5\%$
[possible error = 3% + 0.5% = 3.5%]

REFERENCES

1. Chatfield, C., *Statistics for Technology*, Chapman and Hall, London (1983).

CHAPTER 6

Transmission and Processing of Measurement Signals

Having considered the subject of measurement errors in detail, consideration will now be given in this chapter to the mechanisms available and problems encountered in transmitting measurement signals from the point of measurement to the place where the signals are recorded, analysed and used. For various reasons, signal transmission introduces errors into measurements, and therefore this chapter will go on to discuss the various ways in which measurement signals can be processed to improve their quality. In fact, signal processing is usually necessary even in the absence of transmission-induced errors because, even though the aim is to design measurement systems and transducers so that errors are minimized, it is not always possible, or at least it is not always cost-effective, to remove all errors.

6.1 SIGNAL TRANSMISSION

Measurement sensors and transducers are often located in process plant operating in very hostile conditions, such as high temperatures, flames, fumes, smoke, steam and large volumes of dust. In such environments, it is impossible to locate the other measurement system components which are required to process the sensor outputs, and therefore the signals from the measurement devices have to be transmitted to a different point, away from the harsh environment, where it is more feasible to locate the other measurement system components.

The need to locate measuring system components away from the harsh environment which exists close to most process plant means that the measurement signals from plant sensors must be transmitted at least a moderate distance of a few metres. However, organizational and economic considerations often require that all process plant signals be collected together at one point for analysis, processing and recording of the information received. Where such a central plant control room is provided, this necessitates transmission of the measurement signals over even greater distances than would otherwise be necessary.

It is inevitable that, during transmission from the measurement point to another location, there will be some degradation in the quality of the signals. Hence, the main aim in designing signal transmission systems is to minimize this degradation. The most serious difficulty associated with long-distance signal transmission is attenuation of the signal due to the resistance of the signal wires, and this causes particular difficulties in the case of sensors that have a low-magnitude output. This obviously gets worse as the transmission distance increases. One way to compensate for attenuation is to amplify the signal prior to transmission, although, unless transmission distances are very large, the problem can be avoided altogether if the conductors used to carry the signals are of adequate cross-section. However, if a reasonable signal-to-noise ratio is to be obtained for low-output sensors, signal amplification before transmission still remains necessary.

Contamination of the measurement signal by noise during transmission is a further problem. Many sources of noise exist in industrial environments, such as radiated electromagnetic fields from electrical machinery and power cables, induced fields through wiring loops, and voltage spikes in the mains power supply. Fortunately, the magnitude of this noise contamination can be greatly reduced if the wires carrying the signals are adequately shielded (see section on shielding below).

Many other techniques of improving the quality of signal transmission exist. Most of these consist of converting the measurement signal prior to transmission from its usual varying voltage form into some other form which is less susceptible to contamination during transmission. It is then usually converted back to a varying voltage after transmission. Transmission as a varying current is a common method, which requires both a voltage-to-current converter and then a current-to-voltage converter. Digital transmission, where the varying voltage is converted into a varying frequency signal is also commonly employed. Finally, mention must be made of fibre-optic transmission, which is increasingly common. This converts the measurement signal into a modulated light wave which is transmitted along a fibre-optic cable. All of these techniques are discussed in the sections below.

Signal Amplification

Signal amplification is carried out when the typical output level of a measurement transducer is considered to be too low. Amplification by analogue means is usually carried out by an operational amplifier. This is normally required to have a high input impedance so that its loading effect on the transducer output signal is minimized. In some circumstances, such as when amplifying the output signal from accelerometers and some optical detectors, the amplifier must also have a high frequency response, to avoid distortion of the output reading.

The operational amplifier is an electronic device which has two input terminals and one output terminal, the two inputs being known as the inverting input and non-inverting input respectively. When connected as shown in Figure 6.1, it provides signal amplification. The raw (unprocessed) signal V_i is connected to the inverting input through a resistor R_1 and the non-inverting input is connected to ground (earth). A feedback path is provided from the output terminal through a resistor R_2 to the inverting input terminal. Assuming ideal operational amplifier characteristics, the processed signal V_o at the output terminal is then related to the voltage V_i at the input terminal by the expression:

$$V_o = -\frac{R_2 V_i}{R_1} \tag{6.1}$$

The amount of signal amplification is therefore defined by the relative values of R_1 and R_2. This ratio between R_1 and R_2 in the amplifier configuration is often known as the amplifier gain or closed-loop gain. If, for instance, $R_1 = 1\ \mathrm{M\Omega}$ and $R_2 = 10\ \mathrm{M\Omega}$, an amplification factor of 10 is obtained (i.e. gain = 10). It is important to note that, in this standard way of connecting the operational amplifier (often known as the inverting configuration), the polarity (sign) of the processed signal is inverted. This can be corrected if necessary by feeding the signal through a further amplifier configured for unity gain ($R_1 = R_2$). This inverts the signal again and returns it to its original polarity.

Shielding

Shielding consists of surrounding the signal wires in a cable with a braided metal shield which is connected to earth. This provides a high degree of

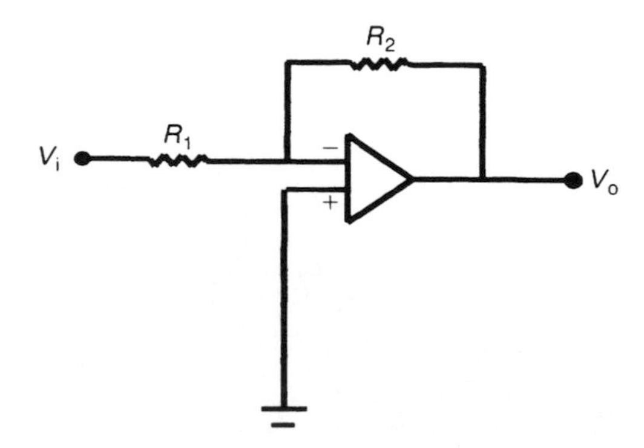

Figure 6.1. Operational amplifier connected for signal amplification

noise protection, especially against capacitive-induced noise due to the proximity of signal wires to high-current power conductors.

Current Loop Transmission (Voltage-to-Current Conversion)

The signal-attenuation effect of conductor resistances can be minimized if varying voltage signals are transmitted as varying current signals. This requires a voltage-to-current converter of the form shown in Figure 6.2, which is commonly known as a 4–20 mA current-loop interface. Two voltage-controlled current sources are used, one (I_1) providing a constant 4 mA output which is used as the power supply current, and the other (I_2) providing a variable 0–16 mA output which is proportional to the input voltage level. The net output current therefore varies between 4 mA and 20 mA. This is a very commonly used means of connecting remote instruments to a central control room.

Current-to-Voltage Conversion

Current-to-voltage conversion is used at the termination of transmission lines to change signals transmitted as varying currents back to varying voltages. An operational amplifier, connected as shown in Figure 6.3, is

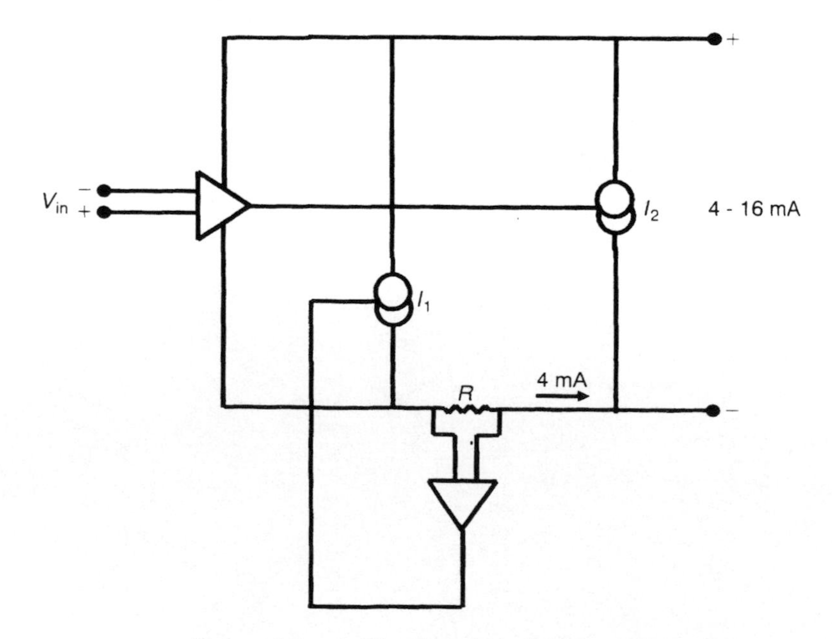

Figure 6.2. 4–20 mA current loop interface

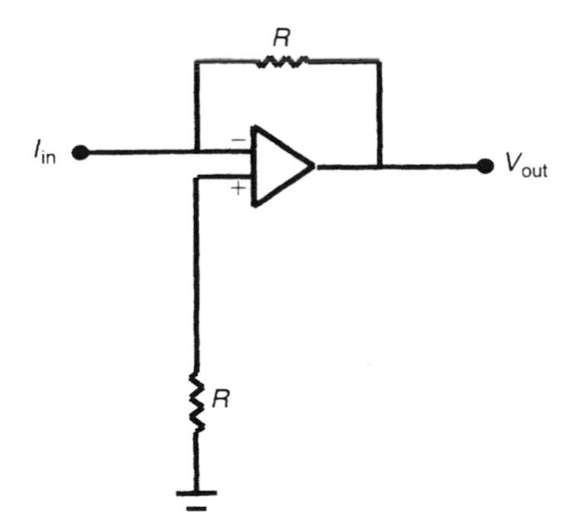

Figure 6.3. Operational amplifier connected for current-to-voltage conversion

suitable for this purpose. The output voltage V_{out} is simply related to the input current I by:

$$V_{out} = IR$$

Digital Transmission (Voltage-to-Frequency Conversion)

High immunity to noise can be obtained in signal transmission if the signal is transmitted in a digital format. This is achieved by applying the input analogue voltage signal to the input of a voltage-to-frequency converter circuit which converts the voltage variations into corresponding frequency variations. Such frequency variations can then be readily transmitted in a digital format. After transmission, reconversion into an analogue voltage signal is possible using a frequency-to-voltage converter.

Fibre-Optic Transmission

Fibre-optic signal transmission involves transforming electrical signals into a modulated light wave which is transmitted along a fibre-optic cable. Then, at the receiving end of the cable, the light is transformed back into electrical form.

Light has a number of advantages over electricity as a medium for transmitting information: it is immune to corruption by neighbouring electromagnetic fields, the attenuation over a given transmitted distance is much less and it is also intrinsically safe. However, there is an associated cost penalty because of the higher cost of a fibre-optic system compared with the cost of metal conductors. The primary reason for this penalty is the high cost of the terminating transducers which perform the signal conversion function at each end of the cable.

The light-transmitting cable contains at least one, but more often a bundle, of glass or plastic fibres. This is terminated at each end by a transducer, as shown in Figure 6.4. At the input end, the transducer converts the signal from the electrical form, in which most signals originate, into light. At the output end, the transducer converts the transmitted light back into an electrical form suitable for use by data recording, manipulation and display systems. These two transducers are often known as the transmitter and receiver respectively.

Signals are normally transmitted along a fibre-optic cable in digital format, although analogue transmission is sometimes used. If there is a requirement to transmit more than one signal, it is more economic to multiplex the signals onto a single cable rather than transmit the signals separately on multiple cables. *Time-division multiplexing* involves switching the analogue signals in turn, in a synchronized sequential manner, into an analogue-to-digital converter which outputs onto the transmission line. At the other end of the transmission line, a digital-to-analogue converter transforms the digital signal back into analogue form and it is then switched in turn onto separate analogue signal lines.

6.2 SIGNAL PROCESSING

As observed in the introduction to this chapter, some form of signal processing is usually necessary whether or not a significant transmission distance is involved between the measurement sensors and the rest of the measurement system. Its purpose is to improve the quality of the reading or signal at the output of the measurement system. Signal processing is designed to compensate both for errors in the raw (unprocessed) data captured by the sensor and also for transmission-induced errors.

The form that signal processing takes depends on the nature of the raw output signals from the measurement transducers. Procedures of signal amplification, signal attenuation, signal linearization, bias removal and signal filtering are all particular forms of signal processing which are applied according to the form of correction required in the raw signal. Mention will also be made in the following sections of certain other special-purpose devices and circuits that are used to manipulate signals for particular purposes. Signal processing functions are normally carried out

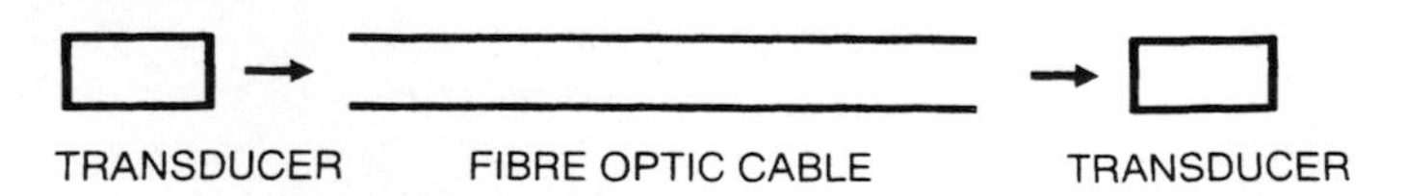

Figure 6.4. Fibre-optic signal transmission

after transmission of the measurement signal from the sensor, except for amplification, which is often needed before transmission.

The implementation of signal processing procedures can be carried out either by analogue techniques or by digital computation on a computer. The choice between these is largely determined by the degree of accuracy required in the signal processing procedure. Analogue signal processing involves the use of various electronic circuits, usually built around the operational amplifier, whereas digital signal processing uses software modules on a digital computer to condition the input measurement data. Digital signal processing is inherently more accurate than analogue techniques, but this advantage is greatly reduced in the case of measurements coming from analogue transducers, which have to be converted by an analogue-to-digital converter prior to digital processing, thereby introducing conversion errors. Digital processing is also slower than analogue processing. Therefore, it is common practice to use analogue techniques for all signal processing tasks except where the accuracy of this is insufficient. It should be noted also that in some measurement situations where a physical quantity is measured by an inherently inaccurate transducer, the extra accuracy provided by digital signal processing is insignificant and therefore inappropriate.

For the purpose of explaining the procedures involved, this chapter concentrates on analogue signal processing and concludes with a relatively brief discussion of the equivalent digital signal processing techniques. One particular reason for this method of treatment is that some prior analogue signal conditioning is often necessary even when the major part of the signal processing is carried out digitally.

6.2.1 Analogue Signal Processing

Signal Amplification

Signal amplification is frequently needed, both for increasing the signal level of transducers that have a low-magnitude output, such as thermocouples, and also for compensating for attenuation of signals during transmission from remote sensors. The standard means of amplifying measurement signals has already been covered in Section 6.1. Therefore, the opportunity will be taken here to introduce a special type of amplifier, known as an *instrumentation amplifier*, which is specially designed and used for the amplification of very low-level signals. This consists of a circuit containing three standard operational amplifiers, as shown in Figure 6.5. The advantage of the instrumentation amplifier compared with a standard operational amplifier is that its differential input impedance is much higher. In consequence, its common mode rejection capability† is much better. This means that, if a

† Common mode rejection ratio describes the ability of the amplifier to reject equal-magnitude signals that appear on both of its inputs.

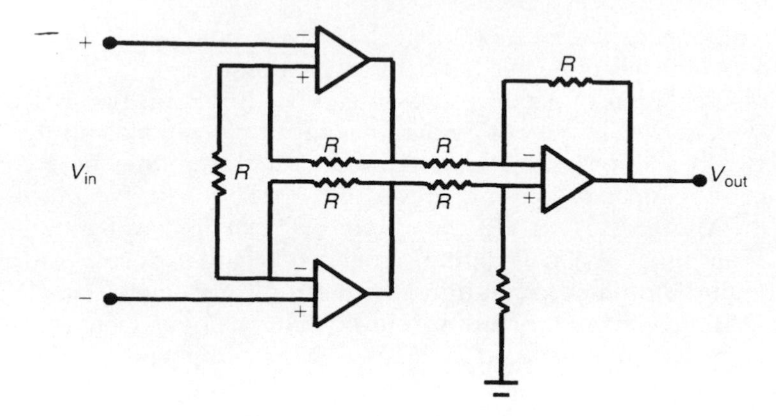

Figure 6.5. Instrumentation amplifier

twisted-wire pair is used to connect a transducer to the differential inputs of the amplifier, any induced noise will contaminate each wire equally and will be rejected by the common mode rejection capacity of the amplifier.

Signal Attenuation

One method of attenuating signals by analogue means is to use a potentiometer connected in a voltage-dividing circuit, as shown in Figure 6.6. For the potentiometer wiper positioned a distance X_w along the resistance element of total length X_t, the voltage level of the processed signal V_o is related to the voltage level of the input signal V_i by the expression:

$$V_o = \frac{X_w V_i}{X_t} \tag{6.2}$$

A major problem with potentiometers in many circuits is that the output can be affected by the impedance of the device (or circuit) connected to its output terminals. An alternative device to the potentiometer for signal attenuation is the operational amplifier. This is connected in the same way

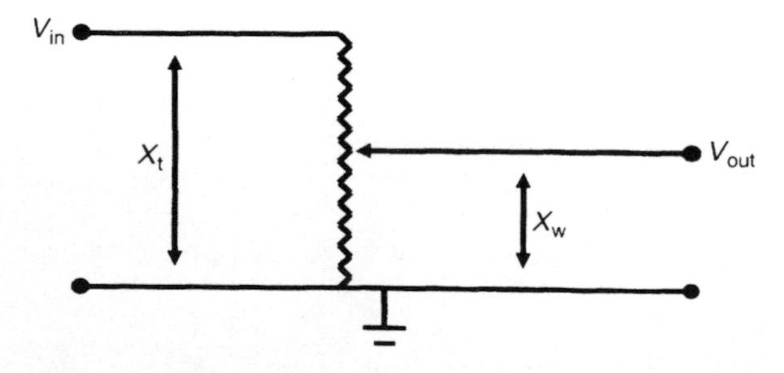

Figure 6.6. Potentiometer in voltage-dividing circuit

as the amplifier shown in Figure 6.1, but R_1 is chosen to be greater than R_2. Equation 6.1 is still valid and therefore, if R_1 is chosen to be 10 MΩ and R_2 as 1 MΩ, an attenuation factor of ten is achieved (gain = 0.1). Use of an operational amplifier as an attenuating device is a more expensive solution than using a potentiometer, particularly because it is an active device and needs a power supply. However, unlike the potentiometer, it is relatively unaffected by the device or circuit connected to its output terminals.

Signal Linearization

Several types of transducer used in measuring instruments have an output which is a non-linear function of the measured input quantity. In many cases, this non-linear signal can be converted to a linear one by special operational amplifier configurations which have an equal and opposite non-linear relationship between the amplifier input and output terminals.

For example, light intensity transducers typically have an exponential relationship between the output signal and the input light intensity of the form:

$$V_o = Ke^{-\alpha Q} \tag{6.3}$$

where Q is the light intensity, V_o is the voltage level of the output signal, and K and α are constants.

If a diode is placed in the feedback path between the input and output terminals of the amplifier as shown in Figure 6.7, the relationship between the amplifier output voltage V_0 and input voltage V_1 is given by:

$$V_0 = C \log_e (V_1) \tag{6.4}$$

If the output of the light intensity transducer with a characteristic given by equation (6.3) is conditioned by an amplifier with a characteristic given by equation (6.4), the voltage level of the processed signal is given by:

$$V_0 = C \log_e (K) - \alpha C Q \tag{6.5}$$

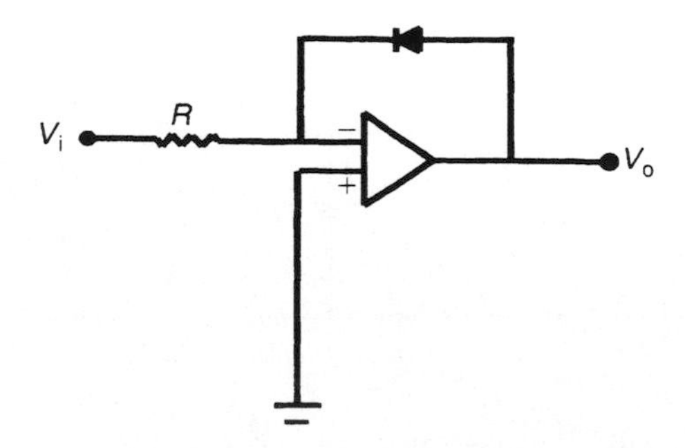

Figure 6.7. Operational amplifier connected for signal linearization

Expression (6.5) shows that the output signal now varies linearly with light intensity Q but with an offset of $C \log_e (K)$. This offset or bias would normally be removed by further signal conditioning, as described below.

Bias Removal

Sometimes, either because of the nature of the measurement transducer itself, or as a result of other signal conditioning operations such as linearization as just described, a bias exists in the output signal. This can be expressed mathematically for a physical quantity x and measurement signal y as:

$$y = Kx + C \tag{6.6}$$

where C represents a bias in the output signal which needs to be removed by signal processing. Analogue processing consists of using an operational amplifier connected in a differential amplification mode, as shown in Figure 6.8. Referring to this circuit, for $R_1 = R_2$ and $R_3 = R_4$, the output V_o is given by:

$$V_o = (R_3/R_1)(V_c - V_i) \tag{6.7}$$

where V_i is the unprocessed measurement signal equal to $(Kx + C)$ and V_c is the output voltage from a potentiometer supplied by a known reference voltage V_{ref}, which is set such that $V_c = C$. Now, substituting these values for V_i and V_c into equation (6.7), y can be written as:

$$y = K'x \tag{6.8}$$

where the new constant K' is related to K according to $K' = -K(R_3/R_1)$. It is clear that the bias has been successfully removed and equation (6.8) is now

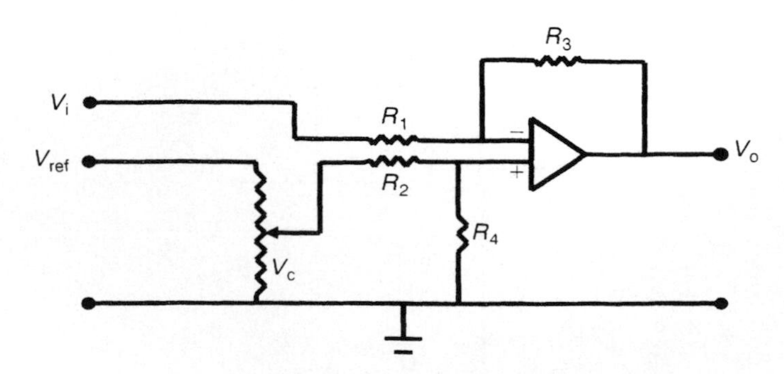

Figure 6.8. Operational amplifier connected in differential amplification mode for bias removal

a linear relationship between the measurement signal y and the measured quantity x.

Signal Filtering

Signal filtering consists of processing a signal to remove a certain band of frequencies within it. The band of frequencies removed can be either at the low-frequency end of the frequency spectrum, at the high-frequency end, at both ends, or in the middle of the spectrum. Filters to perform each of these operations are known respectively as low-pass filters, high-pass filters, band-pass filters and band-stop filters. All such filtering operations can be carried out by either analogue or digital methods.

The result of filtering can be readily understood if the analogy with a procedure such as sieving soil particles is considered. Suppose that a sample of soil A is passed through a system of two sieves of differing meshes such that the soil is divided into three parts, B, C and D, consisting of large, medium and small particles, as shown in Figure 6.9 . Suppose that the system also has a mechanism for delivering one or more of the separated parts, B, C and D, as the system output. If the graded soil output consists of parts C and D, the system is behaving as a low-pass filter (rejecting large particles), whereas if it consists of parts B and C, the system is behaving as a high-pass filter (rejecting small particles). Other options are to deliver just

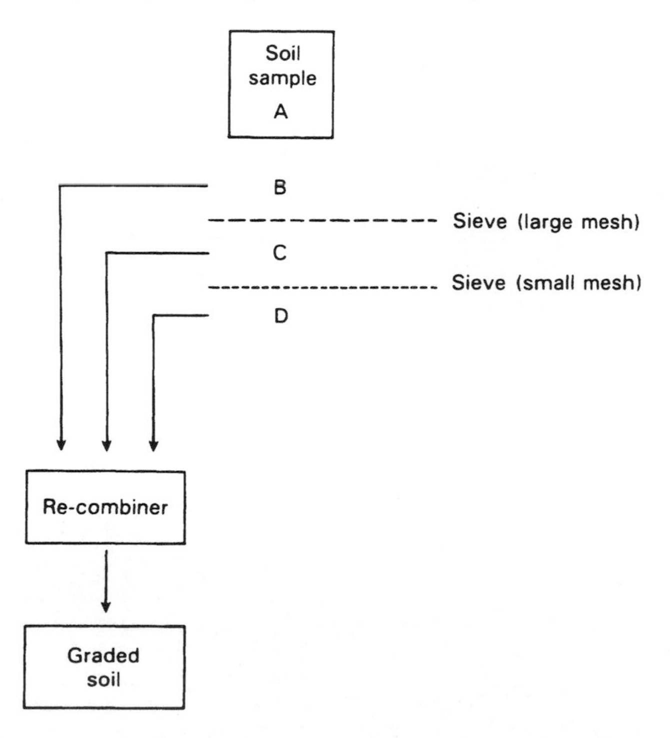

Figure 6.9. Soil sieving analogy of signal filtering

part C (band-pass filter mode) or parts B and D together (band-stop filter mode). As any gardener knows, however, such perfect sieving is not achieved in practice and any form of graded soil output always contains a few particles of the wrong size.

Signal filtering consists of selectively passing or rejecting low-, medium- and high-frequency signals from the frequency spectrum of a general signal. The range of frequencies passed by a filter is known as the *pass band*, the range not passed is known as the *stop band*, and the boundary between the two ranges is known as the *cut-off frequency*. To illustrate this, consider a signal whose frequency spectrum is such that all frequency components in the frequency range from zero to infinity have equal magnitude. If this signal is applied to an ideal filter, then the outputs for a low-pass filter, high-pass filter, band-pass filter and band-stop filter respectively are as shown in Figure 6.10. Note that for the last two types, the bands are defined by a pair of frequencies rather than by a single cut-off frequency.

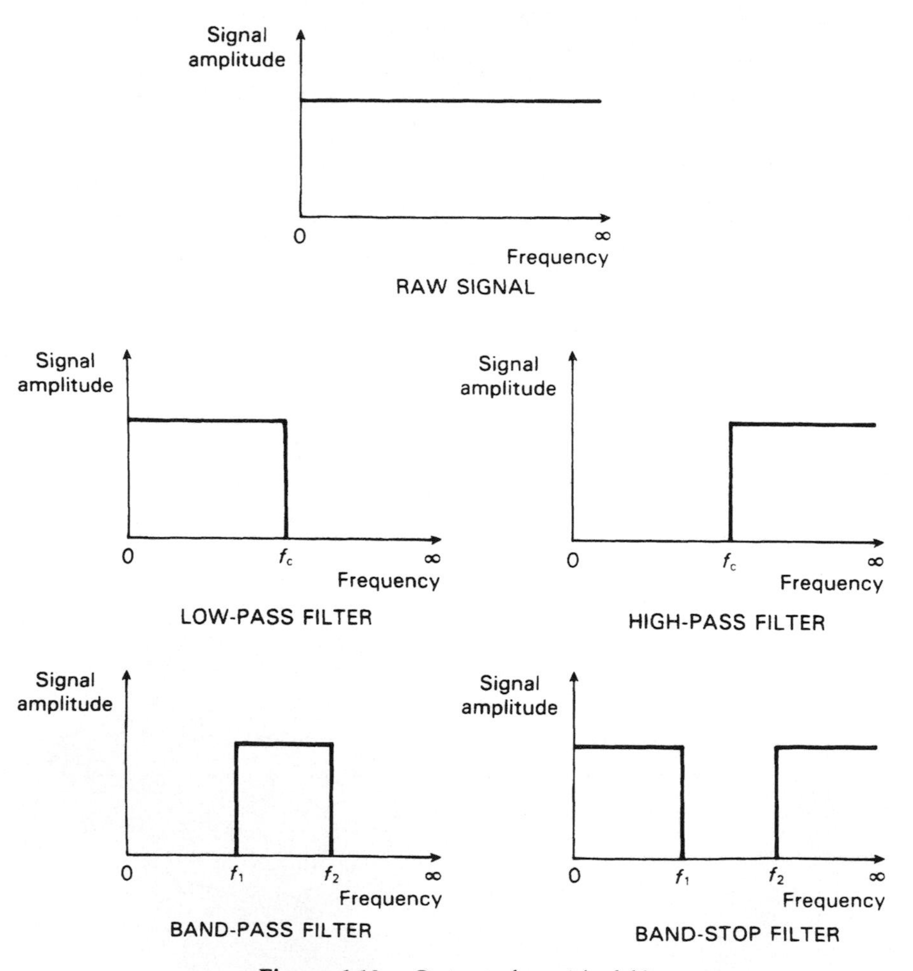

Figure 6.10. Outputs from ideal filters

Just as in the case of the soil sieving analogy presented above, the signal filtering mechanism is not perfect, with unwanted frequency components not being erased completely but only attenuated by varying degrees instead, i.e. the filtered signal always retains some components (of relatively low magnitude) in the unwanted frequency range. There is also a small amount of attenuation of frequencies within the pass-band which increases as the cut-off frequency is approached. Figure 6.11 shows the typical output characteristics of a practical constant-k† filter designed respectively for high-pass, low-pass, band-pass and band-stop filtering. Filter design is concerned with trying to obtain frequency rejection characteristics that are as close to the ideal as possible. However, improvement in characteristics is only achieved at the expense of greater complexity in the design. The filter chosen for any given situation is therefore a compromise between performance, complexity and cost.

In the majority of measurement situations, the physical quantity being measured has a value which is either constant or changing only slowly with time. In these circumstances, the most common types of signal corruption are high-frequency noise components, and the type of signal processing element required is a low-pass filter. In a few cases, the measured signal

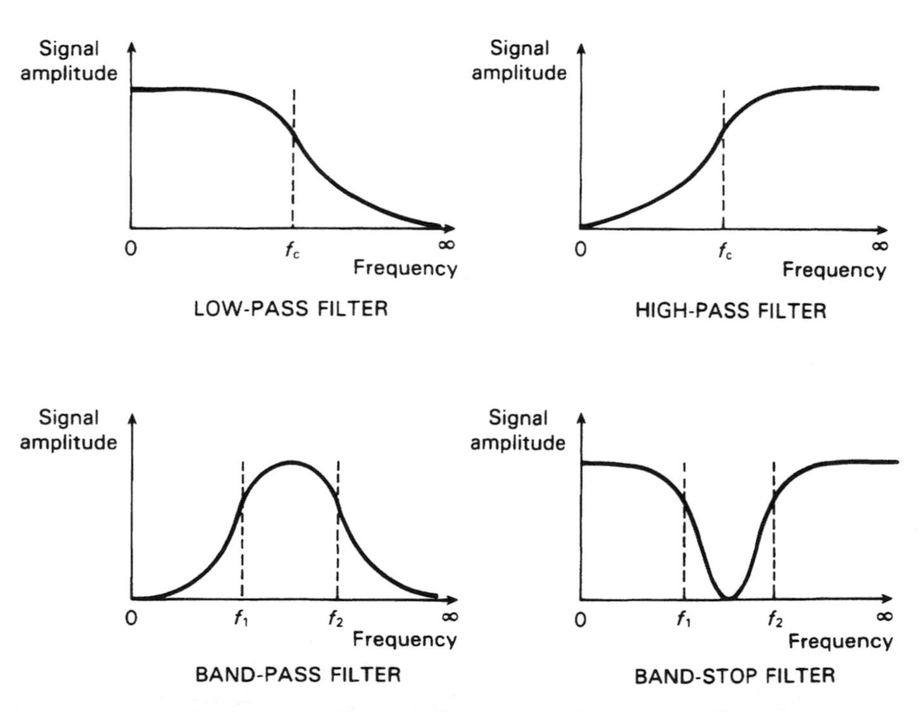

Figure 6.11. Outputs from practical constant-k filters

† 'constant-k' is a term used to describe a common class of passive filters, as discussed in the following section.

itself has a high frequency, for instance when mechanical vibrations are being monitored, and the signal processing required is the application of a high-pass filter to attenuate low-frequency noise components. Band-stop filters can be used where a measurement signal is corrupted by noise at a particular frequency. Such noise is frequently due to mechanical vibrations or proximity of the measurement circuit to other electrical apparatus.

Both passive and active implementations exist for analogue filters. However, the detailed design of these is somewhat complicated and outside the scope of this book. Interested readers are referred to specialist texts on the subject, references [1, 2, 3] for passive filters and [4, 5, 6] for active ones.

Signal Integration

This can be achieved with an operational amplifier connected in the configuration shown in Figure 6.12. This integrates the input signal V_i so that the output signal V_0 is given by:

$$V_o = -\frac{1}{RC}\int V_i \, dt$$

This circuit is used whenever there is a requirement to integrate the output signal from a transducer.

Preamplifier (Voltage Follower)

The preamplifier, also known as a voltage follower, is a unity gain amplifier circuit with a short circuit in the feedback path, as shown in Figure 6.13, such that:

$$V_o = V_i$$

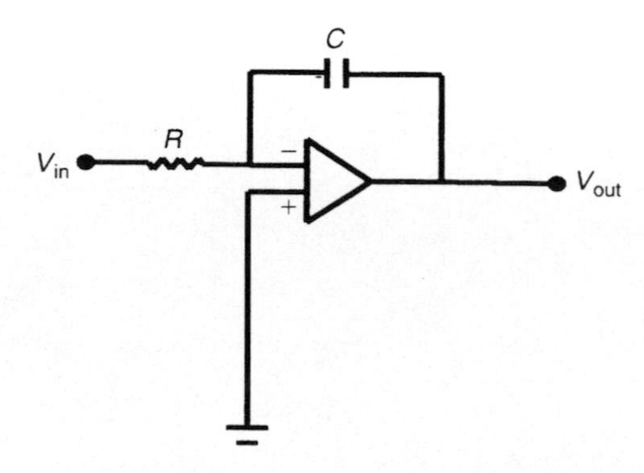

Figure 6.12. Operational amplifier connected as integrating element

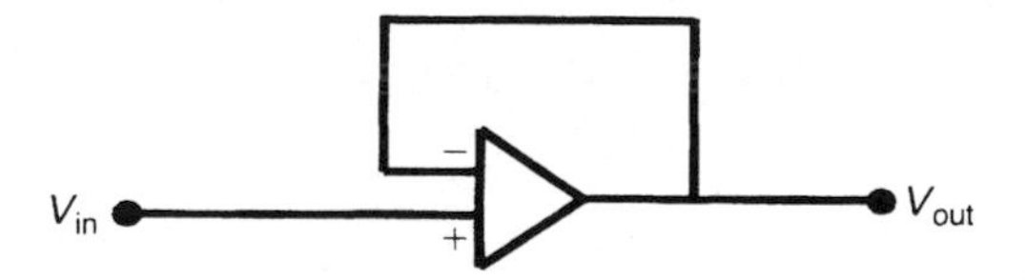

Figure 6.13. Operational amplifier connected as preamplifier (voltage follower)

It has a very high input impedance and its main application is to reduce the load on the measured system. It also has a very low output impedance which is very useful in some impedance-matching applications.

Voltage Comparator

The output of a voltage comparator switches between positive and negative values according to whether the difference between the two input signals to it is positive or negative. An operational amplifier connected as shown in Figure 6.14 gives an output which switches between positive and negative saturation levels according to whether ($V_1 - V_2$) is greater than or less than zero. Alternatively, the voltage of a single input signal can be compared against positive and negative reference levels with the circuit shown in Figure 6.15.

In practice, operational amplifiers have drawbacks as voltage comparators for several reasons. These include propagation delays, and incompatibility of output voltage levels and industry-standard logic circuits. In

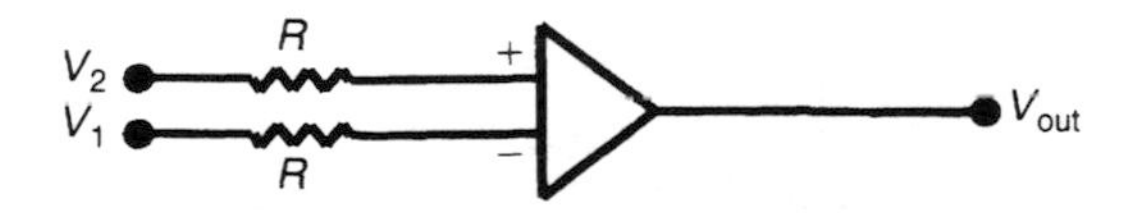

Figure 6.14. Comparison of two voltage signals

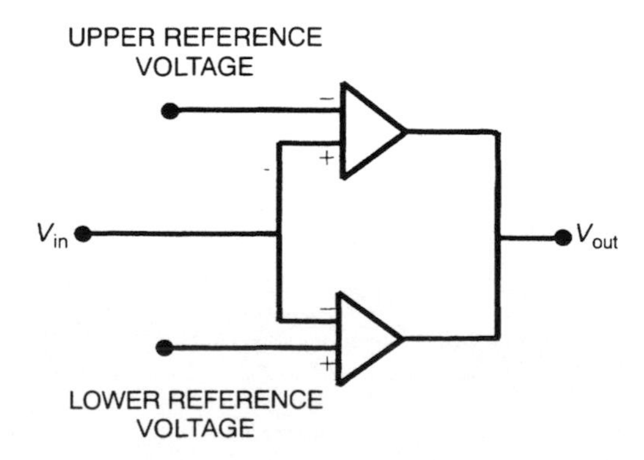

Figure 6.15. Comparison of input signal against reference value

consequence, various other special purpose integrated circuits have been developed for voltage comparison.

Phase-Locked Loop

The phase-locked loop, shown schematically in Figure 6.16, is primarily a circuit for measuring the frequency of a signal. However, because the output waveform is a pure (i.e. perfectly clean) square wave at the same frequency as the input signal, irrespective of the amount of noise, modulation or distortion on the input signal, the phase-locked loop also finds application as a signal processing element for cleaning up poor quality signals.

Signal Addition

The most common mechanism for summing two or more input signals is the use of an operational amplifier connected in signal-inversion mode, as shown in Figure 6.17. For input signal voltages V_1, V_2 and V_3, the output voltage V_o is given by:

$$V_o = -(V_1 + V_2 + V_3)$$

Signal Multiplication

Great care must be taken when choosing a signal multiplier because, whilst many circuits exist for multiplying two analogue signals together, most of them are two-quadrant types which only work for signals of a single polarity, i.e. both positive or both negative. Such schemes are unsuitable for

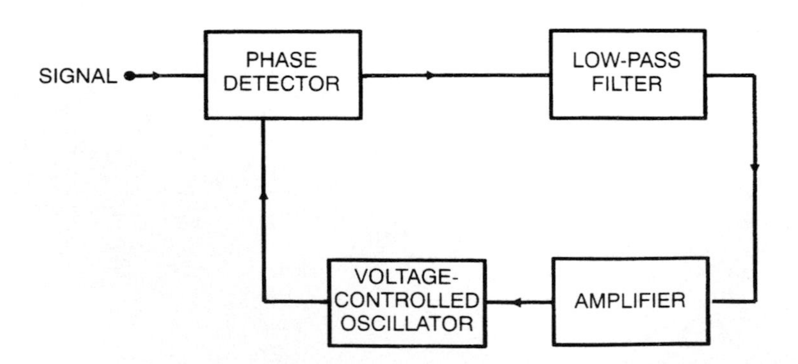

Figure 6.16. Phase-locked loop

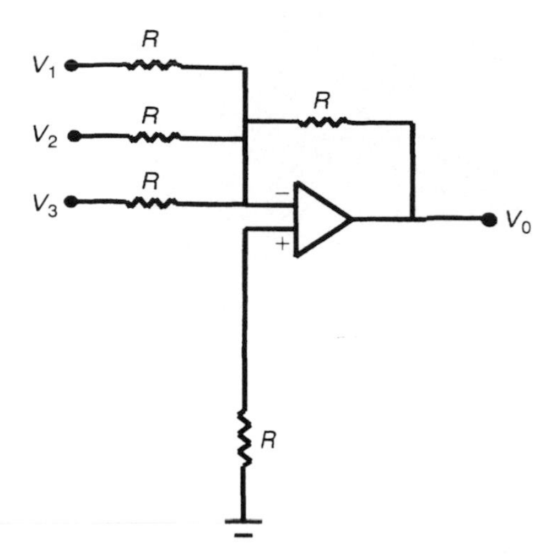

Figure 6.17. Operational amplifier connected for signal addition

general analogue signal processing, where the signals to be multiplied may be of changing polarity.

For analogue signal processing, a four-quadrant multiplier is required. Two forms of such a multiplier are easily available, the Hall-effect multiplier and the translinear multiplier.

Sample and Hold circuit

A sample and hold circuit is often an essential element at the interface between an analogue instrument/transducer and an analogue-to-digital converter. It holds the input signal at a constant level whilst the analogue-to-digital conversion process is taking place and prevents the conversion errors which would probably result if variations in the measured signal were allowed to pass through to the converter. The operational amplifier circuit shown in Figure 6.18 provides this sample and hold function. The input signal is applied to the circuit for a very short time duration with switch S_1 closed and S_2 open, after which S_1 is opened and the signal level is then held until, when the next sample is required, the circuit is reset by closing S_2.

Analogue-to-Digital Conversion

If the analogue measurement signals are going to be input to a computer, there is a fundamental mismatch between the analogue form of output data from sensors and transducers and the digital form of data required by a digital computer. This problem is solved by the provision of an analogue-to-digital converter in the computer input interface. If the computer then

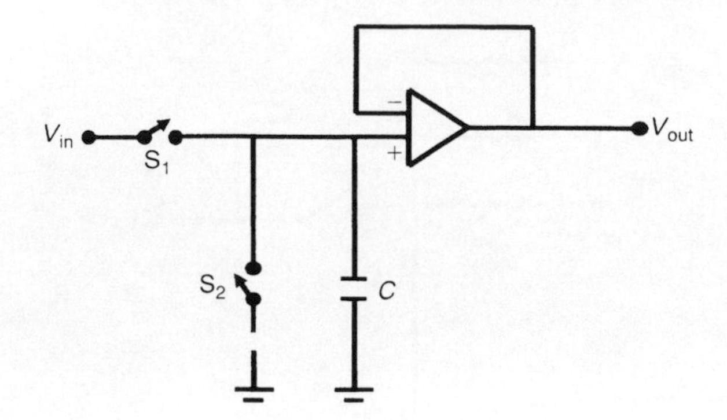

Figure 6.18. Operational amplifier connected as 'sample and hold' circuit

computes a control signal which needs to be output to an actuator which requires a signal in analogue form, a *digital-to-analogue converter* will also be required in the computer output interface.

6.2.2 Digital Signal Processing

Digital techniques achieve much greater levels of accuracy in signal processing than equivalent analogue methods. However, the time taken to process a signal digitally is longer than that required to carry out the same operation by analogue techniques, and the equipment required is more expensive. Some care is therefore needed in making the correct choice between digital and analogue methods for a particular signal processing application.

Whilst digital signal processing elements in a measurement system can exist as separate units, it is more usual to find them as an integral part of an intelligent instrument. However, their construction and mode of operation are the same irrespective of whether they exist physically as separate boxes or within an intelligent instrument.

The hardware aspect of a digital signal processing element consists of a digital computer and analogue interface boards. The actual form that signal processing takes depends on the software program executed by the processor. However, before consideration is given to this, some theoretical aspects of signal sampling need to be discussed.

As mentioned earlier, in Section 6.2.1, digital computers require signals to be in digital form whereas most instrumentation transducers have an output signal in analogue form. Analogue-to-digital conversion is therefore required at the interface between analogue transducers and a digital computer. The procedure followed is to sample the analogue signal at a particular moment in time and then convert the analogue value to an equivalent digital one. This conversion takes a certain finite time, during which the analogue signal can be changing in value. The next sample of the analogue signal cannot be taken until the conversion of the last sample to digital form is completed. The representation within a digital computer of a continuous analogue signal is therefore a sequence of samples whose pattern only approximately follows the shape of the original signal. This pattern of samples taken at successive, equal intervals of time is known as a discrete signal. The process of conversion from a continuous analogue signal to a discrete digital one is illustrated for a sine wave in Figure 6.19.

The raw analogue signal in Figure 6.19 has a frequency of approximately 0.75 cycles per second. With the rate of sampling shown, which is approximately 11 samples per second, reconstruction of the samples matches the original analogue signal very well. If the rate of sampling were decreased, the fit between the reconstructed samples and the original signal would be less accurate. If the rate of sampling were very much less than the frequency of the raw analogue signal, such as 1 sample per second, only the samples marked 'X' in Figure 6.19 would be obtained. Fitting a line through these 'X's incorrectly estimates a signal whose frequency is approximately 0.25 cycles per second. This phenomenon, whereby the process of sampling transmutes a high-frequency signal into a lower frequency one, is known as *aliasing*. To avoid aliasing, it is necessary theoretically for the sampling rate to be at least twice the highest frequency in the analogue signal sampled. In practice, sampling rates of between 5 and 10 times the highest frequency signal are normally chosen so that the discrete sampled signal is a close approximation to the original analogue signal in amplitude as well as frequency.

Problems can arise in sampling when the raw analogue signal is corrupted by high-frequency noise of unknown characteristics. It would be normal practice to choose the sampling interval as say a ten-times multiple of the frequency of the measurement component in the raw signal. If such a sampling interval is chosen, aliasing can in certain circumstances transmute high-frequency noise components into the same frequency range as the measurement component in the signal, thus giving erroneous results. This

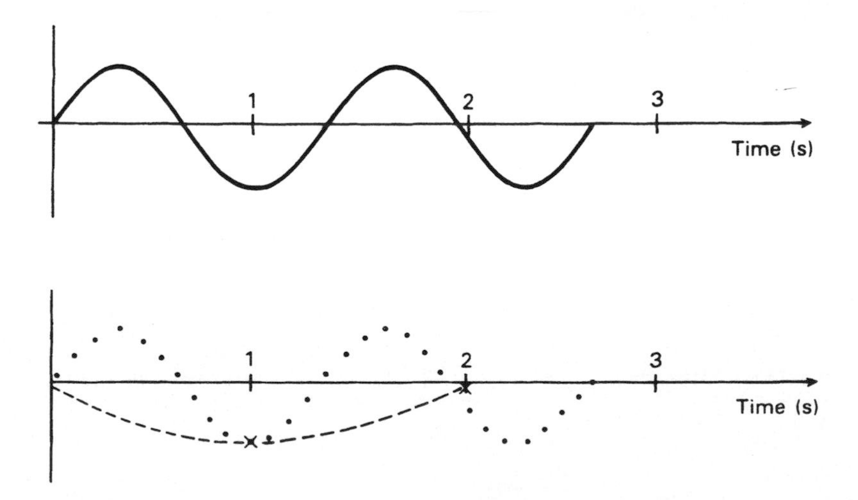

Figure 6.19. Conversion of continuous analogue signal to discrete sampled signal

is one of the circumstances mentioned earlier, where prior analogue signal conditioning using a low-pass filter must be carried out before processing the signal digitally.

One further factor which affects the quality of a signal when it is converted from analogue to digital form is *quantization*. Quantization describes the procedure whereby the continuous analogue signal is converted into a number of discrete levels. At any particular value of the analogue signal, the digital representation is either the discrete level immediately above this value or the discrete level immediately below this value. If the difference between two successive discrete levels is represented by the parameter Q, then the maximum error in each digital sample of the raw analogue signal is $\pm Q/2$. This error is known as the quantization error and is clearly proportional to the resolution of the analogue-to-digital converter, i.e. to the number of bits used to represent the samples in digital form.

Once a satisfactory digital representation in discrete form of an analogue signal has been obtained, the procedures of signal amplification, signal attenuation and bias removal become trivial. For signal amplification and attenuation, all samples have to be multiplied or divided by a fixed constant. Bias removal involves simply adding or subtracting a fixed constant from each sample of the signal.

Signal linearization requires a priori knowledge of the type of nonlinearity involved, in the form of a mathematical equation which expresses the relationship between the output measurements from an instrument and the value of the physical quantity being measured. This can be obtained either theoretically through knowledge of the physical laws governing the system or empirically using input–output data obtained from the measurement system under controlled conditions. Once this relationship has been obtained, it is used to calculate the value of the measured physical quantity corresponding to each discrete sample of the measurement signal. Whilst the amount of computation involved in this is greater than in the trivial cases of signal amplification, etc., already mentioned, the computational burden is still relatively small in most measurement situations.

Digital signal processing can also perform all of the filtering functions mentioned earlier in respect of analogue filters, i.e. low-pass, high-pass, band-pass and band-stop. However, the design of digital filters requires a high level of theoretical knowledge, including the use of z-transform theory, which is outside the range of this book. The reader is therefore referred to specialist texts which describe digital filter design [6, 7].

REFERENCES

1. Blinchikoff, H. J., *Filtering in the time and frequency domains*, John Wiley & Sons, New York (1976).
2. Skilling, H. H., *Electrical Engineering Circuits*, John Wiley & Sons, New York (1967).

3. Williams, E., *Electric filter circuits*, Pitman, London (1963).
4. Hilburn, J. L. and Johnson, D. E., *Manual of active filter design*, McGraw-Hill, New York (1973).
5. Stephenson, F. W., *Active filter design handbook*, John Wiley & Sons, New York (1985).
6. Huelsman, L. P., *Active Filters: Lumped, Distributed, Integrated, Digital and Parametric*, McGraw-Hill, New York (1970).
7. Lynn, P. A., *The Analysis and Processing of Signals*, Macmillan, London (1973).

CHAPTER 7

Measurement Signal Recording and Data Presentation

7.1 INTRODUCTION

Earlier chapters have considered the various factors which are relevant to obtaining good quality measurements. However, there has been little consideration so far about how measurements are used. In many situations, measurements are used immediately in the control of manufacturing systems or processes. Such use typically takes the form of a plant operator reading a measurement of a process parameter and then taking appropriate control action. For example, an operator controlling a furnace will read a thermometer and then either turn up or turn down the heat source according to the measured temperature. Alternatively, measurement signals can be fed directly into an automatic control system which applies on-line control to the controlled process.

In contrast to these examples where measurement data is used immediately, there are many other situations where there is a need to save measurements for later analysis. Such analysis is an essential part of quality control systems, because it can often predict the onset of production problems before they become serious enough to affect the quality of a product significantly. If data is to be analysed, it must first be recorded and stored in some permanent form, and after analysis, the results must be displayed in a convenient and usable form. This chapter therefore starts with a discussion of the various means available for recording measurement signals. Following this, the various ways of displaying and presenting data and the results of analysis are considered.

7.2 MEASUREMENT SIGNAL RECORDING

Many types of instrument are available for recording data, including chart recorders, ultraviolet recorders, fibre-optic recorders and magnetic tape recorders. In addition to these instruments, computer data logging can also be used to record measurement signals, and this is an increasingly common method.

Chart recorders are probably the most widely used method of making permanent records of measurement signals, and they carry out this task simply, cheaply and reliably. However, their dynamic response is poor and they cannot record signals with frequencies greater than about 30 Hz. Ultraviolet recorders can cope with much higher frequencies up to about 13 kHz, but they are very delicate and easily damaged instruments which are also quite expensive. Fibre-optic recorders are even more expensive than ultraviolet recorders but have a higher bandwidth still. Finally, magnetic tape recorders have a bandwidth up to 80 KHz.

7.2.1 Galvanometric Chart Recorders

These are the most common form of chart recorder and account for about 80 percent of those sold. They work on the same principle as a moving-coil voltmeter except that the pointer draws an ink trace on paper, as illustrated in Figure 7.1, instead of merely moving against a scale. The measured signal is applied to the coil, and the angular deflection of this and its attached pointer is proportional to the magnitude of the signal applied. The pointer normally carries a pen, and, by using a motor running at constant speed to drive the chart paper that the pointer is in contact with, the pen produces a time history of the measured signal.

Inspection of Figure 7.2 shows that the displacement y of the pen across the chart recorder is given by $y = R\sin\theta$. This sine relationship between the input signal and the displacement y is non-linear, and results in an error of 0.7 percent for deflections of $\pm 10°$. However, in practice, the error is greater than this because, with the pen moving in an arc, it is difficult to relate the magnitude of deflection to the time axis accurately. One way of overcoming this is to print a grid on the chart paper in the form of circular arcs, as illustrated in Figure 7.3. Unfortunately, measurement errors still occur in reading this type of chart because interpolation for points drawn between the curved grid lines is difficult.

An alternative solution is to use heat-sensitive chart paper directed over a knife-edge, and to replace the pen by a heated stylus, as illustrated in Figure 7.4. The input–output relationship is still non-linear, with the deflection y being proportional to $\tan\theta$ as shown in Figure 7.5, and the reading error for excursions of $\pm 10°$ is still 0.7 percent. However, the rectilinearly scaled chart paper now required, as shown in Figure 7.6, allows much easier and hence more accurate interpolation between the grid lines.

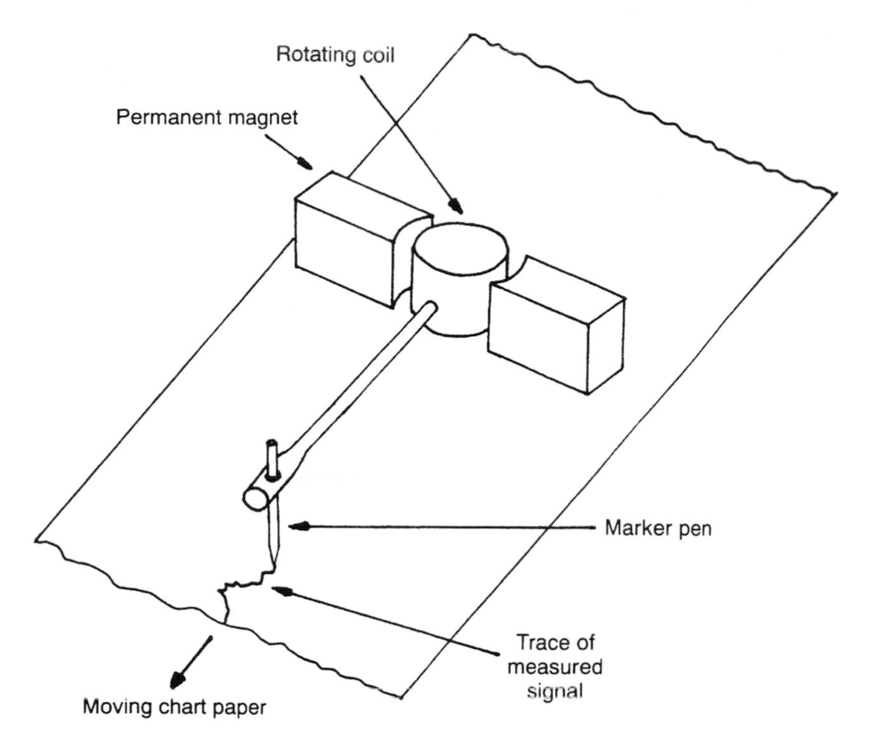

Figure 7.1. Simple galvanometric chart recorder

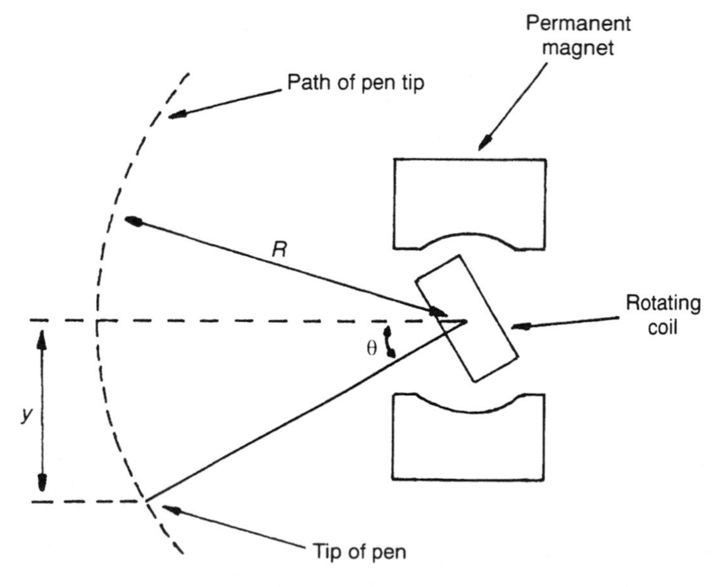

Figure 7.2. y versus θ relationship for standard recorder

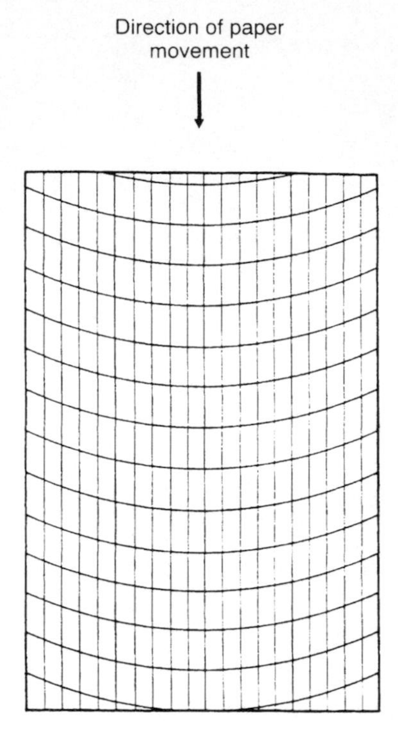

Figure 7.3. Curvilinear paper for standard recorder

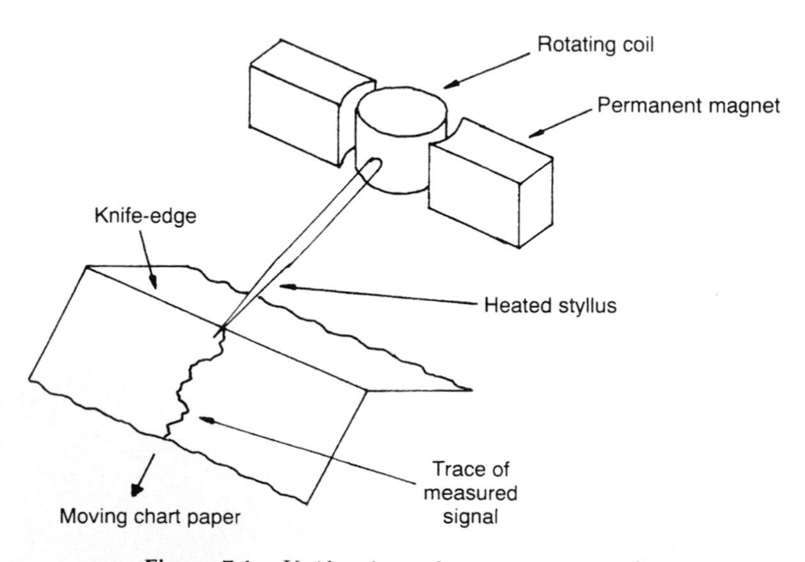

Figure 7.4. Knife-edge galvanometric recorder

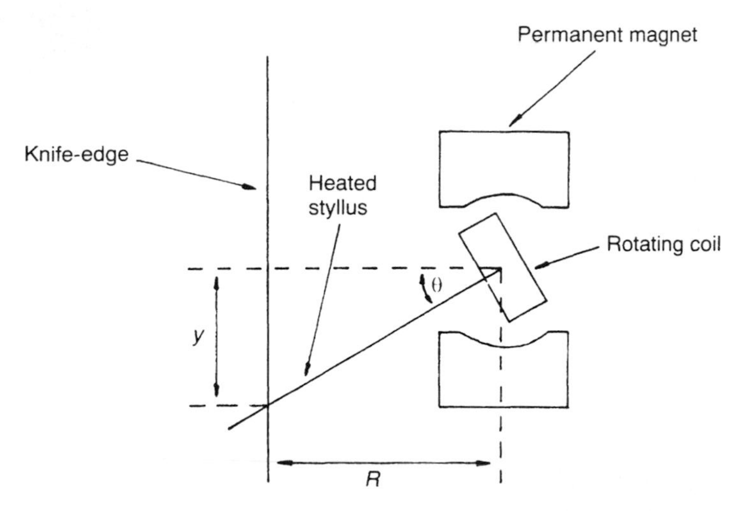

Figure 7.5. y versus θ relationship for knife-edge recorder

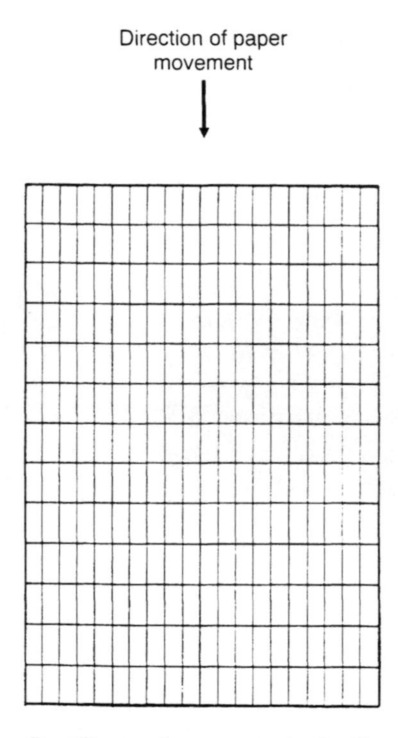

Figure 7.6. Rectilinear chart paper for knife-edge recorder

When all error sources in galvanometric recorders are taken into account, the typical uncertainty in measurements is likely to be about ±2 percent. Measurement resolution is normally 1 percent of full-scale reading and the maximum frequency allowable for input signals is 30 Hz.

7.2.2 Potentiometric Chart Recorders

Potentiometric recorders are more expensive than galvanometric types but have much better static characteristics. They can achieve an inaccuracy of ±0.1 percent or less of full scale and measurement resolution of 0.2 percent f.s., which are both much better than the corresponding figures for galvanometric recorders. They employ a servo system, as shown in Figure 7.7, in which the pen is driven by a servo-motor, and a potentiometer on the pen feeds back a signal proportional to the pen position. This position signal is compared with the measured signal, and the difference is applied as an error signal which drives the motor. However, a consequence of this electromechanical balancing mechanism is to give the instrument a slow response time in the range of 0.2–2.0 s. This means that potentiometric recorders are only suitable for measuring either d.c. signals or signals that vary only very slowly with time.

7.2.3 Ultraviolet Recorders

Ultraviolet recorders work on very similar principles to standard galvanometric chart recorders, but achieve a very significant reduction in system inertia by mounting a narrow mirror rather that a pen system on the moving coil. This mirror reflects a beam of ultraviolet light onto ultraviolet-sensitive paper. The arrangement enables signals at frequencies up to 13 kHz to be recorded with a typical inaccuracy of ±2 percent of full-scale deflection.

It is usual to find several of such mirror-galvanometer systems mounted in parallel within one instrument to provide a multichannel recording capability, as illustrated in Figure 7.8. Whilst it is possible to obtain satisfactory permanent signal recordings by this method, special precautions are necessary to protect the UV-sensitive paper from light before use and to spray a fixing lacquer on it after recording. Such instruments must also be handled with extreme care, because the mirror galvanometers and their delicate mounting systems are easily damaged by relatively small shocks.

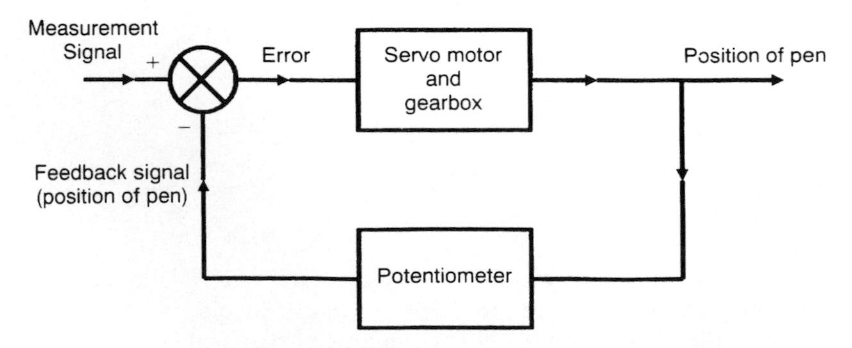

Figure 7.7. Servo system of potentiometric recorder

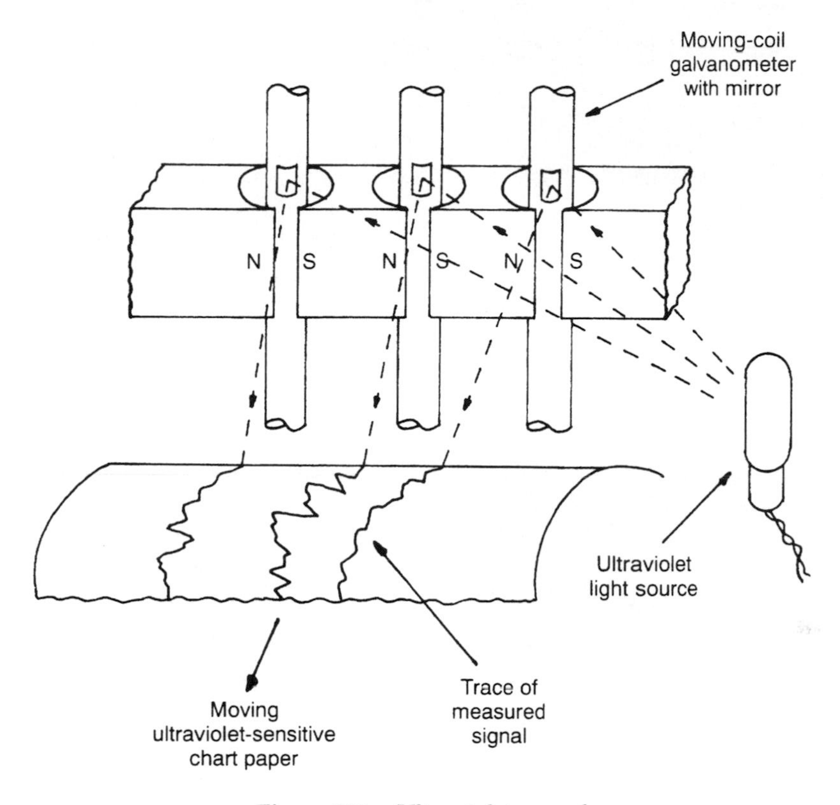

Figure 7.8. Ultraviolet recorder

7.2.4 Fibre-Optic Recorders

The fibre-optic recorder is a recent development which uses a fibre-optic system to direct light onto light-sensitive paper. Fibre-optic recorders are similar to oscilloscopes in construction, in so far as they have an electron gun and focusing system which directs a stream of electrons onto one point on a fluorescent screen. The screen is usually a long thin one instead of the square type found in an oscilloscope and only one set of deflection plates is provided. The signal to be recorded is applied to the deflection plates and the movement of the focused spot of electrons on the screen is proportional to the signal amplitude. A narrow strip of fibre optics in contact with the fluorescent screen transmits the motion of the spot to photosensitive paper held in close proximity to the other end of the fibre-optic strip. By driving the photosensitive paper at a constant speed past the fibre-optic strip, a time history of the measured signal is obtained. Such recorders are much more expensive than ultraviolet recorders but have an even higher bandwidth.

Whilst the construction above is the more common in fibre-optic recorders, a second type also exists which uses a conventional square screen instead of a long thin one. This has a square faceplate attached to the screen housing a square array of fibre optics. The other side of the fibre-optic system is in contact with chart paper. The effect of this is to provide a hard copy of the typical form of display obtainable on a cathode ray oscilloscope.

7.2.5 Magnetic Tape Recorders

Magnetic tape recorders can record analogue signals up to 80 kHz in frequency. As the speed of the tape transport can be switched between several values, signals can be recorded at high speed and replayed at a lower speed. Such time-scaling of the recorded information allows a hard copy of the signal behaviour to be obtained from instruments such as ultraviolet and galvanometric recorders whose bandwidth is insufficient to allow direct signal recording. A 200 Hz signal cannot be recorded directly on a chart recorder, but if it is recorded on a magnetic tape recorder running at high speed and then replayed at a speed ten times lower, its frequency will be time-scaled to 20 Hz which is then within the frequency range that a chart recorder can handle. Instrumentation tape recorders typically have between four and ten channels, allowing many signals to be recorded simultaneously.

The two basic types of analogue tape recording technique are direct recording and frequency-modulated recording. Direct recording offers the best frequency bandwidth but the accuracy of signal amplitude recording is quite poor, and this seriously limits the usefulness of this technique in most applications. The frequency-modulated technique offers much better amplitude recording accuracy, with an inaccuracy of only ± 5 percent at signal frequencies of 80 kHz. In consequence, this technique is very much more common than direct recording.

7.2.6 Computer Data Logging

The only technique available for recording signals at frequencies higher than 80 kHz involves using a digital computer. As the signals to be recorded are usually in analogue form, a prerequisite for this is an analogue-to-digital converter board within the computer to sample the analogue signals and convert them to the digital form in which the computer operates. Careful choice of the sampling interval is necessary so that an accurate digital record of the signal is obtained as explained in Chapter 6, and problems of aliasing, etc., are not encountered. Some prior analogue signal conditioning may also be required in some circumstances when a computer is used for data logging, again as mentioned in Chapter 6.

7.3 DATA PRESENTATION

The magnitudes of recorded signals, and results derived from them, can be presented on paper in either tabular or graphical form. The decision about which form of presentation to use is not always clear-cut. In some circumstances, one of the two forms is clearly best. However, in many data collection exercises, it is appropriate to use a combination of both forms, so making best use of the merits of each technique. It should be mentioned that a large number of computer packages are now available which present data in tabular and graphical forms, and these are much less laborious than 'paper' presentations produced by human hand.

7.3.1 Tabular Data Presentation

A tabular presentation allows data values to be recorded in a precise way which maintains exactly the accuracy to which the data values were measured. In other words, the data values are written down exactly as measured. Besides recording the raw data values as measured, tables often contain further values calculated from the raw data.

An example of a tabular data presentation is given in Table 7.1. This records the results of an experiment to determine the strain induced in a bar of material under a range of stresses. Data was obtained by applying a sequence of forces to the end of the bar and using an extensometer to measure the change in length. Values of the stress and strain in the bar are calculated from these measurements and are also included in the table. The final row, which is of crucial importance in any tabular presentation, is the estimate of possible error in each calculated result.

A table of measurements and calculations should conform to several rules as illustrated in Table 7.1.

Table 7.1. Sample tabular presentation of measured applied forces and extensometer readings and calculations of stress and strain

	Force applied (kN)	Extensometer reading (divisions)	Stress (N/m^2)	Strain
	0	0	0	0
	1	2.8	7.7	1.40×10^{-6}
	2	4.0	15.5	1.98×10^{-6}
	3	4.9	23.2	2.42×10^{-6}
	4	5.8	31.0	2.86×10^{-6}
	5	6.6	38.7	3.26×10^{-6}
	6	7.4	46.5	3.66×10^{-6}
	7	8.2	54.2	4.05×10^{-6}
	8	9.0	62.0	4.44×10^{-6}
	9	9.8	69.7	4.84×10^{-6}
	10	10.6	77.5	5.24×10^{-6}
Possible error in measurements (%)	±0.2	±0.2	±1.5	$\pm 0.1 \times 10^{-6}$

1. The table should have a title which explains what data are being presented within the table.

2. Each column of figures in the table should refer to the measurements or calculations associated with one quantity only.

3. Each column of figures should be headed by a title which identifies the data values contained in the column.

4. The units in which quantities in each column are measured should be stated at the top of the column.

5. All headings and columns should be separated by bold horizontal (and sometimes vertical) lines.

6. The errors associated with each data value quoted in the table should be given. The form shown in Table 7.1 is a suitable way to do this when the error level is the same for all data values in a particular column. However, if error levels vary, then it is preferable to write the error boundaries alongside each entry in the table.

7.3.2 Graphical Presentation of Data

Presentation of data in graphical form involves some compromise in the accuracy to which the data are recorded, as the exact values of measurements are lost. However, graphical presentation has two important advantages over tabular presentation.

1. Graphs provide a pictorial representation of results which is more readily comprehended than a set of tabular results.

2. Graphs are particularly useful for expressing the quantitative significance of results and showing whether a linear relationship exists between two variables.

Figure 7.9 shows a graph drawn from the stress and strain values given in the Table 7.1. Construction of the graph involves first of all marking the points corresponding to the stress and strain values. The next step is to draw some line through those data points which best represents the relationship between the two variables. This line will normally be either a straight one or a smooth curve. The data points will not usually lie exactly on this line but instead will lie on either side of it. The magnitude of the excursions of the data points from the line drawn will depend on the magnitude of the random measurement errors associated with the data. As for tables, certain rules exist for the proper representation of data in graphical form.

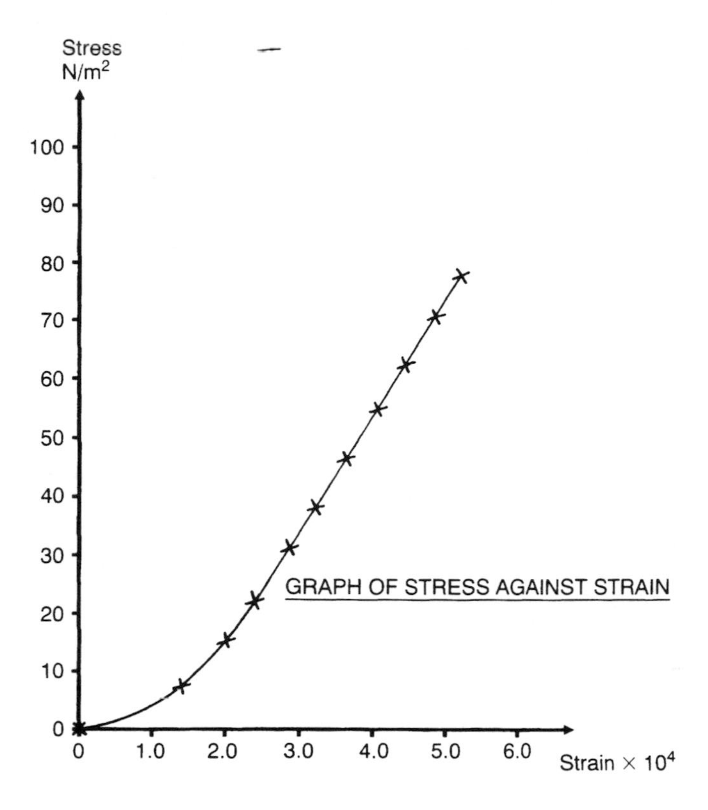

Figure 7.9. Sample graphical presentation of data: graph of stress against strain

1. The graph should have a title or caption which explains what data are being presented in the graph.

2. Both axes of the graph should be labelled to express clearly what variable is associated with each axis and to define the units in which the variables are expressed.

3. The number of points marked along each axis should be kept reasonably small – about five divisions is often a suitable number.

4. No attempt should be made to draw the graph outside the boundaries corresponding to the maximum and minimum data values measured, i.e. in Figure 7.9, the graph stops at a point corresponding to the highest measured stress value of 77.5.

7.3.3 Fitting Curves to Data Points on a Graph

The procedure of drawing a straight line or smooth curve as appropriate which passes close to all data points on a graph, rather than joining the data points by a jagged line which passes through each data point, is justified on account of the random errors which are known to affect measurements. Any line between the data points is mathematically acceptable as a graphical representation of the data if the maximum deviation of any data point from the line is within the boundaries of the identified level of possible measurement error. However, within the range of possible lines that could be drawn, only one will be optimal. This optimal line is where the sum of negative errors in data points on one side of the line is balanced by the sum of positive errors in data points on the other side of the line. The nature of the data points is often such that a perfectly acceptable approximation to the optimum can be obtained by drawing a line through the data points by eye. In other cases, however, it is necessary to fit a line by regression techniques (see Appendix 5 for further details of regression techniques).

CHAPTER 8

Practical Implementation of Measurement and Calibration Procedures

Earlier chapters have been concerned with setting up the theoretical framework of the measurement and calibration procedures necessary for operation of a quality control and assurance system. Having laid down these theoretical foundations, it is now possible to go on to examine the practical aspects involved in implementing the measurement and calibration procedures required.

The procedures to be described in this chapter represent the best possible practice attainable. However, it must be understood that what is sensible, practical, achievable and affordable in any given situation may differ in substantial respects from these ideals. The most appropriate person to give advice about what standards of measurement accuracy and calibration are appropriate and acceptable in any given situation is a consultant who has a lot of experience in the particular industry involved.

8.1 CHOICE OF PRIMARY MEASURING INSTRUMENTS

The initial step in setting up a measurement and calibration system associated with a quality control system is to specify what physical parameters in the manufacturing process have to be measured. In addition, the accuracy with which each parameter must be measured and the environmental conditions in which the measuring instruments must operate are further vital pieces of information. Provision of this information clearly requires expert knowledge from personnel who are intimately acquainted with the operation of the manufacturing plant in question.

Once this information has been acquired, the next stage is to choose suitable instruments for measuring each of the relevant process parameters in the manufacturing system. This again requires the use of an expert instrumentation engineer who has knowledge of all the instruments available for measuring each physical quantity and is able to evaluate their accuracy, cost and suitability for operation under the environmental conditions pertaining. As far as possible, measurement systems and instruments should be chosen which are as insensitive as possible to the operating environment, although this requirement is often difficult to meet because of cost and other performance considerations. Maintainability and constancy of performance are also important criteria in this evaluation because the instrument chosen should be capable of operating for long periods without performance degradation and a need for costly maintenance. In consequence of this, the initial cost of an instrument often has a low weighting in the evaluation exercise. The later chapters in this book offer assistance in this evaluation by providing lists and data for all the instruments available for measuring a range of physical quantities. However, new techniques and instruments are being developed all the time and therefore instrumentation engineers must keep abreast of the latest developments by reading appropriate technical journals regularly.

8.2 CHOICE OF SECONDARY MEASURING INSTRUMENTS

Once all the primary measuring instruments have been specified as described above, the instrumentation engineer must look carefully at their operating characteristics and especially at their behaviour under the environmental conditions relevant to the manufacturing operation. Any susceptibility of their performance characteristics to variations in the environmental conditions existing must be identified and quantified. Appropriate instruments must then be chosen to measure the relevant environmental parameters so that appropriate corrections can be made to the measurements obtained from the primary measuring instruments. The choice of secondary instruments is carried out following principles similar to those for primary instruments, by choosing an instrument with suitable accuracy, characteristics and cost from the range available.

8.3 CALCULATION OF MEASUREMENT ACCURACY

Quality control requires that all manufacturing-process-related measurements should conform to specified and documented measurement limits. This means that the accuracy of each individual measurement and also the overall accuracy of measurement systems consisting of several components must all be quantified.

The starting point for calculating measurement accuracy in a system is to identify and quantify the various sources of measurement error present, as described in Chapter 5. In situations where there are significant environmental inputs in the system, knowledge of the operational characteristics of the measuring instruments used, as discussed in Chapter 4, is also important. This then allows the readings obtained from secondary transducers to be used to correct the outputs from the primary measuring instruments. Other processing of the output signals from the primary measuring instruments is also carried out, as described in Chapter 6, to remove bias, noise etc., and ensure that the accuracy of each measurement is the best that is obtainable. Having calculated the separate accuracies of measuring system components in this manner, the overall accuracy of a measurement system containing several components can then be calculated according to the formulae presented at the end of Chapter 5.

When measurement accuracy figures are written down in the quality system documentation, the figures quoted must reflect the maximum errors which may exist in measurements. Such maximum errors will occur when instruments have drifted furthest from their quoted characteristics just prior to recalibration. Knowledge of the way in which instrument characteristics drift and the calibration frequency set is obviously required before this maximum error can be defined. This is then the magnitude of measurement error which the calibration system guarantees will not be exceeded provided that the calibration procedures and frequency are followed correctly.

To summarize therefore, the quoted accuracy level for a measurement system is derived by the following steps.

(a) Establish the characteristics of all instruments under controlled environmental conditions by calibrating them against standards instruments traceable back to reference standards.

(b) Establish the maximum variation in instrument error levels at the extremes of possible operating conditions and just prior to the instrument being calibrated, thus setting the error for each measurement system component.

(c) Calculate the error band for the whole measurement system from the error bands of individual system components.

8.4 CALIBRATION PROCEDURES

A fundamental requirement in all quality control and assurance systems is that calibration procedures be implemented to monitor the accuracy of all measurement systems employed in order that the specified quality level be maintained. Such periodic calibration of instruments ensures that their output reading will be within stated measurement-limit bounds when they are used under the environmental conditions specified for the calibration exercise.

Every instrument in the workplace has a designated person who is responsible for it. That person must ensure that all instruments for which he/she is responsible are calibrated at the correct times by approved personnel. Approved personnel means either staff within the company who have attended all the necessary courses relevant to the calibration duties or subcontractors outside the company who are verified as being able to provide calibration services satisfactorily.

If possible, calibration should be carried out under the controlled standard environmental conditions specified in the instrument data sheet. Unfortunately, it is not always possible to create all of these specified conditions. Therefore, the first step in instrument calibration must be to establish whether any of the environmental parameters pertaining during calibration differ from the required values specified in the data sheet. Any environmental parameters which are not at the correct value represent an environmental input whose effect on the instrument output must be quantified and compensated for. The procedure for such quantification is as follows.

Whilst keeping all other environmental parameters at some constant level, the environmental input under investigation is varied in steps over a range of values, causing the instrument output to vary in steps over a corresponding range of readings. This allows an input–output relationship to be drawn for that particular environmental input.

Following this quantification of the effect of any environmental parameters which are away from the standard values specified for calibration, the instrument calibration exercise can proceed. This consists of applying inputs of varying magnitude to the instrument being calibrated. At the same time, the same inputs are also applied to a 'standard' instrument whose output is compared with the values registered by the instrument being calibrated. The 'standard' instrument referred to is a high quality one kept solely for calibration duties whose accuracy is at least ten times better than the nominal value of the instrument being calibrated. The 'standard' instrument must form part of a calibration chain and be calibrated itself against a yet more accurate standard. A typical calibration chain was shown in Figure 3.2. Adherence to this practice fulfils the requirement that the accuracy of all process instruments and the 'standard' instruments used to calibrate them be *traceable* back to the fundamental standards maintained by the National Physical Laboratory (in the UK), or other appropriate national body responsible for the maintenance of measurement standards.

Effect of Environmental Inputs

It was mentioned above that calibration guarantees that the accuracy of the output reading of a calibrated instrument will be at a certain level when the instrument is used under stated environmental conditions. Outside these conditions, the accuracy of the instrument will vary to a greater or lesser extent according to its susceptibility to the environmental inputs inherent in a varying environment. In all normal measurement situations, it is impossible to control the environmental conditions to be at the levels specified for calibration, and therefore due account must be taken of this by appropriate correction of the primary measuring instrument output readings.

The choice of suitable secondary measuring instruments to monitor the level of environmental inputs has already been discussed in Section 8.2. However, apart from knowing the magnitude of environmental inputs, it is also necessary to quantify their effect on the primary measurement. The effect of environmental inputs is determined in a very similar manner to that of calibration procedure described above. The level of the environmental input being investigated is varied in steps over a certain range whilst the level of the desired input (primary measurement) and all other environmental inputs is held constant, and the corresponding output readings are recorded. This allows an input–output relationship to be drawn for that particular environmental input. The same procedure is then repeated for all other environmental inputs.

This procedure represents a statement of a desired ideal rather than a realistic picture of what is normally carried out in practice. It is usual to disregard environmental inputs that only have a small effect on the output of measuring instruments, as the measurement errors introduced by so doing are insignificant. Changes in atmospheric pressure are typical of such environmental inputs and are generally disregarded.

Rate of Change of Instrument Characteristics

The discussion so far has been concerned with specifying what the characteristics of an instrument are immediately following calibration. These characteristics will change over a period of time, because of factors such as ageing effects, mechanical wear, long-term environmental changes, and the effects of dust, fumes and chemicals in the operating atmosphere. The extent and rate of this change must be determined so that the appropriate time when the instrument will need recalibrating can be determined. Instrument manufacturers can give very useful guidance on this, but their figures can only be approximate, as the rate of variation in characteristics will be affected by the operating environment of the particular application of the measurement system. Some practical investigation of the rate of variation of characteristics will therefore always be necessary.

Determining the Calibration Frequency Required

When installing instruments in a new system, or new instruments in an existing system, it is highly likely that much thought will have been given to the suitability of the instruments chosen for their intended application. Knowledge of the instruments' characteristics and their operating environment will normally allow good approximations to be made of the rate at which their performance will change with time and therefore what frequency of recalibration will be required.

Because good instrument calibration is so essential to the proper functioning of quality control systems, however, it is prudent not to rely too much on such a priori predictions. It is always possible that some very significant factor in the procedure for estimating the required calibration frequency has been inadvertently neglected. Therefore, as long as the circumstances permit, it is sensible to start from basics in deriving the required calibration frequency and not use past-history information about the instrument and its operating environment. Applying this philosophy, the following frequency of checking instrument characteristics might be appropriate (assuming 24 h/d working).

Day 1: Once per hour for first 4 hours, then every 4 hours.

Days 2–3: Every 12 hours

Days 4–7: Once per day

Weeks 2–3: Twice per week

Weeks 3–4: Once per week

Week 6: Once

Week 8: Once

Months 3–6: Once per month

Month 9: Once

Month 12: Once

Month 18: Once

Month 24: Once

Then once per year

The above reducing frequency of calibration checks should be applied until a point is reached where a deterioration in the instrument's accuracy is first detected. Comparison of the amount of performance degradation with the inaccuracy level that is permissible in the instrument will show whether the instrument should be calibrated immediately at this point or whether it can be safely left for a further period. If the above pattern of calibration checks were followed for an instrument and the check at week 8 showed a deterioration in accuracy which was close to the permissible limit, this would determine that the calibration frequency for the instrument should be every eight weeks.

The above method of establishing the optimum calibration frequency is clearly an ideal which cannot always be achieved in practice, and indeed for some types of instrument this level of rigour is unnecessary. When used on many production processes, for instance, it would be unacceptable to interrupt production every hour to recheck instrument calibrations unless a very good case could be made about why this was necessary. Also, the nature of some instruments, for example a mercury-in-glass thermometer, means that calibration checks will only ever be required infrequently. However, for instruments such as unprotected base-metal thermocouples, initial calibration checks after one hour of operation would not be at all inappropriate.

8.5 PROCEDURE FOLLOWING CALIBRATION

When the instrument is calibrated against a standard instrument, its accuracy will be shown to be either inside or outside the required measurement accuracy limits. If the instrument is found to be inside the required measurement limits, the only course of action required is to record the calibration results in the instrument's record sheet and then put it back into use until the next scheduled time for calibration.

The options available if the instrument is found to be outside the required measurement limits depends on whether its characteristics can be adjusted and the extent to which this is possible. If the instrument has adjustment screws, these should be turned until the characteristics of the instrument are within the specified measurement limits. Following this, the adjustment screws must be sealed to prevent tampering during the instrument's subsequent use. The instrument can then be returned to its normal operating position for further use.

The second possible course of action if the instrument is outside measurement limits covers the case where no adjustment is possible or the range of possible adjustment is insufficient to bring the instrument back within measurement limits. In this event, the instrument must be withdrawn from use, and this withdrawal must be marked prominently on it to prevent it from being reused inadvertently. The options available then are to either send the instrument for repair if this is feasible or to scrap it.

8.6 DOCUMENTATION OF MEASUREMENT AND CALIBRATION SYSTEMS

All measurement and calibration systems implemented must be fully documented. In the case of a small company, all the relevant information should be contained within one manual, whereas for a large company it is often more appropriate to have separate volumes covering corporate and division procedures respectively.

The manual (or manuals) documenting the measurement and calibration procedures associated with the quality control systems in operation should contain the following information.

Documentation Revision Level and Distribution

1. The date and revision level of the manual.

2. The number of copies of the manual issued and the names of the authorized personnel holding them.

3. The procedure for amendment of the manual.

Measurement Systems

4. A full description of the measurement requirements throughout the workplace.

5. The measurement limits required for each measurement system

6. The system of serial numbers used to uniquely identify all instruments used in the workplace.

7. A list of the instruments used and the name of the person responsible for each.

8. A description of each type of instrument used.

9. The required calibration frequency for each instrument (where for some reason calibration is not required for any particular instrument, the documentation must specifically state that it is excluded).

10. Instructions on the proper way of using instruments.

11. Information about any environmental control or other special precautions to be taken when using instruments.

12. Training courses (if any) to be attended by persons using instruments.

Calibration Procedures

13. The standard instruments to be used for calibration.

14. The required method of storing and handling standard instruments.

15. The standard environmental conditions required for performance of calibration functions (if a common and well-accepted procedure is published elsewhere, this can be referenced rather than including the procedure in full in the calibration document).

16. The standard format required for the recording of calibration results within the Instrument Manual (including the date of calibration).

17. The standard format required for recording calibration results on the instruments themselves (where appropriate).

18. The procedure to be followed if an instrument is found to be outside the calibration limits.

19. The method of marking instruments withdrawn from use because they are outside calibration limits.

20. The traceability of the calibration system back to national reference standards.

21. Training courses to be attended by personnel performing instrument calibration duties and refresher courses (if any) required..

Calibration Reviews

22. The review procedure for the continued effectiveness of the calibration system being operated.

23. The results of each such effectiveness review.

A typical structure of the quality manual documenting this information is given in Appendix 6.

8.7 INSTRUMENT RECORDS

A separate record for each instrument in the system must be maintained which specifies as a minimum:

1. its serial number
2. the name of the person responsible for calibrating it
3. the required calibration frequency
4. the date of the last calibration
5. the calibration results

These sheets should be bound together as a Site Instrument Manual. A suitable format for recording this information is given in Table 3.1.

CHAPTER 9

Reliability in Manufacturing Systems

9.1 INTRODUCTION

The earlier chapters in this book are concerned with the measurement and error-assessment techniques relevant to quality control procedures during manufacture. However, assurance of the quality of a product immediately after it has been delivered is only one of the parameters in the customer satisfaction equation. Of at least equal importance is the reliability and longevity of the product after manufacture. The benefits accruing from efforts put into quality control procedures during product manufacture will be quickly lost if the product later gains a reputation of unreliability over its subsequent expected working life. Reliability is an essential component in quality assurance and this chapter is therefore concerned with considering the various means available for measuring and maintaining product reliability. A more comprehensive treatment of the subject can be found in [1].

Reliability is formally defined as the ability of a product to perform its required function within the bounds of specified working conditions for a stated period of time. Factors such as manufacturing tolerances, quality variations in raw materials used and differing operating conditions all conspire to make the faultless operating life of a product impossible to predict. Such factors are subject to random variation and chance, and therefore the reliability of a product cannot be defined in absolute terms. The nearest one can get to an absolute quantification of reliability are quasi-absolute terms like the mean-time-between-failures, which expresses the average time a product works without failure. Otherwise, reliability has to be expressed as a statistical parameter which defines the probability that no faults will develop in a product over a specified interval of time.

One immediate difficulty which arises when attempts are made to quantify the reliability of a product is defining what should be counted as a failure. Failures can generally be divided into three categories:

- *Critical failures* Where failure causes total loss of function in the product.
- *Major failures* Where failure causes a major loss of function in a product but it can still continue to be used to some extent.
- *Minor failures* Where failure leaves the product still able to be used to perform is major purpose but with the loss of some convenience function.

In a product such as a transistor radio, failure of the speaker would be regarded as a critical failure because that would render the radio totally useless. Failure of a noise suppression circuit within it would probably be counted as a major failure, as it would make the radio difficult to listen to though some programs might be heard if the user listened carefully. A minor failure might be failure of the tone adjustment control, which would impair the quality of sound reproduction by a small degree but still allow programmes to be heard reasonably well.

It is probably becoming apparent that classification of failures into these three categories is not necessarily straightforward. Many types of failure lie on the borderline between categories and it is a matter of personal judgement which side of the border they are placed. Nevertheless, deciding what should be counted as a serious failure is a prerequisite in reliability analysis. Once this has been decided, reliability can be quantified either in quasi-absolute or probabilistic terms.

9.2 RELIABILITY QUANTIFICATION IN QUASI-ABSOLUTE TERMS

Whilst reliability is essentially probabilistic in nature, it can be quantified in quasi-absolute terms by the mean-time-between-failures and the mean-time-to-failure parameters. It must be emphasized that these two quantities are only average values calculated over a number of identical samples of a product. The actual values for any particular sample of a product may vary substantially from these mean values.

The *mean-time-between-failures* (MTBF) is a parameter which expresses the mean number of failures which occur in a product over a given period of time. For example, suppose that the history of a robot manipulator is logged over a one-year (365-day) period and the time intervals in days between faults occurring which require repair are as follows:

11 23 27 16 19 32 6 24 13 21 26 15 14 33 29 12 17 22

The mean interval is 20 days which is therefore the mean-time-between-failures. A simpler way to calculate the MTBF is to merely record the total number of failures which occur over a given time interval. For N failures over a time interval t, the failure rate θ is given by:

$$\theta = \frac{N}{t}$$

The MTBF is the reciprocal of θ, i.e.

$$\text{MTBF} = \frac{1}{\theta}$$

The mean-time-to-failure (MTTF) is a parameter which is associated with products which are discarded at the first failure because it is either impossible or uneconomic to repair them. It expresses the average time before failure occurs, calculated over a number of identical products. Consumer goods such as cheap transistor radios are examples of products which cannot be repaired economically and are therefore discarded at the first major or critical failure. For example, suppose that a batch of 20 radios is put through an accelerated-wear test (in which a long period of typical use is simulated over a much shorter period) and the simulated life in years before serious failure for each radio is as follows:-

7 9 13 6 10 11 8 9 14 8 8 12 9 15 11 9 10 12 8 11

The mean of these twenty numbers is 10. Therefore the simulated mean-time-to-failure is 10 years.

For many products, MTBF and MTTF figures are both relevant. The nature of a product often means that minor repairable faults will occur at various points in time during its use, and then, after a certain, greater length of time, a catastrophic failure will occur. Over its working lifetime, the frequency of faults is quantified by the MTBF parameter, and the estimated working life before irredeemable failure occurs is quantified by the MTTF parameter.

A further reliability-associated term of importance is the *mean-time-to-repair* (MTTR). This expresses the average time needed for repair of a product, calculated over a number of typical faults which are likely to occur in it. Returning to the example of the robot manipulator, suppose that the time taken in hours to repair each of eighteen faults was as follows:

4 1 3 2 1 9 2 1 7 2 3 4 1 3 2 4 4 1

The mean of these values is 3: therefore the mean-time-to-repair was 3 h.

The relative importance of the mean-time-to-repair parameter varies according to the product to which it is applied. When applied to a critical element in a production process, the mean-time-to-repair is of equal importance to the mean-time-between-failures. What really matters in this case is the proportion of the total available production time which is lost whilst the critical element is inoperative. Clearly, repair time is of equal importance to the frequency of fault occurrence. Often, an element whose MTBF is low but whose MTTR is also low will be preferable to an alternative element where the MTBF is a little higher but the MTTR is a lot higher.

This argument about the relative importance of the MTBF and MTTR parameters assumes, however, that elements have to be repaired immediately on failure. Where critical elements in production processes do not have a great cost, it is normal practice to keep spare elements on stand-by to replace elements on the production process as they fail. In this case, the MTTR is of very little importance. The amount of lost production depends on how long it takes to remove and replace a failed element. In this mode of working, the MTBF parameter assumes very much greater importance because the total production time lost in a given interval of time is clearly proportional to the number of faults occurring over the interval. The MTTR would only become important if it became very large and interrupted the supply of stand-by elements.

The MTBF and MTTR parameters are often expressed in terms of a combined quantity known as the *availability* figure. This measures the proportion of the total time that a product is working, i.e. the proportion of the total time that it is in an unfailed state. The availability of a product is defined as the following ratio:

$$\text{AVAILABILITY} = \frac{\text{uptime}}{\text{total time}} = \frac{\text{MTBF}}{\text{MTBF} + \text{MTTR}}$$

Alternatively, the availability can be expressed as:

$$\text{AVAILABILITY} = \frac{1}{1 + \theta \cdot \text{MTTR}}$$

where θ is the failure rate.

In terms of customer satisfaction, the aim must always be to maximize the MTBF figures and minimise the MTTR figure, thus maximizing the availability. As far as the MTBF (and MTTF) figures are concerned, good design and high quality control standards during manufacture are the appropriate means of maximizing these figures. Good design procedures which mean that faults are easy to repair are also an important factor in reducing the MTTR figure. However, many factors affecting the MTTR are outside the manufacturers control to a large extent because they are strongly influenced by customer practices. If a customer chooses to do his own repairs, the time taken to effect each repair is governed by the skill of the personnel he employs to do the work and the stocks of spare parts he keeps. If, on the other hand, repair work is entrusted to the manufacturer, then the manufacturer can do much to reduce the MTTR by ensuring that their maintenance staff are well-trained and motivated and respond quickly to breakdown calls. The manufacturer must also maintain an adequate stock of spare parts and have a means of ensuring that parts are delivered speedily as soon as they are requested. The existence of an efficient parts-delivery service is an important contribution to reducing MTTR figures even when the customer does his own maintenance.

9.3 FAILURE PATTERNS

The failure rate of a product may increase, stay the same or decrease over its life. Material fatigue is a typical reason for the failure rate to increase over the life of a product. In the early part of their life, when all components are relatively new, many products exhibit a very low incidence of faults. Then, at a later stage, when fatigue and other ageing processes start to have a significant effect, the rate of faults increases and continues to increase thereafter.

Electronic components are typical of products whose rate of fault incidence decreases over a period of time. Most manufacturing defects, such as substandard components, poor sealing against contamination, faulty assembly or bad connections, show up very early in the life of such products. Thereafter, the rate of fault-incidence remains at a low and approximately constant level for a substantial period of time. Where such a failure pattern is known to exist, it is normal practice to 'burn in' components until all of the components that are likely to fail in this early period of their life have failed. This should mean that by the time a product is delivered to the customer, it has reached a stage where the rate of fault-incidence is constant. If this screening process is controlled correctly, customers will never experience this failure pattern where the rate of fault-incidence decreases with time.

A constant pattern of failure over the whole lifetime of a product is typically exhibited by complex systems containing many different components. The various components within such systems each have their own failure pattern where the failure rate is increasing or decreasing with time. The greater the number of such components within a system, the greater is the tendency for the failure patterns in the individual components to cancel out and the rate of fault-incidence to assume a constant value.

It is very important to stress at this stage that the MTBF, MTTF and MTTR figures discussed are average values. The greater the number of products that the average is calculated over, the greater will be the accuracy of the figure derived. Any particular sample of a product, however, may have MTBF etc. figures which differ significantly from the mean value.

Many types of product exhibit all of the three patterns of failure above over some period of their working lives, and have a failure-rate/age curve of the form shown in Figure 9.1(a). This is frequently referred to as the '*bath-tub curve*'. Manufacturers would normally burn in such products before delivery to the customer, to move the failure pattern beyond the left-hand shaded region in Figure 9.1(a) in which the failure rate is decreasing from a high initial value. The failure pattern of the product after delivery, as perceived by the customer, would therefore be as shown in Figure 9.1(b).

Normal practice would be to replace the product either when its

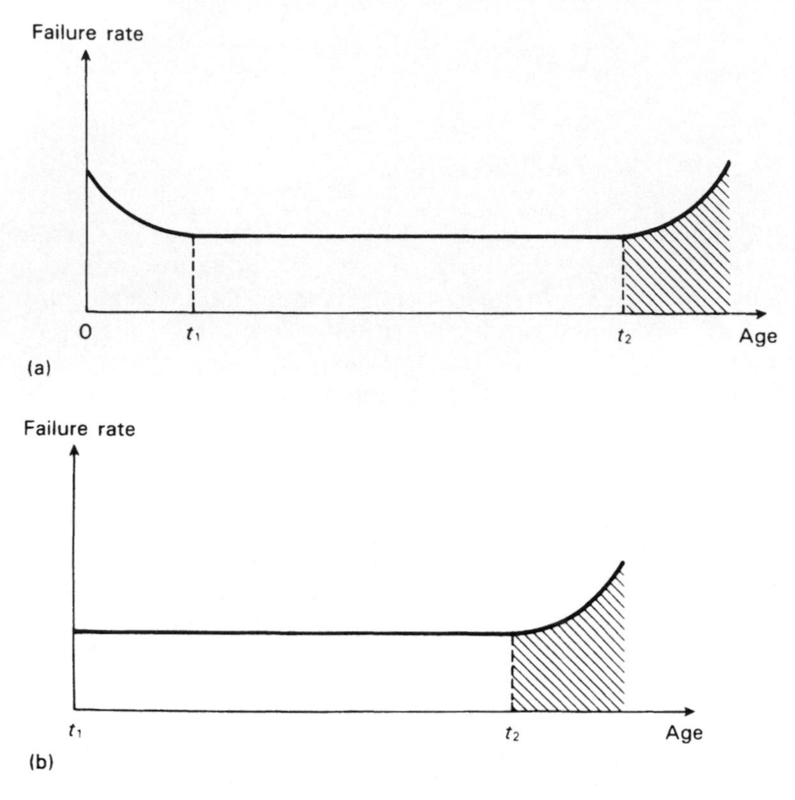

Figure 9.1. 'Bath-tub' curves: (a) typical failure-rate/age characteristic for product after manufacture; (b) target failure-rate/age characteristic for product after delivery to customer

reliability reaches the right-hand shaded region in Figure 9.1 or shortly afterwards. This ensures that products are replaced before the rate of fault-incidence reaches a high level.

9.4 RELIABILITY QUANTIFICATION IN PROBABILISTIC TERMS

It has already been explained how reliability can be expressed in quasi-absolute terms as the mean-time-between-failures. If the average number of failures in a given time for a certain product is θ, the MTBF can be expressed as:

$$\text{MTBF} = \frac{1}{\theta} \tag{9.1}$$

In probabilistic terms, the *reliability* R_X of a product X is defined as the probability that the product will not fail within a certain period of time. The *unreliability* U_X is a corresponding term which expresses the probability that the product will fail within a certain time interval. U_X and R_X are related by the expression:

$$U_X = 1 - R_X \tag{9.2}$$

For a given time interval t, U_X is related to θ (and hence to the MTBF) by the expression:

$$U_X = 1 - e^{-\theta t} \tag{9.3}$$

(It must be noted that this expression is only valid if the failure pattern is in the centre region of the bath-tub curve, i.e. the failure rate is a constant. Also, the product must be in a working state at time $t = 0$, i.e. it must be delivered without faults.)

Examination of equation (9.3) shows that at time $t = 0$ the unreliability is zero. Also, as t tends to ∞, the unreliability tends to a value of 1. This agrees with intuitive expectations that the value of unreliability should lie between values of 0 and 1.

Another point of interest in equation (9.3) is to consider the unreliability when t = MTBF, i.e. $t = 1/\theta$. Then

$$U_X = 1 - e^{-1} = 0.63$$

i.e. the probability of a product failing after it has been operating for a length of time equal to the MTBF is 63 percent.

Further analysis of equation (9.3) shows that, for $\theta t \leqslant 0.1$

$$U_X \approx \theta t, \qquad [\theta t \leqslant 0.1] \tag{9.4}$$

This is a useful formula for calculating (approximately) the reliability of a critical product which is only used for a time which is a small proportion of its MTBF.

9.5 LAWS OF RELIABILITY IN COMPLEX SYSTEMS

If a piece of equipment contains 100 integrated circuits which each have a mean failure rate of one failure per 10^6 hours, will the equipment operate for 1000 hours without failure? This is clearly a question that a customer would like to see answered, and it is equally clear that it is impossible to give an answer in absolute terms. All that can be done is to express the likelihood or probability that the equipment will work for 1000 hours without failure. The basic principles of probability (see Appendix 4) can be extended and applied to express the reliability of a system containing multiple components which are either in series or in parallel.

Reliability of Components in Series

In many systems containing multiple components, the whole system fails if any one component within it fails. A good example of this is a system of fairy lights connected in series on a Christmas tree. One way in which the reliability of such a system of series components can be quantified is in terms of the probability that none of the components will fail within a given interval of time.

Applying the joint probability rule (equation A4.6), the reliability R_S of a system of n series components can be expressed as the product of the separate reliabilities of the individual components:

$$R_S = R_1 R_2 R_3 \ldots R_n \qquad (9.5)$$

In the case of n identical system components, equation (9.5) simplifies to

$$R_S = (R_X)^n, \qquad (9.6)$$

where R_X is the reliability of each component.

In the case of the integrated circuits mentioned at the start of this section, the probability of any individual chip failing within 1000 h of operation is 0.1 percent. The reliability of each component is therefore (from equation 9.2):

$$1.0 - 0.001 = 0.999$$

As all of the transistors in the system are nominally identical, equation (9.6) is applicable and the system reliability can be expressed as

$$R_S = (0.999)^{100} = 0.905$$

Thus, there is a 90.5 percent probability that the system will operate for 1000 hours without failure.

Reliability of Components in Parallel

In many systems containing components connected in parallel, total system failure only occurs when all elements fail. Street lights are an example of such a system where all lamps are connected in parallel onto the mains electrical supply. The street is only totally dark if all the lamps fail. For such systems, the system reliability R_S is given by:

$$R_S = 1 - U_S, \qquad (9.7)$$

where U_S is the unreliability of the system. U_S is calculated in a similar manner to that of equations (9.5) and (9.6):

$$U_S = U_1 U_2 U_3 \ldots U_n \tag{9.8}$$

or for identical system components:

$$U_S = (U_X)^n \tag{9.9}$$

From equation (9.9),

$$R_S = (1 - U_S) = 1 - U_X{}^n = 1 - (1 - R_X)^n \tag{9.10}$$

As an example of the use of these equations, consider a lighting system containing four lamps connected in parallel. What is the probability of total system failure within the first 1000 hours of operation, given that the reliability of a single lamp over 1000 hours is 90 percent?

All components in the system are identical, and therefore the system reliability can be calculated from equations (9.7) and (9.9)

$$U_X = 1 - R_X = 0.1$$

Hence, $U_S = (0.1)^4 = 0.0001$

Thus, the probability of total system failure (no light at all) over 1000 hours of operation is 0.01 percent or, expressed otherwise, the system reliability is 99.99 percent.

Binomial Law of Reliability

This law is applicable to systems of parallel components where some component failure can be tolerated, but system failure occurs before all individual components within it have failed. If R and U are the reliability and unreliability of n identical components connected in parallel in a system, then, by the binomial law of reliability:

$$(R + U)^n = 1 \tag{9.11}$$

Expanding equation (9.11):

$$R^n + nR^{n-1}U + \frac{n(n-1)}{1 \cdot 2} R^{n-2}U^2 + \ldots + \frac{n(n-1)(n-2)\ldots(n-r)}{1 \cdot 2 \cdot 3 \ldots r} R^{n-r}U^r + U^n = 1 \tag{9.12}$$

If system does not fail unless all n components fail, then:

$$R_S = 1 - U^n \tag{9.13}$$

If system fails when r components fail, then:

$$R_S = R^n + nR^{n-1}U + \ldots + \frac{n(n-1)(n-2)\ldots(n-r-1)}{1{\cdot}2{\cdot}3\ldots(r-1)}U^{n-1} \qquad (9.14)$$

and since $R_S = 1 - U_S$,

$$U_S = \frac{n(n-1)(n-2)\ldots(n-\mathrm{r})}{1{\cdot}2{\cdot}3\ldots r}R^{n-r}U^r + \ldots + U^n \qquad (9.15)$$

9.6 AVAILABILITY OF COMPLEX SYSTEMS

The behaviour of complex systems containing many components is frequently described in terms of their availability rather than their reliability. This is because the availability figure is more directly relevant to production cost calculations. The equations for calculating the availabilities of systems containing series or parallel components have an identical form to those for calculating the corresponding reliabilities. Thus, for two series components:

$$A_S = A_1 \times A_2$$

and, for two parallel components:

$$A_S = A_1 + A_2 - (A_1 \times A_2)$$

where A_S is the net system reliability and A_1 and A_2 are the reliabilities of the separate system components.

9.7 RELIABILITY CALCULATIONS FOR SAMPLE SYSTEMS

Example 9.1
The environment in a computer room is controlled by three identical air-conditioning units connected in parallel, as shown in Figure 9.2. The computer can continue to function if only one of the three units is working.

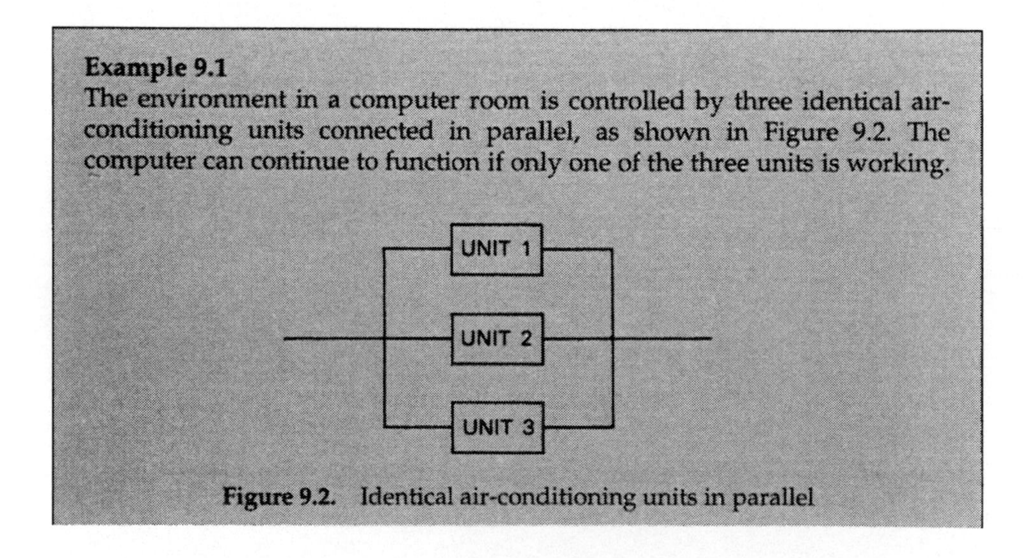

Figure 9.2. Identical air-conditioning units in parallel

If the reliability of each air-conditioning unit is 90 percent over 10 000 hours of operation, what is the overall reliability of the system?

For each unit, $R = 0.9$ and $U = 0.1$. Applying equation (9.14), with $n = 3$ and $r = 1$:

$$R_S = (0.9)^3 + 3{\cdot}(0.9)^2{\cdot}(0.1) + 3{\cdot}(0.9){\cdot}(0.1)^2 = 0.999$$

Thus, the overall system reliability is 99.9 percent, i.e. the probability that all three air-conditioning units will break down and leave the computer unable to run is only 0.01 percent.

Example 9.2

In a machine shop, finished goods can be produced by either of two parallel routes A and B as shown in Figure 9.3. If the reliabilities of the machines are R_1, R_2, R_3, R_4, what is the total system reliability? (The system is only regarded as failed if neither of the two routes A and B is operational.)

From equation (9.5): $R_A = R_1 R_2$
$R_B = R_3 R_4$

From equation (9.2): $U_A = 1 - R_A = (1 - R_1 R_2)$
$U_B = 1 - R_B = (1 - R_3 R_4)$

From equation (9.7): $R_S = 1 - U_S$

From equation (9.8): $U_S = U_A U_B = (1 - R_1 R_2)(1 - R_3 R_4)$

Hence, $R_s = 1 - (1 - R_1 R_2)(1 - R_3 R_4)$

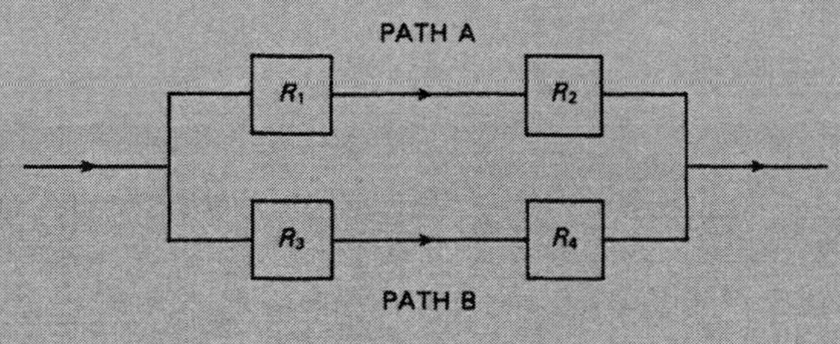

Figure 9.3. Production routes in parallel

Example 9.3

Figure 9.4 shows a block diagram of some units in a production system which are connected in series. The reliabilities of the units for 1000 hours of operation are as follows:

$$R_1 = R_3 = R_5 = 0.99; \qquad R_2 = R_4 = 0.90$$

What is the overall system reliability?

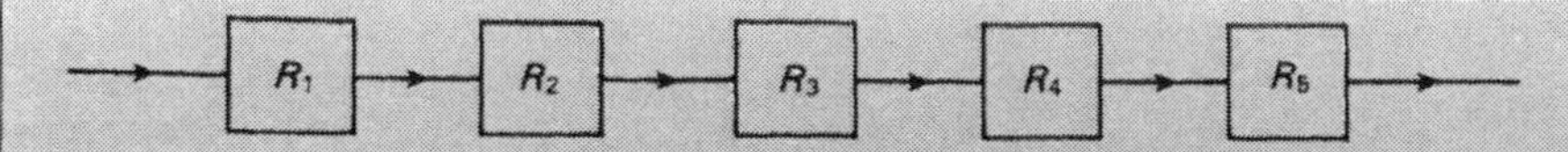

Figure 9.4. Production routes in series

Applying equation (9.5):

$$R_S = (0.99)^3(0.90)^2 = 0.786$$

Thus, the overall system reliability is 78.6 percent.

Example 9.4
Figure 9.5(a) shows a block diagram of the same production system as in Example 9.3 but where the least reliable units in the system are duplicated. A switching system automatically transfers production to the stand-by unit when one of the duplicated units fails. If the values of R for each unit are the same as in Example 9.3, what is the new system reliability?

The system is equivalent to the system shown in Figure 9.5(b), where R_A and R_B represent the net reliabilities of the pairs of identical production units. From equation (9.10)

$$R_A = R_B = 1 - (1 - 0.90)^2 = 1 - 0.01 = 0.99$$

Thus

$$R_S = R_1 R_A R_3 R_B R_5 = (0.99)^5 = 0.951$$

Therefore, the overall system reliability is now 95.1 percent.

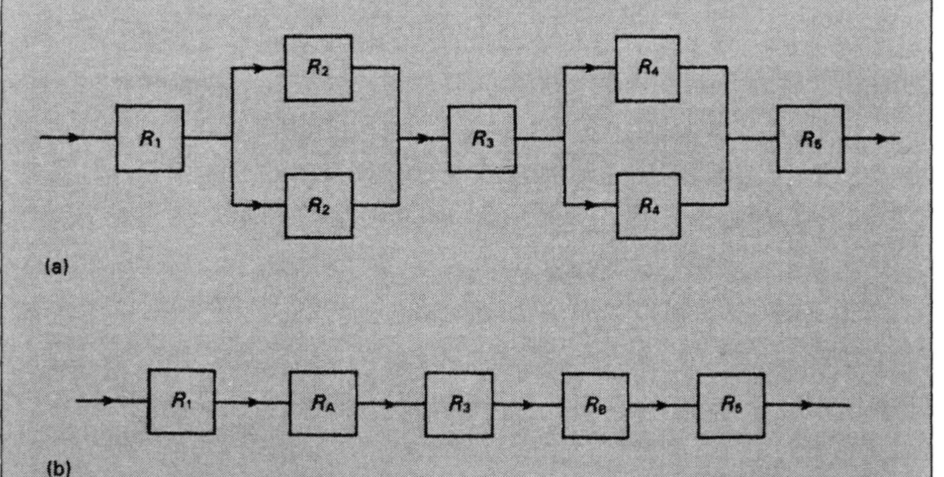

Figure 9.5. Improving reliability: (a) production units in series with duplicated units; (b) alternative diagram for calculating net efficiency of system (a) above

9.8 ACHIEVING HIGH RELIABILITY

Reliability considerations must be at the forefront of the designer's mind whenever he/she is developing a new product or modifying an existing one. The design of the product should be such that manufacturing defects are minimized and the incidence of faults during subsequent use of the product is kept to as low a level as possible. However, there is always an overriding constraint that, whilst the finished product must perform the task required of it reliably, the cost of producing it must be sensible with respect to the prevailing market conditions. Thus, features designed to enhance reliability can only be included within a product if their cost does not make the product too expensive for the market it has to compete in. The arguments put forward in Chapter 2 about specifying the most cost-effective level of quality control apply equally well to reliability considerations.

Reliability and quality control are strongly correlated. Good quality control during the manufacture of a product is of fundamental importance in achieving high reliability when the product is put into use. The more tightly the production parameters in the manufacture of a product can be constrained to their target values, the less will weaknesses be inherent in the product and the less likely it will be to fail. Additionally, if the product being manufactured is identified as having a failure pattern which initially decreases with time, a burn-in procedure should be followed so that its failure pattern has reached steady state before it is delivered to the customer.

Sound analysis at the design stage, careful quality control during manufacture and burn-in periods after manufacture where appropriate will ensure that the reliability of any particular product is the best that can practically be attained. In their subsequent use, however, many products form part of a larger system, and the requirement then is to maximize the reliability of the whole system at minimum cost.

Several general rules about enhancing the reliability of a product containing several components become apparent if the examples in Section 9.7 are studied. Firstly, it follows from Example 9.3 (and also from equation 9.5) that the greater the number of series components in a system, the less the overall system reliability tends to be. Therefore, attempts must be made at the design stage of such a system to include as few separate series components as possible. A system having been designed with as few separate series components as possible, Example 9.4 shows that, by identifying the critical components in the system that have the worst reliability and duplicating them with identical components connected in parallel, the system reliability can be improved substantially. If this practice is followed, provision must be provided for replacing failed components by stand-by units. The most efficient way of doing this is via an automatic-acting switching system, but manual methods of replacement can also often work reasonably well. Similar principles of duplication apply in such areas as the design of lighting systems. In a particular room, calculations may show that a single 80 W fluorescent tube mounted centrally will provide adequate lighting. If reliability considerations are brought in, however, the designer will specify two 40 W fluorescent tubes rather than a single 80 W one because this still provides some lighting even if one tube fails. The gain in system reliability is very much greater than the cost of achieving it because the energy consumption will be the same in either case and the only increased cost is the difference in cost between two 40 W tubes and one 80 W one. The general principle ensuing from this is that a system containing two or more identical components in parallel is usually more reliable than one containing only a single component.

The principle of increasing reliability by placing components in parallel is often extended by deliberately putting more components in a system than it needs to function at 100 percent efficiency. This practice is known by the term *redundancy*. It is commonly applied in electrical circuits where bad connections are a frequent cause of malfunctions. Where connections are made by plugs and sockets, it is often arranged that the same connection is made by two separate pairs of plugs and sockets. The second pair is redundant, i.e. it is not normally needed and the system can function at 100 percent efficiency without it, but it becomes useful if the first pair fails. A common example of this is the multiple earth connections to be found in a car, where, in good conditions, only one connection is theoretically necessary.

To summarize then, reliability considerations should be dominant at the design stage of a product. The design should be such that the product is as easy to manufacture as possible. This will ensure that the incidence of faults created in products during manufacture will be minimized. High standards of quality control are of course a necessary condition for this as well. Finally, the general rule applies that, in a complex system, reliability is maximized when the number of series components is minimized and the number of parallel components which each perform the same function is as large as possible.

REFERENCES

1. Dhillon, B. S. and Singh, C. *Engineering Reliability: New techniques and applications*, John Wiley & Sons (1981).

CHAPTER 10

Software Quality Metrics

10.1 INTRODUCTION

With the continual growth in computerized elements within manufacturing systems, the performance of software programs within such systems has become a potentially weak link in quality control and quality assurance procedures. Although the discipline of software engineering has established many sound practices in software design, quality assurance rules demand that the quality of the software produced is verified. Therefore, rigorous procedures are needed that can measure the performance, reliability and freedom from error of software components in production systems.

Software engineering is now a well-established discipline that provides sets of design rules that try to ensure that software development proceeds according to rigorously designed procedures so that the software produced has the attributes of high reliability and maintainability whilst the costs of code design and writing are minimized. Formal procedures for achieving high quality in software are set out in BS 7165 [1] and ISO 9000–3 [2]. Any academic library or bookshop also offers a large number of texts on software design procedures. These differ significantly in their style of approach, but all have the common aim of encouraging the production of error-free software which conforms to the design specification. It is not the intention in this text to enter into arguments about which of these software design approaches is best. The choice between the different software design techniques on offer is largely a matter of personal preference, and this is therefore left to the reader. However, it is assumed in this text that any quality-sensitive software contributing to a manufacturing system will have been produced according to good software engineering principles, using one of the alternative design regimes promoted in the relevant texts.

For quality assurance purposes, software design according to such rigorous requirements is a necessary, but not sufficient, requirement. Adopting design procedures that should achieve good quality is not enough: the actual achievement of good quality must also be demonstrated. Thus, in addition to rigorous design, metrics must also be applied to test the success of the applied design procedures in producing error-free software that meets all the design specifications and produces the results intended.

The science of measuring various attributes of software is widely known in the software industry as *software metrics* and it is applied in practice to many different aspects of software and its production. These applications of software metrics can be divided into two classes: measurement of internal attributes and measurement of external attributes. Internal attributes are those aspects of software which are only concerned with the structure and features of the software itself, such as the length of code, modularity, structuredness and internal code documentation. External attributes are concerned with the effect on, and interaction with, the external environment that the software is associated with, and include things like reliability, portability, code-efficiency, maintainability, fitness for purpose and ability of the software to check for data errors. These internal and external attributes are not independent; there is strong coupling between them. For instance, the size and structure of a code, and the extent to which it is produced according to good software engineering principles, is strongly correlated with its ultimate reliability, maintainability and fitness of purpose.

Besides these direct measurements of the internal and external attributes of the software itself, the term 'software metrics' is often extended to include the measurement of various aspects of the software production process. These include measurements of the times for software specification, design, coding and testing in person-hours and the productivity per person. All of these factors have cost implications and there are clearly strong economic arguments for ensuring, for instance, that the productivity of programmers is maximized and software production times are minimized, whilst ensuring that the quality of the software product produced is also maximized.

In common with software engineering in general, there are now many texts devoted to the subject of software metrics (e.g. [3]), and it is not the intention to repeat coverage of this same ground in general software metrics in this chapter. It is an a priori assumption here that good design principles have been applied in producing software that is part of a production process and that appropriate software metrics have been applied as part of the design process. However, having produced the software according to procedures that are designed to ensure its quality, the remaining question is: 'Have these careful design procedures succeeded in producing good quality software?' This is the key question in quality assurance procedures, and the purpose of this chapter is to present the various mechanisms available for answering this question by measuring the quality of software.

In many texts covering software metrics, the issue of software quality metrics is given little or no attention. There is a frequent assumption that adherence to good software design and production practices will ensure that good quality is achieved. Good design is certainly a necessary ingredient in achieving good quality, but quality assurance standards such as ISO 9000 demand that metrics are applied that actually measure the quality of the software product. This chapter is therefore devoted to a review of the various possible approaches to software quality metrics.

10.2 GENERAL QUALITY METRICS

The very first test of software quality is to determine whether the code satisfies the specified requirements, i.e. whether it performs the tasks and produces the results that it is supposed to. It is essential that there is traceability between the specified requirements and the design components of the software. The process of confirming this traceability, which is often known as the *verification and validation phase* of software development, is described in various texts on software design (e.g. [4]). This phase of testing is very much a binary assessment of quality: it determines only whether software *does* or *does not* satisfy the various requirements specified. There is no quantification of how well the requirements are satisfied. Therefore, following this initial verification and validation stage, the next requirement is to assess the software quality quantitatively.

The quality of software can be measured in a number of different ways. These ways include a count of the number of errors within it, its efficiency, its maintainability, its reliability, and the degree to which it is fault-tolerant. Which of these is the most appropriate measure depends on what the measurement is being made for.

The first possible quality measure, that of *counting the number of errors* found per thousand lines of code, is one which is quoted in several software engineering texts. This is of very little real value. In practice, testing will fail to detect all the errors in any given piece of software and, from a quality assurance point of view, the important issue is how many more errors are likely to turn up and how soon they will emerge. Thus, for the present purposes, the error per thousand lines of code metric can immediately be discarded.

Software efficiency is concerned with the speed of execution and memory occupancy of the software code. The time-efficiency of a software program is measured in terms of the number of machine operations required to execute some algorithm of given size. Clearly, reducing redundant computational operations increases the speed of execution and thus the time-efficiency of the program. Memory occupancy is measured in similar terms, and efficiency is improved by reducing the amount of unnecessary use of memory. Execution speed and memory occupancy are inversely proportional to some extent, and some compromise between them in program design is usually necessary. The issue of software efficiency is considered in more detail in various software metrics texts (e.g. [3]). However, it is not considered further here because it is largely irrelevant to the measurement of the quality of software in terms of its effect on the quality of products in an industrial manufacturing system that it is part of, except in a few special circumstances where variations in the values of input parameters may cause variations in the execution time of the program which might in turn affect the availability of the results at the time that the production system needs them.

Software maintainability is defined in some texts in terms of 'the mean time to repair a bug'. However, maintainability must encompass more activities than this and, in particular, should include the process of modifying software either to change the functions that it performs or to add new functions. The need for such modification indicates, to some extent, poor requirements specifications in the first place, before the software was produced. However, with changing manufacturing methods and requirements, such post-production software modification is often inevitable. Estimating the mean-time-to-repair (MTTR) bugs is a problem which is solvable once the cause of the bugs is identified, but estimating the time necessary to effect modifications in response to new or changed user requirements is an altogether more difficult task. In fact, it is impossible to predict a mean time for completing modifications to software, because this depends entirely on the number and extent of the modifications that are called for. Thus, whilst maintainability is a desirable goal in software design, the impossibility of measuring this in absolute terms means that it cannot be quantified for quality assurance purposes. In consequence, the software quality metrics proposed in this chapter will only include *reliability prediction* and *measurement of fault-tolerance.*

Fault-tolerance is an important aspect of software reliability. The software must respond in an acceptable way to any unexpected conditions arising or errors occurring in the input data (e.g. because of sensor failure). Any errors in the inputs to software algorithms must not be transmitted through to the output, where they might cause quality problems in the manufacturing system that the software is part of. This is most easily achieved by setting limits on the acceptable range of each input variable to the software, and then testing all input data against these limits. Whenever any input parameter exceeds the boundaries specified as acceptable, the software must respond by flagging an error condition to the production system controller. This enables appropriate action to be taken, which will either involve shutting down production until the fault is rectified, or at least will require that extra checks are made to ensure that the fault is not affecting product quality.

10.3 QUALITY METRICS IN TERMS OF RELIABILITY

In any testing of software to assess its reliability, it is important to draw a clear distinction between failures of the computer system as a whole and the emergence of errors in the functioning of a software program running within the computer hardware. From a quality point of view, total failures of the computer system which result in a total loss of output are less serious, because the fact that there is a fault is obvious. What is much more serious is the possibility of software errors causing incorrect, though plausible, results. This may lead to problems occurring, without any warning signs, in the product quality of manufacturing systems that the software is part of.

Software reliability is defined in many texts as the mean-time-between-failures (MTBF), i.e. the mean time between errors being found, by analogy to the general engineering definition of reliability. However, applying the general engineering definition of reliability to software is not appropriate because the characteristics of the error mechanisms in software and in engineering hardware systems are fundamentally different. Hardware systems which work correctly when first introduced can develop faults at any time in the future, and so the MTBF is a sensible measure of reliability. However, software does not change with time: if it starts off being error free, then it will remain so. Thus, specifying an MTBF value for software is of little use from a quality assurance point of view. What a customer wants to know, in advance of its use, is that the software is not going to fail. Therefore, software reliability must somehow be expressed as a probability that errors will not occur after the software has been put into use.

A fundamental problem in predicting that errors will not occur in software is that, however exhaustive the testing is, it is impossible to say with certainty that all bugs have been found and eliminated. Software errors can be quantified by two parameters, P and Q, where

P is the number of errors discovered by testing the code (i.e. the number of known errors)

Q is the total number of errors in the code (both known and unknown, i.e. both discovered and undiscovered)

The problem of calculating Q for a program is known to be an intractable one (except for trivially small programs). Therefore, the best that can be done is to develop a rigorous software testing procedure that establishes a measurement of P which can be used to predict Q to a high level of probability.

Measurement of software quality in terms of its reliability is not currently widespread in manufacturing industry. However, even where it is measured, it is common to find that only the parameter P is measured, and it is this alone which is used as the quality metric. This is grossly unsatisfactory. Some means must be found of predicting the quantity Q from the measurement of P, to an acceptable level of probability that the value predicted is a close approximation to the true value.

10.3.1 Software Testing and Correction

The purpose of testing is to detect the errors which exist in software. Whilst it is never possible to detect all the errors which might exist, the aim must always be to find as many errors as possible. To achieve this aim, testing must be carried out in a rigorous manner which is fully documented.

The classical approach to software testing is known as *bottom-up testing*. In this scheme of testing, each software unit is tested with the full range of inputs that it is designed to handle, checking in each case that the output is as expected. Then the units are assembled and checked as a whole, to determine whether there is any problem in the way that the separate units interact. Referring to the software structure shown in Figure 10.1(a), with program modules A, B, C, D, E, F, this means testing in the order: F, then E, then D, then C, then B, then A, and finally all modules A–F together.

An alternative approach, known as top-down testing, involves testing the top-level structure of the program first, then gradually bringing in subordinate program units and testing each new partial structure until the whole program has been assembled and tested. Referring to Figure 10.1(a) again, this would involve testing the program modules in the order: A, then A with B, then A and B with C, then A–C with D, then A–D with E, and finally A–E with F. The claimed advantage of the top-down approach is that any structural program errors are found more quickly. However, it is more difficult to implement because, until each subprogram is included in the partially-assembled software, any results that it should provide to the top-level module have to be emulated in some way. Thus, whilst serious structural errors are detected sooner, the overall time required for program testing is increased compared with the bottom-up approach.

In the case of more complex programs, neither top-down nor bottom-up approaches to testing are sufficient by themselves. Where a program has

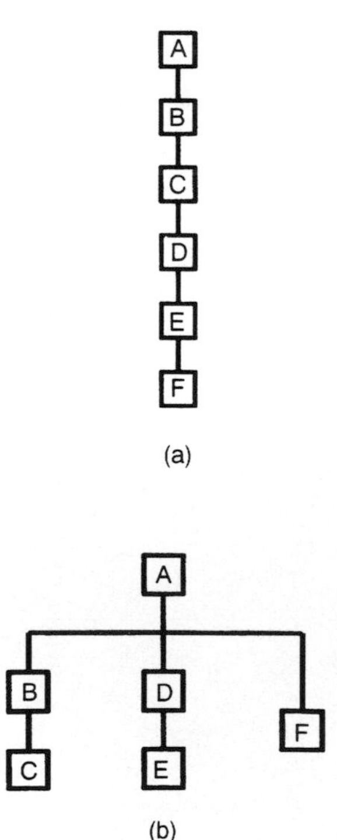

Figure 10.1. Software structures: (a) program structure suitable for either bottom-up or top-down testing; (b) program structure requiring sideways integration in testing

two or more possible vertical paths, with each path containing several program modules, some form of *sideways integration* in the software testing is necessary. Sideways integration involves taking each vertical path in turn and testing the modules within it by either a top-down or bottom-up approach. An approach which could be called *bottom-up-with-sideways integration* involves testing each separate vertical path in isolation and then test them all together with the top-level program module. Referring to the program structure shown in Figure 10.1(b), this would involve testing in the order: C, then B, then B with C, then E, then D, then D with E, then F, then A, and finally all modules A–F together. Alternatively, in an approach which could be called *top-down-with-sideways integration*, testing could start by including just one vertical path with the top-level program module, and then adding other vertical paths one by one, retesting of course after each stage of addition. For the program structure shown in Figure 10.1(b), this would involve testing the program modules in the order: A, then A with B, then A with B and C, then A–C with D, then A–C with D and E, and finally A–E with F.

Whichever approach to software testing is adopted, it is important to ensure that the procedure tests all possible paths through, and loops within, each program segment and the program as a whole. This requires a systematic and thorough approach, but it is a relatively easy and straightforward task to accomplish as long as the testing is well managed and methodical. However, the process of testing the software for the full range of possible inputs poses a much greater difficulty. In the case of several input parameters, testing the program exhaustively for errors with every possible combination of input values is clearly impractical, as this would involve almost infinite time and expense. Hence, the requirement is to devise a suitable testing procedure which tests a representative selection of input data value combinations. Choosing this representative set of test inputs is a problem of considerable complexity. It would not be appropriate to discuss solutions to this problem here, as this chapter seeks only to give an overview of the necessary considerations in software testing. However, suitable procedures are described in many dedicated software engineering texts (e.g. [6], [7]).

Whenever the testing procedure identifies a program error, the cause of the problem has to be found and then rectified by making appropriate corrections to the software code. Having done this, the full testing procedure must be repeated from the start, unless it is very clear that the changes made to the code within a particular module of software cannot have any effects outside the module. Unless this condition clearly holds, full retesting of the whole software is necessary to avoid the possibility of the changes made in one part of the software code having unexpected effects elsewhere in the program.

The existence of documented testing procedures, and evidence that they are properly carried out, is a necessary part of quality assurance. It is also the starting point for the final aspect of quality assurance: that of predicting that the software will not fail during use.

10.3.2 Quantifying Reliability

The classical approach to quantifying software reliability, as described in some texts on software metrics, e.g. [3], [7], establishes what the rate of error discovery is during testing and then extrapolates this into an estimate of the mean-time-between-failures (MTBF) for the software once it has been put into use as part of the manufacturing system. However, as explained at the start of Section 10.3, this is unsatisfactory from a quality assurance point of view. This approach accepts that errors in the software exist but makes a prediction that they will not emerge very frequently. Part of this approach is to continue software testing until the predicted MTBF is greater than the projected time-horizon of use of the software in the manufacturing system, i.e. such that there is an expectation that no software errors will occur during the lifetime of the manufacturing system. However, from a quality assurance point of view, it is far more convincing to be able to say, 'There is a high probability that there are zero errors in the software', than to say, 'There are a finite number of errors in the software but they are unlikely to emerge within the expected lifetime of use of the software'. Thus the quantification of software reliability adopted in this text is the probability of zero errors rather than the MTBF.

Establishing the probability that there are no errors in a piece of software is based on calculating the quantity Q defined earlier in Section 10.3. It was suggested in that section that Q could be estimated from the quantity of errors P in the software detected during rigorous software testing. It was observed that P will usually be less than Q because of the expectation that, however exhaustive the testing regime is, it will not successfully detect all the errors that exist. However, as there is a stated, quality-assurance-directed requirement for zero errors in the software, the above calculation of Q is only the starting point of a procedure which, having estimated a value for Q, must involve further software testing until the predicted value of Q is zero.

A suitable mechanism for estimating the quantity Q and then reducing it to zero is known as *error-seeding* [5]. In this method, the programmer responsible for producing the software deliberately puts a number of errors Z into the program. The software is then given to a different programmer to test. This testing will identify a number of errors given by $P + Y$, where P is the number of errors identified that the original programmer did not deliberately put into the program and Y is the number of deliberately inserted errors that are detected. Normally, P will be less than Q and Y will be less than Z. However, on the assumption that the ratio of seeded errors detected to the total number of seeded errors present will be the same as the ratio of the unseeded errors detected to the total number of unseeded errors in the software, the following expression can be written:

$$\frac{P}{Q} = \frac{Y}{Z} \tag{10.1}$$

In this expression, P and Y are measured by the program tester and Z is the known number of deliberate errors introduced by the program's author. Hence, it should be possible to calculate the total number of errors Q from:

$$Q = \frac{ZP}{Y} \tag{10.2}$$

Unfortunately, expression (10.1) makes assumptions that are not necessarily true in practice. In particular, it assumes that the seeded errors are representative of all the other unintentional (non-seeded) errors that are present in the software, both in proportion and character. This representativeness can never be completely achieved in practice because, if errors are unknown, then their characteristics and the distribution between different types is also unknown. Thus, whilst this approach may be able to give an approximate indication of the value of Q, it can never predict its actual value with certainty.

The performance of this approach can be improved if two independent program testers are used to test the same program [6]. Suppose that the number of errors detected by each tester is P_1 and P_2 respectively. Normally, the errors found by the two testers will be the same in some cases, but each will also find errors that the other did not find. Let C be the number of errors that both testers find. If S_1 and S_2 represent the error-detection success of each tester, these successes can be quantified as:

$$S_1 = P_1/Q; \qquad S_2 = P_2/Q \tag{10.3}$$

where Q is the estimate of the total number of errors in the program, as before.

It is reasonable to assume that the proportion of errors P_1 that tester 1 finds out of the total number of errors Q is the same proportion as the number of errors C that he/she finds out of the number P_2 found by tester 2, i.e.

$$\frac{P_1}{Q} = \frac{C}{P_2} = S_1 \qquad \text{and hence,} \qquad P_2 = \frac{C}{S_1} \tag{10.4}$$

From equation (10.3), $Q = P_2/S_2$, and substituting the value of P_2 obtained from (10.4), the following expression for Q is obtained:

$$Q = \frac{C}{S_1 S_2} \tag{10.5}$$

From (10.4), $S_1 = C/P_2$ and from (10.3), $S_2 = P_2 S_1/P_1 = C/P_1$. Thus, substituting for S_1 and S_2 in (10.5),

$$Q = \frac{P_1 P_2}{C} \tag{10.6}$$

Example 10.1
Let the number of program errors found by two testers be 30 and 36 respectively, and let the number of common errors (i.e. the instances where a particular error is found by both testers) be 20. Then, $P_1 = 30$, $P_2 = 36$ and $C = 20$. Evaluating equation (10.6) with these values gives $Q = 54$.

For quality assurance purposes, program testing should continue until the number of errors which have been found is equal to the predicted total number of errors Q. In the case of the above example, this means continuing testing until 54 errors have been found. However, the problem with doing this is that Q is only an estimated quantity. For the program mentioned in this example, there may only be, in fact, 52 or 53 errors in the program. Thus to continue testing until 54 errors have been found would mean testing for ever! Thus, once 52 or 53 errors have been found and testing for a significant time after this detects no more errors, the testing procedure should be terminated, even though the program may still contain one or two errors. The approximate nature of the calculated value of Q also means that its true value could be 55 or 56, and if testing is stopped once 54 errors have been found, this means that the software certainly still contains errors. Thus, the fact that Q is only an estimated value means the statement that a program is error free once the number of errors detected is equal to Q can only be expressed in probabilistic terms.

To quantify this probability, further testing of the program is necessary [6]. The starting point for this further testing is the stage when the total number of errors Q that are predicted have been found (or when the number found is slightly less than Q but further testing does not seem to be finding any more errors). The next step is to seed the program with W new errors and then test it until all W seeded errors have been found. Provided that no new errors have been found during this further testing phase, the probability that the program is error free can then be expressed as:

$$P_{\text{no error}} = \frac{W}{W+1} \tag{10.7}$$

For $W = 10$, $P_{\text{no error}} = 0.91$ or 91 percent. Thus, the probability that the program is error free after this final stage of testing depends on the number of seeded errors which are introduced and detected without any new unknown errors being found. If of course a new error is found during this procedure, it must be corrected and then the seeding and testing procedure must be repeated. Any specified level of probability that the program is error free can be achieved by an appropriate choice of W in equation (10.7). Thus, for example, if the probability of no error is required to be 99.5 percent, the value of W required is 199.

10.3.3 Enhancement of Reliability by *N*-version Programming

Reliability can be greatly improved by developing three different versions of the same software, with each version being produced by a different programmer but according to a common requirements specification. Then, assuming that there are no errors in the specification itself, the three outputs can be compared. Any event that occurs when the outputs from two of the programs are the same but the third shows a different value indicates an error in the third program. Sometimes, for applications that are very critical, even five versions of the software are created. Then, if a 'voting' system is used, up to two out of the five versions can be faulty without incorrect outputs being generated. However, even two versions of a program are sufficient to successfully indicate the existence of errors if the outputs differ. This last scheme does not identify which of the outputs is correct, but it is often adequate from a quality assurance viewpoint, provided that either production is stopped until the error source is found and rectified, or extra tests are made to ensure that product quality has not been affected.

Whilst this approach reduces the chance of software errors affecting product quality, it is not foolproof because the degree of independence between programs cannot be guaranteed. Different programmers, who may be trained in the same place and use the same design techniques, may generate different programs which have the same errors. Thus, this method has the best chance of success if the programmers are trained independently and use different design techniques. Ideally, they should also work for different companies and not communicate with each other at all whilst writing their programs. However, whatever measures are taken to ensure the independence of program development, it is a fact of life that determined programmers will always find ways of circumventing the controls in place and will collaborate to share the effort of development of some software components and therefore reduce their workload. The best defence against this is to ensure that the procedures specified by ISO 9000 to ensure product quality, as explained earlier, are fully implemented. Of particular importance is the need to train all programmers involved in the need for quality control and to establish a corporate empathy towards achieving high quality. Programmers must fully understand why having different versions of a program is necessary and how this will contribute towards improved reliability. If they are sympathetic towards this need, they will co-operate with it and not try to 'cheat' in their independent production of program versions.

REFERENCES

1. BS 7165: *Recommendations for the achievement of quality in software*, British Standards Institute (1991).
2. ISO 9000-3: *International Organization for Standards*, Geneva (1991).

3. Fenton, N. E., *Software metrics: a rigorous approach*, Chapman and Hall (1991).
4. Mather, R. N., Methodology for business system development, *IEEE Trans. Software Eng.*, 13(5), 593–601, (1987).
5. Mills, H. D. *On the statistical validation of computer programs*, IBM Federal Systems Division, Maryland (1972).
6. Pfleeger, S. L., *Software engineering: the production of quality software*, Macmillan (1987).
7. Shooman, M. L., *Software engineering: design, reliability and management*, McGraw-Hill (1983).

CHAPTER 11

Statistical Process Control

It has already been established that an important part of quality control is monitoring the values of all process variables and parameters of a product that have an effect on quality. This appears at first sight to be relatively simple: all it would seem to require is to measure all the relevant quantities with calibrated instruments (whilst correcting for any environmental errors etc.) and make sure that each one remains at its correct value. Unfortunately, a serious problem arises in the case of parameters that are known to suffer random variations. In this case, almost every measurement will be away from the target value, and some mechanism is therefore needed to determine whether or not the deviations are within the range that can reasonably be expected.

Statistical process control provides this mechanism. It determines whether the variation in measurements of a process or product parameter is within the range to be expected according to the known random deviation characteristics of the parameter, or whether the deviation is outside the expected range and indicative of some problem in the production process. When applied correctly, statistical process control is an extremely valuable technique in manufacturing systems. It is able to determine when things are starting to go wrong before the stage is reached where the problem has grown to a sufficient magnitude to significantly effect the quality of the product being produced. This early warning of the development of production problems usually leads to significant production cost savings, apart from being an important element in the quality system operated.

In a book of this nature, it is inevitable that only an overview of the main principles of statistical process control can be covered. If required, more detailed information can be obtained from specialist texts on the subject, such as [1].

11.1 CONDITIONS FOR APPLICATION OF STATISTICAL PROCESS CONTROL

Before statistical process control can be applied to a process variable, three conditions must be satisfied. Firstly, any systematic errors in the measurements must have been identified and either eliminated or compensated for. Secondly, the deviations of the measurements from the mean value must be truly random and Gaussian. Thirdly, the process must be under *statistical control*, which means that the mean and standard deviation of the measurements must remain constant. If, and only if, these three conditions are satisfied can statistical process control be applied.

The condition regarding *systematic errors* is fairly easy to comply with. The various sources of systematic error were discussed at length in Chapter 5, and appropriate mechanisms were presented there for either eliminating them or compensating for them.

The condition regarding *Gaussian distribution* is rather difficult to test in a mathematically rigorous way. However, it is normally sufficient to plot some measurements on a histogram and confirm that they are symmetrically distributed about the mean value in a manner similar to that shown in Figure 5.1.

The final condition specified was about the process being under *statistical control*. This is verified by taking several sets of successive samples of a process parameter measurement and calculating the mean and standard deviation of each set. These will not be identically equal going from one set of measurements to the next because each set only consists of a finite number of measurements out of an infinite data set (see the discussion on 'standard error of the mean' in Section 5.2.2 for further explanation), but differences from one data set to the next should only be small and they should not show any trend such as steadily increasing or decreasing values. The existence of statistical control is verified mathematically if the means of successive sets of measurements do not go outside the boundaries equal to three times the standard error of the mean. This is best explained by considering an example.

In a particular chemical production process involving an exothermic reaction, water cooling is used to control the temperature so that the correct products are obtained from the reaction. Suppose that several successive sets of 20 temperature measurements are taken and the mean and standard deviation (σ) are calculated for each, giving the following values

Set 1: mean = 120.0; $\sigma = 2.05$

Set 2: mean = 119.5; $\sigma = 2.13$

Set 3: mean = 121.3; $\sigma = 2.02$

Set 4: mean = 120.4; $\sigma = 1.98$

Set 5: mean = 119.8; $\sigma = 2.08$

The first point to note is that the values of mean and standard deviation are not showing any increasing or decreasing trends. The standard error of the mean should now be calculated (see Section 5.2.2)

$$\alpha = \sigma/\sqrt{n} = 2.13/\sqrt{20} = 0.48$$

(where n is the number of measurements in each data set). Hence,

$$3\alpha = 1.44$$

The mean of the five successive data-set means can be calculated as

$$\text{mean} = (120.0 + 119.5 + 121.3 + 120.4 + 119.8)/5 = 120.2$$

The maximum deviation of any of the five data-set means from this overall mean is $(121.3 - 120.2) = 1.1$. This is less than the calculated value for 3α of 1.44, showing that the process is under statistical control.

11.2 PRINCIPLES OF STATISTICAL PROCESS CONTROL

Before proceeding further it is useful to review some of the material on random errors presented in Section 5.2. Figure 5.2 shows that, for measurements subject to random variation within a Gaussian distribution, both very large and very small values extending towards plus and minus infinity are possible. However, the nature of the distribution is such that such very large and small values only occur rarely, and most measurements will fall reasonably close to the mean value. Statistical process control relies on this fact that most values will be close to the mean, and it is able to set a criterion for determining whether any particular measurement is inside or outside what could reasonably be expected according to random variation.

It was established in Section 5.2.2 that 99.7 percent of measurements lie within boundaries of $\pm 3\sigma$ either side of the mean of a set of measurements which are subject to random errors and part of a Gaussian distribution. This is normally used as the criterion for determining whether a particular measurement is inside the boundary of values that can reasonably be expected according to the known random variation characteristics. If any measurement falls outside the boundaries of $\pm 3\sigma$, there is only a 0.3 percent chance that such a large deviation is due just to random effects, and this is used to indicate that there is some abnormal condition in the production process which should be investigated and corrected.

Measurements that go outside the boundaries of $\pm 3\sigma$ usually indicate that some problem has already developed. However, in many production situations, it is desirable to have some prior warning that things are starting to go wrong before this stage is reached. Such prior warning is usually provided by establishing an additional check for values that go outside boundaries of $\pm 2\sigma$. For a set of measurements subject to random errors, 95.4 percent (as shown in Section 5.2.2) will be inside these boundaries and thus only 4.6 percent will be outside. This is not a high enough probability to say that there is definitely a problem developing, but it can be used to initiate checks, especially if more than one measurement falls outside the $\pm 2\sigma$ boundaries within a short space of time.

Statistical process control of a production parameter is normally carried out in practice by plotting successive measurements on special charts. These are often known as *Shewhart charts*, after the American, Shewhart, who invented them in the 1920s. Three common types of chart are the *XBAR chart*, the *CUSUM chart* and the *RANGE chart*.

11.3 XBAR CHART (OR MEAN CHART)

The *XBAR chart* is the one which is probably used most extensively. It is also sometimes referred to as a MEAN chart. The purpose of an XBAR chart is to show up measurements that differ from the mean value of the measurements by an abnormal amount in relation to what could reasonably be expected due to random errors. The mean of the measurements is commonly represented as $\overline{X}$ or XBAR, explaining why the chart is given the name 'XBAR chart'.

In its most common form of implementation, an XBAR chart has the mean parameter value ($\overline{X}$) and the $\pm 2\sigma$ and $\pm 3\sigma$ boundaries drawn on it†. The $\pm 2\sigma$ boundaries are normally labelled as the upper and lower warning limits (UWL and LWL) and the $\pm 3\sigma$ ones as the upper and lower action limits (UAL and LAL). Referring back to the set of mass measurements given in Table 5.1, suppose that these are measurements in grams of the mass of small loaves produced in a bakery. The mean of the measurements is 81.18 g and the standard deviation is 0.30. The $\pm 2\sigma$ and $\pm 3\sigma$ boundaries are therefore ± 0.60 and ± 0.90 respectively and the warning and action limits can be calculated as follows:

$$\text{UWL} = 81.18 + 0.60 = 81.78; \qquad \text{LWL} = 81.18 - 0.60 = 80.58$$

$$\text{UAL} = 81.18 + 0.90 = 82.08; \qquad \text{LAL} = 81.18 - 0.90 = 80.28$$

An XBAR chart can be drawn for these mass measurements as shown in Figure 11.1.

The basic principle of using an XBAR chart is to plot a succession of measurements of the process variable under control on the chart. If the

† The formal standards on statistical process control described in ISO 8258 [2] and BS 7785 [3] specify slightly different boundaries of $\pm 1.96\sigma$ and $\pm 3.09\sigma$, which encompass 95.0 percent and 99.8 percent respectively of measurements that are only varying owing to random effects.

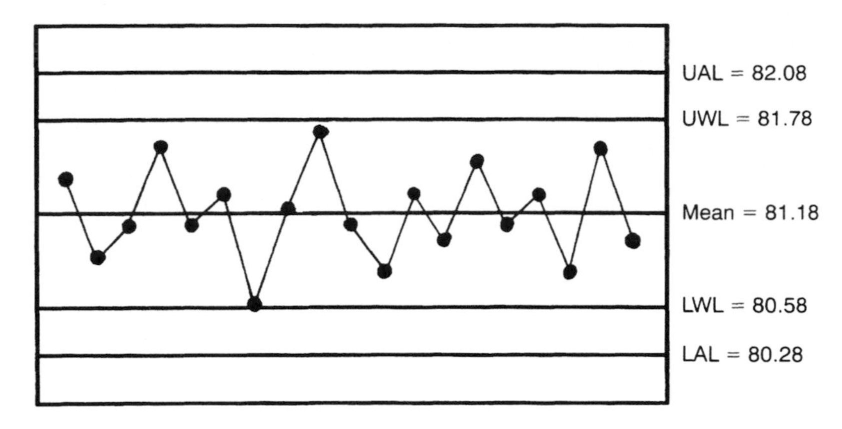

Figure 11.1. XBAR chart for the 19 mass measurements given in Table 5.1

process is properly under control, then all measurements plotted will remain within the boundaries marked by the LWL and UWL lines. This much is common to all statistical process control schemes. However, practice varies with regard to the response generated if measurements go outside the warning or action limits. The basic rules say that a measurement going outside the $\pm 2\sigma$ boundaries, as shown in Figure 11.2(a), should be taken as a warning that a fault may be developing in the production process, initiating checks that the sensors are working properly and carrying out other observations according to the nature of the process. Measurements going outside $\pm 3\sigma$ limits, as shown in Figure 11.2(b), indicate that a definite fault has occurred, triggering remedial action which may include shutting the plant down if the fault cannot be readily identified and corrected.

However, in practice, these basic rules of statistical process control are often modified in various ways. To some extent, these differences in practice arise from differences in the sampling rate, which may vary from a few seconds for fast processes to as slow as one sample per day for petrochemical processes. The sections below attempt to cover some of the ways in which the basic rules for responding to out-of-limit measurements are altered.

11.3.1 Sample Averaging

If a single measurement goes outside the $\pm 2\sigma$ or $\pm 3\sigma$ boundaries on an XBAR chart, there is a finite probability (of 4.6 percent and 0.3 percent respectively) that the cause is a random deviation of the measurement and instrument producing it rather than a fault in the controlled process. The incidence of false alarms when measurements exceed these defined $\pm 2\sigma$ and $\pm 3\sigma$ boundaries can be greatly reduced if several successive measurements are averaged and the mean plotted on the control chart. If just two

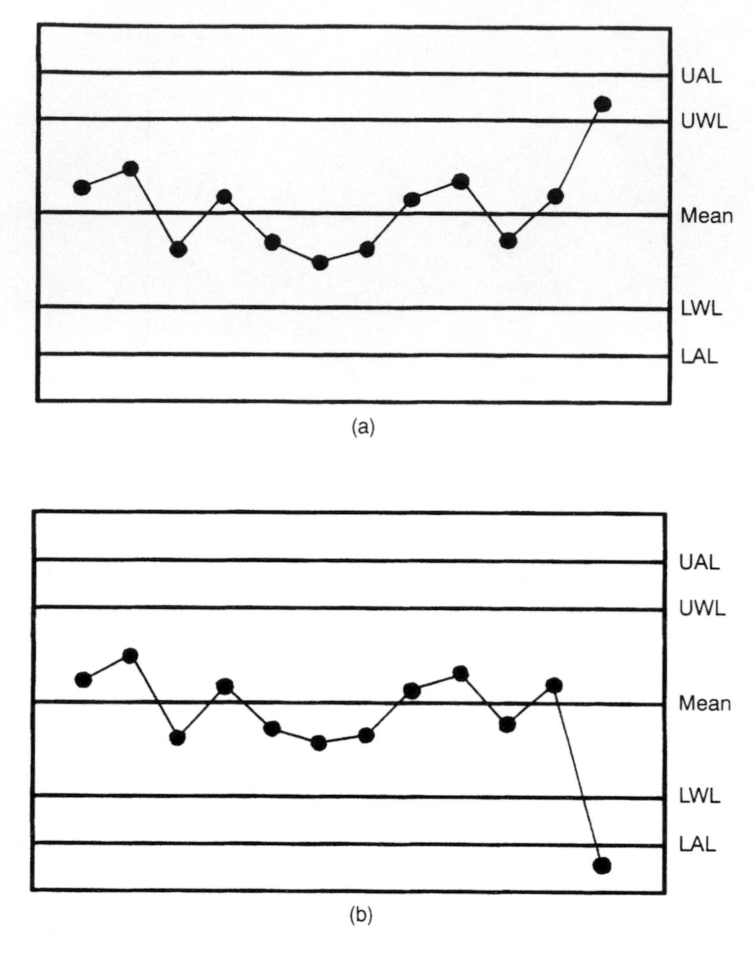

Figure 11.2. Use of XBAR chart to indicate process problems: (a) measurements going outside $\pm 2\sigma$ limits ; (b) measurements going outside $\pm 3\sigma$ limits

successive measurements are averaged, the probability of them both exceeding the warning limit of $\pm 2\sigma$ owing only to random effects can be expressed as $0.046 \times 0.046 = 0.0021$ or 0.21 percent. If the average of three successive measurements is plotted, the probability falls to $0.046^3 = 0.000\,097$ or 0.01 percent. Figure 11.3(a) shows what happens if the means of successive sets of two measurements in Table 5.1 are plotted, and Figure 11.3(b) shows a similar plot of the means of successive sets of three measurements. If Figure 11.3 is compared with Figure 11.1, it is apparent that averaging has the effect of moving the points plotted closer to the mean line.

Such plotting of the average of several successive measurements is clearly useful when a fast process is involved and samples are taken every few seconds. However, in the case of a slow process which is sampled infrequently, the technique is not of much use because, if a real fault has

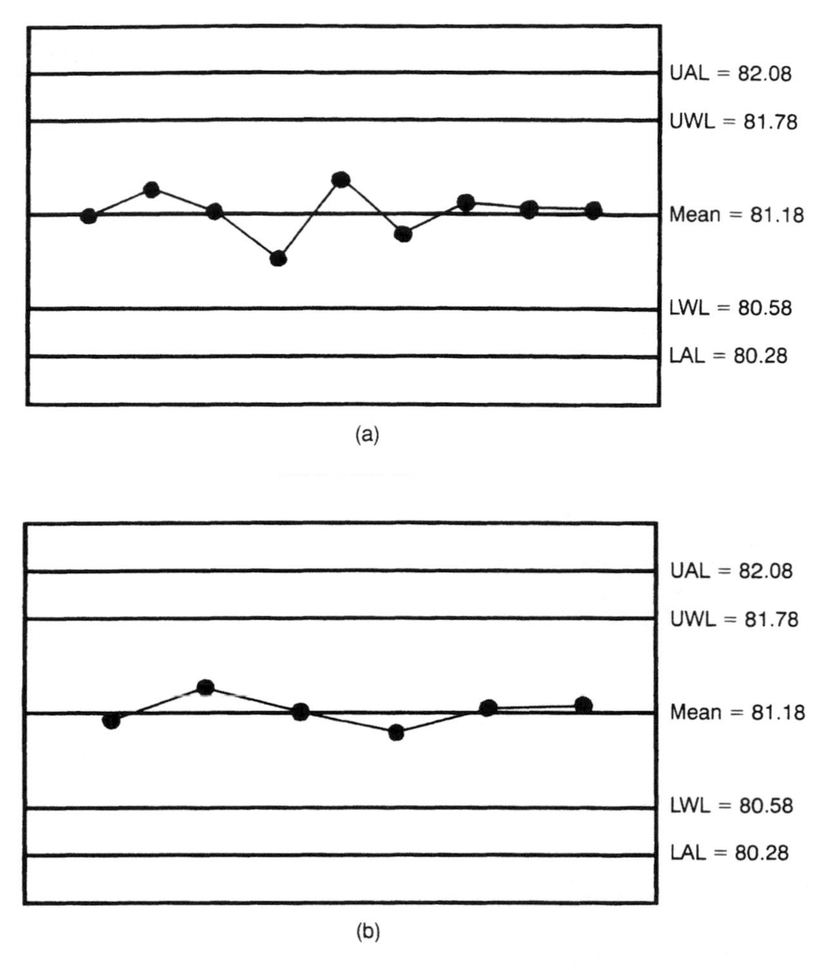

Figure 11.3. Plotting the mean of successive measurements on XBAR chart: (a) means of two measurements; (b) means of three measurements

developed in the production process, serious damage will result before the problem is identified and corrective action is taken.

11.3.2 Trends

The basic rules of statistical process control treat excursions of measurements outside $\pm 2\sigma$ limits as a warning which generates investigative action. However, if the process is changing rapidly, it may go on to exceed the unacceptable $\pm 3\sigma$ limits before there has been sufficient time to respond properly to such a warning. This difficulty may be avoided if an additional rule of looking for trends is introduced into the control algorithm. A trend is a sequence of values which are either increasing or decreasing in value,

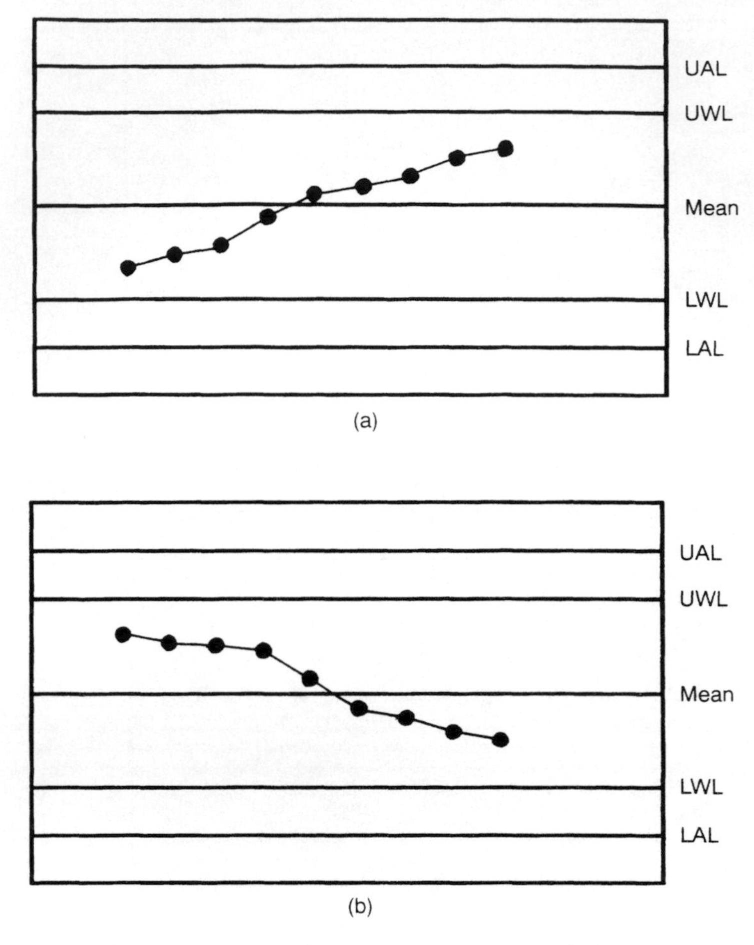

Figure 11.4. Trends on XBAR chart: (a) increasing trend; (b) decreasing trend

owing to events like a tool wearing. Figure 11.4 shows examples of such increasing and decreasing trends.

What has not been established so far is how many points there should be in a trend before action is taken. It is very important to get this correct, otherwise false alarms may be generated. Starting from any measurement P, the probability of the next value being greater than P is represented by the parameter A in Figure 11.5 and the probability of it being smaller is represented by the parameter B. In general, A is not equal to B, and neither probability is easy to quantify. This makes it difficult to calculate the probability A^n or B^n that a succession of n values will increase or decrease consistently owing only to random effects. However, because the sum $A + B$ is the total distance between the $+2\sigma$ and -2σ limits, i.e. equal to 0.954, there is a particular point where A and B are equal with a value of 0.477. This condition is approximately satisfied when the starting measurement P is close to the mean value. Therefore, starting from any particular

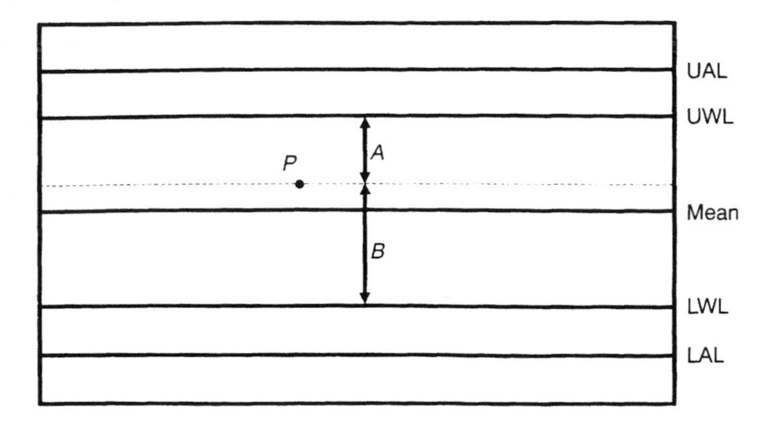

Figure 11.5. Probability of next measurement being greater or smaller

point on the chart which is close to the mean value, there is approximately a 0.477 probability that the next measurement will be larger. The chance that the third measurement will be larger again is 0.477×0.477, i.e. 0.2275. The probability that this rising trend will continue just because of random variations to the fifth measurement is 0.052 or 5.2 percent. The probability of this trend of five measurements occurring because of random variations is similar to the chance (4.6 percent) of one measurement exceeding the $\pm 2\sigma$ boundaries. Such a trend containing five measurements is therefore taken as a warning of possible process problems. Continuing further, a trend continuing to the ninth measurement only has a 0.27 percent chance of occurring naturally with random variations. This is similar to the chance (0.3 percent) of a measurement exceeding the $\pm 3\sigma$ boundaries, and such a trend of nine measurements is therefore taken to indicate a problem with the process, which must be investigated and corrected.

11.3.3 Runs

Like trends, runs can very usefully predict the onset of process problems before measurements start to approach the $\pm 2\sigma$ or $\pm 3\sigma$ boundaries. A run is a sequence of measurements that are consistently above or below the mean value, as shown in Figure 11.6. The probability of any particular measurement lying between the mean and the $+2\sigma$ boundary is 0.477 or 47.7 percent (because 95.4 percent of the measurements lie between $+2\sigma$ and -2σ). The probability of five successive values lying on the same side of the mean due to random effects is $0.477^4 = 0.052$ or 5.2 percent. This is almost the same as the 4.6 percent chance of a measurement lying outside the $\pm 2\sigma$ limits, and it is taken as a possible warning of process problems. The probability of nine successive measurements lying on the same side of the mean is only $0.477^8 = 0.0027$ or 0.27 percent. This is similar to the 0.3 percent probability of a single measurement lying outside the $\pm 3\sigma$ boundaries, and is therefore

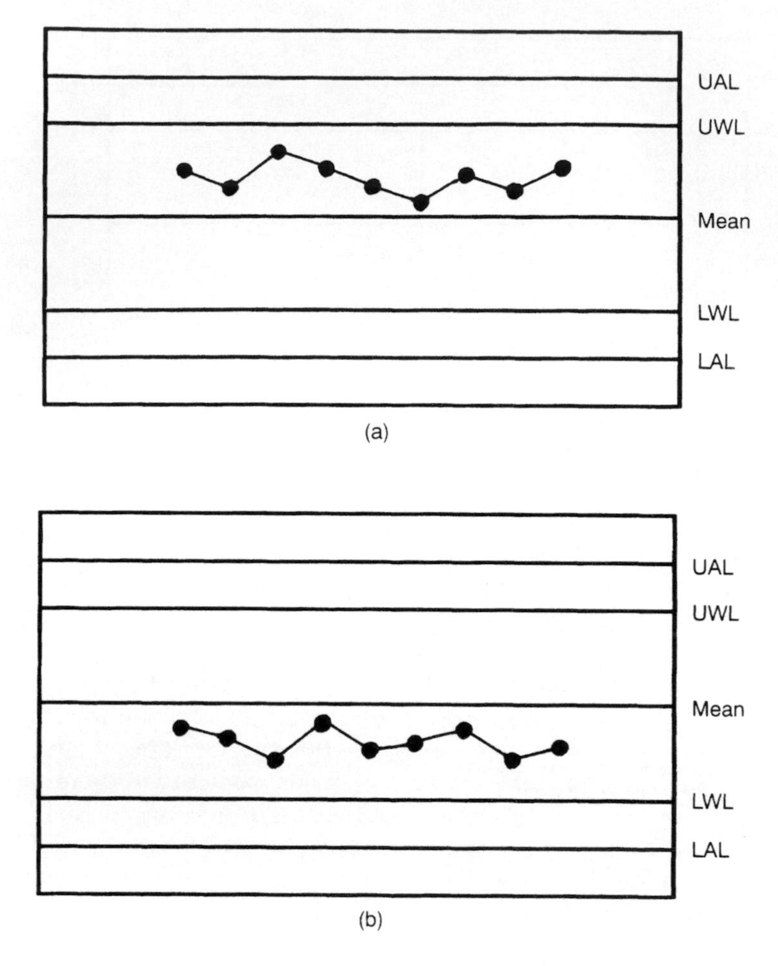

Figure 11.6. Runs on XBAR chart: (a) measurements increasing; (b) measurements decreasing

taken to indicate a problem with the process which must be investigated and corrected.

11.3.4 Simplified XBAR Chart

If the statistical process control algorithm bases the decision about control action on two or more successive measurements, then it becomes unnecessary to have both warning and action limits. For example, if the decision is based on plots of the mean of two successive measurements, or on a run/trend of nine increasing or decreasing measurements, then a simpler control chart with just $\pm 2\sigma$ action boundaries and no warning boundaries is sufficient, as shown in Figure 11.7. If the mean of more than two measurements, or runs/trends with a greater number of measurements than nine, is

Figure 11.7. Simplified XBAR control chart

used as the control criterion, then action limits lower than $\pm 2\sigma$ can be used. The appropriate action limits for plots of the mean of n successive measurements which are equivalent to the $\pm 3\sigma$ limits for single measurements are given by

$$\text{action limits} = \pm \frac{3\sigma}{\sqrt{n}}$$

11.4 CUSUM CHARTS (CUMULATIVE SUM CHARTS)

If the mean value of a process parameter changes, but only by a small amount, an XBAR chart may fail to show that the change has occurred because all measurements may continue to fall within the $\pm 2\sigma$ warning boundaries for some time after the change. Plotting the mean of several successive measurements may be of some help in this situation, as such means are likely to show a run of mean values which are consistently either above or below the mean value drawn on the XBAR chart. However, a CUSUM chart is generally much better for indicating small changes in the mean value of a parameter.

A CUSUM chart plots the cumulative sum of the deviations of each measurement from the mean value. For a series of n measurements, x_i, $i = 1, \ldots n$, the cumulative sum is given by

$$\text{CUSUM} = \sum_{i=1}^{n} d_i$$

Referring back to the set of 19 successive mass measurements given in Table 5.1, the CUSUM can be formed as shown in Table 11.1. Whilst the process is under statistical control, the CUSUM should remain close to zero and only make small excursions either side of zero. Whilst only random variations are taking place, negative deviations should occur in similar numbers to

Table 11.1. CUSUM calculation for mass measurements given in Table 5.1

Measurement	d_i	CUSUM ($\sum d_i$)
81.4	0.22	0.22
80.9	−0.28	−0.06
81.1	−0.08	−0.14
81.6	0.42	0.28
81.1	−0.08	0.2
81.3	0.12	0.32
80.6	−0.58	−0.26
81.2	0.02	−0.24
81.7	0.52	0.28
81.1	−0.08	0.2
80.8	−0.38	−0.18
81.3	0.12	−0.06
81.0	−0.18	−0.24
81.5	0.32	0.08
81.1	−0.08	0.00
81.3	0.12	0.12
80.8	−0.38	−0.26
81.6	0.42	0.16
81.0	−0.18	−0.02

positive ones and so the sum should be approximately zero. However, a potential problem can arise because of error in the mean value used to calculate the CUSUM. Referring back to the 19 mass measurements in Table 5.1, the mean is 81.178 947 if it is calculated to eight significant digits. This was approximated to a value of 81.18 for calculating the CUSUM. This means that there is an error of 0.001 053 at every step in the CUSUM calculation. Thus, after calculating the CUSUM for two measurements, the error is 0.002 106, after three measurements it is 0.003 159 and after 19 measurements it is 0.02. Clearly, the errors are cumulative and result in a significantly biased value of the CUSUM after a number of measurements. This necessitates reinitializing the CUSUM to zero periodically.

To demonstrate the value of CUSUM calculation, suppose that some change occurs in the production process after these 19 measurements have been taken and plotted, so that the mean increases from 81.18 to 81.28, and that a further 10 measurements are then taken after the change. Table 11.2 shows these further ten measurements and the CUSUM calculation. If the XBAR chart in Figure 11.1 is continued by plotting these next 10 measurements, as shown in Figure 11.8, there is little indication that there has been a change in the process. However, if the full set of mass measurements is plotted on a CUSUM plot, as shown in Figure 11.9, the cumulative sum is seen to move randomly either side of zero for the first 19 measurements and then to move steadily upwards over the next 10 measurements. This

Table 11.2. Data values and CUSUM calculation for 10 mass measurements following process change (starting from CUSUM value after the 19 measurements in Table 5.1 as calculated in Table 11.1)

Measurement	d_i	CUSUM ($\sum d_i$)
81.4	0.22	0.20
80.9	−0.28	−0.08
81.5	0.32	0.24
81.7	0.52	0.76
81.1	−0.08	0.68
81.3	0.12	0.80
81.6	0.42	1.22
80.9	−0.28	0.94
81.7	0.52	1.46
80.7	−0.48	0.98

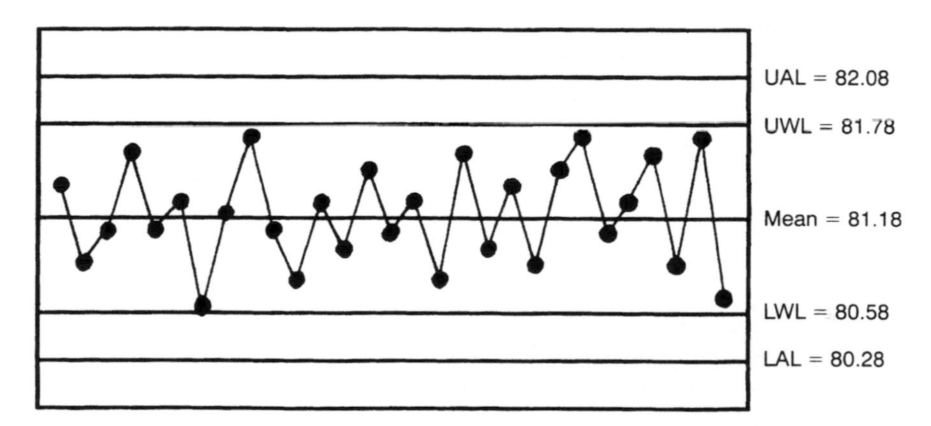

Figure 11.8. XBAR chart for 29 mass measurements

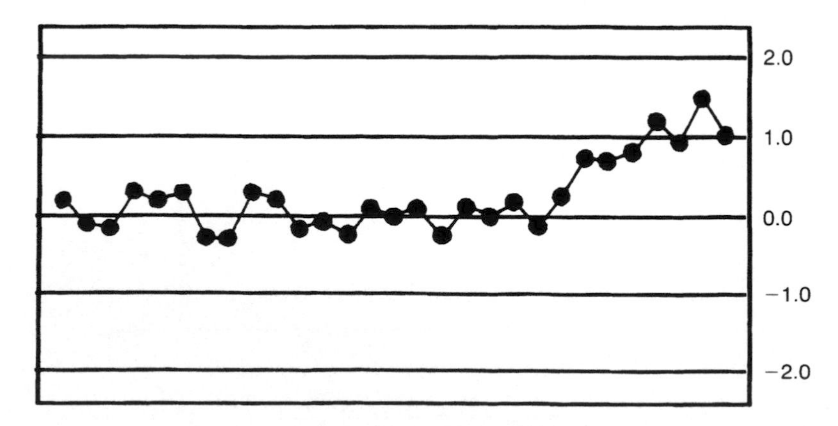

Figure 11.9. CUSUM chart for 29 mass measurements

upwards behaviour of the CUSUM plot is indicative of a change having taken place in the process.

The CUSUM chart is therefore much better than the XBAR chart for detecting small process changes. However, if a large change suddenly occurs which takes a measurement outside the $\pm 2\sigma$ or $\pm 3\sigma$ boundaries, this is indicated more quickly by an XBAR chart. This can be demonstrated if we return to the original set of nineteen measurements in Table 5.1 and consider what happens if these are followed by a measurement of 81.8, which is just outside the $+2\sigma$ boundary of 81.78. The XBAR chart for the $19+1$ measurements shown in Figure 11.10(a) immediately shows this excursion outside the upper warning limit. However, the corresponding CUSUM plot in Figure 11.10(b) fails to show this clearly. Thus, in practice, there is a strong case for plotting both XBAR and CUSUM charts in parallel.

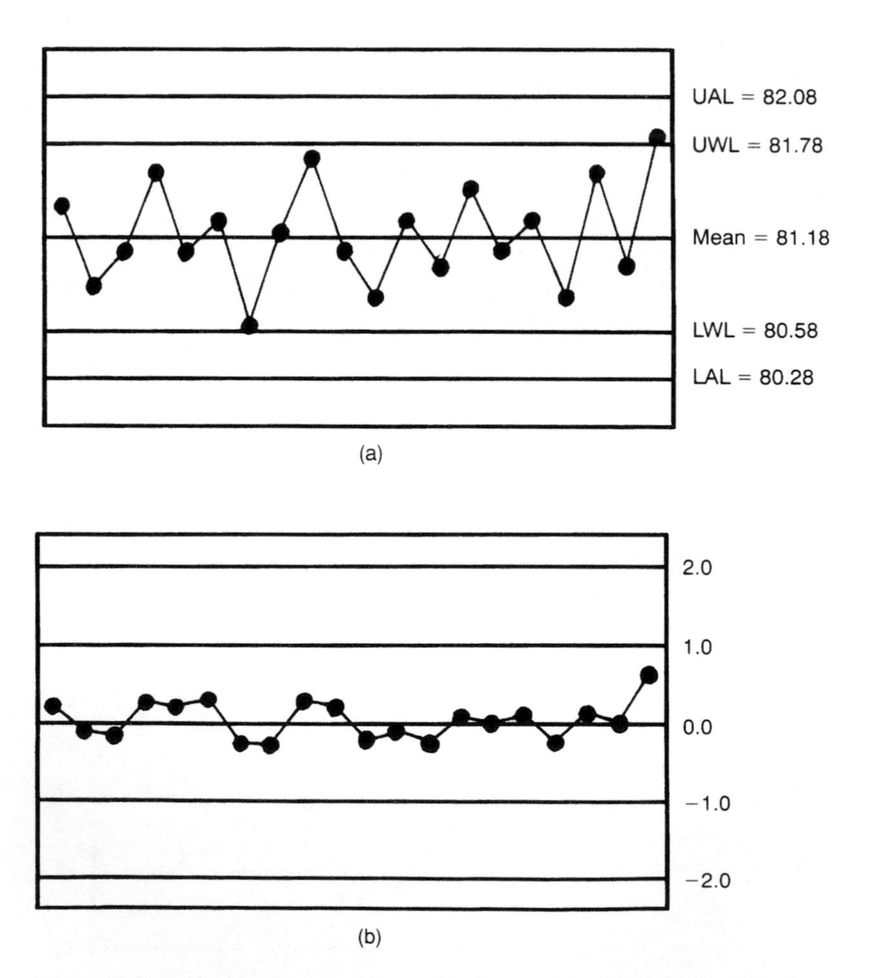

Figure 11.10. Measurement exceeding $+2\sigma$ limit: (a) plotted on XBAR chart; (b) plotted on CUSUM chart

11.5 RANGE CHARTS (R CHARTS)

The RANGE chart is often called an 'R' chart for simplicity. The chart plots the range of a set of several successive measurements of a parameter. The range of the measurements is simply the difference between the largest value and the smallest value in the set. Such samples of several successive measurements are taken at regular intervals of time and the range of each is plotted on the chart. The sample size for measurements to be plotted on an R chart can be as small as two, but it is more usual for the sample size to be at least four.

Table 11.3 shows some measurements of the thickness of paper produced in a paper mill. Four measurements are taken at a time and this procedure is repeated every hour. The range of each set of four measurements is calculated in Table 11.3 and Figure 11.11 shows the corresponding R chart plotted.

The main use of an R chart is in identifying long-term changes in a manufacturing process which cause the spread of measurements to change. The usual observation is an increase in spread, but a decrease can also sometimes occur, usually due to some malfunction in a measuring instrument. Whilst the process continues to operate correctly, the points plotted on the chart will remain within a constant band of values. However, any long-term trend for the points plotted on the chart to increase in magnitude is indicative of some change having taken place in the process which should be investigated. Likewise, any significant decrease in spread usually indicates that a measuring instrument is not working properly.

11.5.1 Control Limits for R Chart

Determining the control limits for an R chart is nothing like as straightforward as it is for XBAR and CUSUM charts because the distribution of the ranges of samples is positively skewed as shown in Figure 11.12. This means that the control limits are asymmetrical about the mean range.

The warning and action control limits are normally based on 97.5 percent and 99.9 percent probability limits (i.e. so that there is only a 2.5 percent and 0.1 percent probability respectively that any sample range outside these limits is due only to random effects). The actual values of the limits are calculated as

$$\text{Limit} = K\overline{R} \tag{11.1}$$

where $\overline{R}$ is the mean range value and K is a constant. Values for K depend on the size of each sample and are derived from published tables, such as those in [1].

For a sample size (n) of 4, the values of K are

$$K_{\text{UAL}} = 2.57; \quad K_{\text{LAL}} = 0.10; \quad K_{\text{UWL}} = 1.93; \quad K_{\text{LWL}} = 0.29$$

Table 11.3. Paper thickness measurement and range calculation for successive samples

Measurment (μm)	Maximum value x_1	Minimum value x_2	Range $(x_1 - x_2)$
97			
106			
104	106	97	9
101			
95			
99			
107	107	95	12
98			
102			
96			
101	102	96	6
99			
94			
105			
102	105	94	11
95			
100			
103			
97	103	97	6
101			
96			
102			
104	104	96	8
101			
93			
97			
103	103	93	10
99			
106			
100			
102	106	94	12
94			
97			
101			
98	102	97	5
102			
106			
99			
96	106	96	10
102			

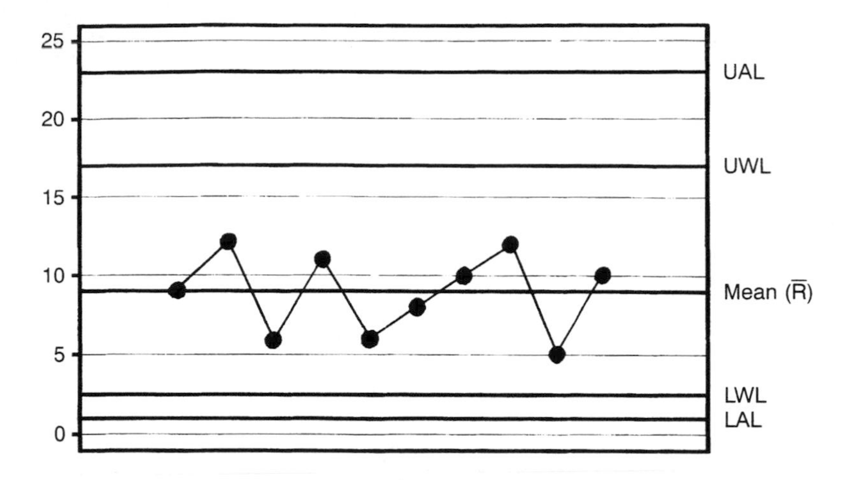

Figure 11.11. Range (R) chart for paper thickness measurements

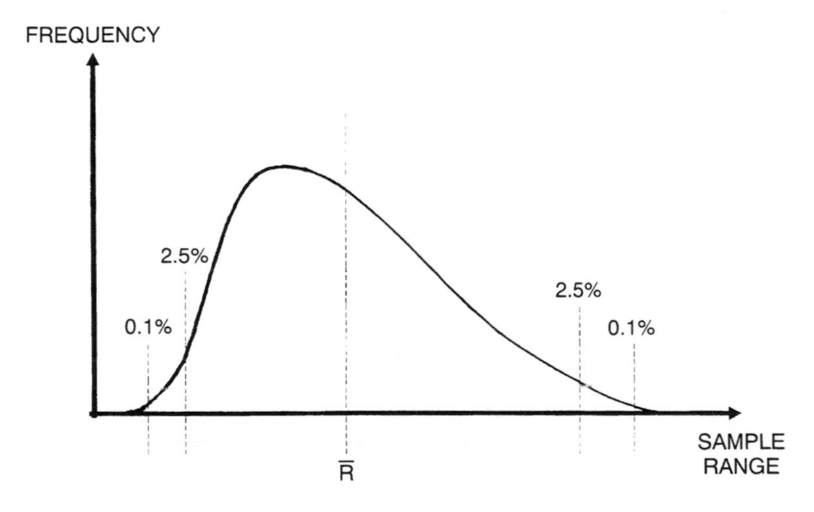

Figure 11.12. Positively skewed control limits on R chart

The mean range value is calculated from:

$$\overline{R} = \left\{ \sum (R_1 + R_2 + \ldots + R_s) \right\} / s$$

where R_1, R_2 etc. are the means of each successive sample and s is the total number of sample range values used to calculate the mean sample range. Thus, from Table 11.3:

$$\overline{R} = (9 + 12 + 6 + 11 + 6 + 8 + 10 + 12 + 5 + 10)/10 = 8.9$$

Hence, using equation (11.1) above, the control limit values are:

$$UAL = 22.9; \quad LAL = 0.9; \quad UWL = 17.2; \quad LWL = 2.6$$

11.6 SUMMARY OF CONTROL CHARTS

To operate statistical process control to the best advantage, it is usually necessary to plot all three types of control chart simultaneously in order to gain the best advance warning of all the manufacturing process problems that may arise. When sudden changes occur, such as something breaking, the XBAR chart indicates this most rapidly. However, the other two types of chart are much better at showing up small changes in a process over a long period of time. Such small changes can either take the form of changes in the mean value of one or more process parameters or alternatively take the form of a change in the spread of the measurements about the mean value. In some cases, both the mean value and spread can change together. For detecting changes in the mean, the CUSUM chart is best, whereas the RANGE chart is best for detecting changes in the spread of parameter values.

REFERENCES

1. Oakland, J. S., *Statistical process control* (3rd edn), Butterworth-Heinemann (1996).
2. BS 7785, *Shewhart control charts*, British Standards Institution (1994).
3. ISO 8258, *Shewhart control charts*, International Standards Organization, Geneva (1991).

CHAPTER 12

Product Sampling and Testing

12.1 INTRODUCTION

In the days prior to the development of comprehensive quality systems which established quality control procedures at all stages in the design and manufacture of a product, final product inspection and test procedures were the principal quality control mechanism in manufacturing systems. Although the development of quality control systems has done much to reduce the incidence of defects in products, the main beneficiary has been the financial health of companies as production costs have fallen in line with the reduction in product reject levels. Product testing still remains necessary on quality assurance grounds, in spite of the existence of quality control systems, because it is the only foolproof way of assuring customers of the quality of products. However, the amount of testing necessary to provide the necessary evidence of quality is often less than it used to be in the days before quality systems became common.

The purpose of designing and implementing a quality system in a production process is to make the final product conform consistently to some desired specification. If the quality system achieved this infallibly, there would not in fact be any need at all for testing of the final product, because it would always be as it should be. Unfortunately, this ideal situation is never met in practice for three main reasons.

Firstly, human beings are always involved in the operation of the quality system, at least to some extent. The problem with this is that human beings are prone to making mistakes, particularly when they are tired or when they are interrupted by some other event happening whilst they are performing a quality-related procedure. Therefore, whatever quality procedures are supposed to be performed, there is no guarantee that they will actually be carried out. Also, the procedures that are performed will not necessarily be done to the standard required or carried out with the diligence necessary.

Secondly, the statistical variation of process parameters means that it is often impossible to ensure that particular parameters of a product will be within their design specification. Statistical process control aims to ensure that random variations in each process parameter remain within the range expected. Unfortunately, variations in the value of process parameters may interact in an unpredictable way with regard to their effect on the parameters of the product. One of the limitations of statistical process control is that it can only be applied to monitor the values of process parameters one at a time, and thus it cannot predict the cumulative effect on a product of simultaneous variations in two or more process parameters. Thus, whilst the product parameters will probably be correct if the process parameters remain exactly at their design values, there is no guarantee that the product parameters will be within an acceptable tolerance even if statistical process control proves that the separate process parameters have remained within allowable bounds.

Thirdly, many quality-related control parameters are specified in statistical terms, and so there is always a finite probability that they will be outside acceptable bounds. In consequence, some form of testing of the final product in a manufacturing system is always necessary to ensure that its quality is acceptable.

12.1.1 What is Acceptable Quality?

The way in which 'acceptable quality' is defined and measured varies from product to product. In some cases, the definition is very simple and involves just one product characteristic, for example, whether a casting is whole or whether it has a piece missing. In such a case, the quality criterion is that the number of defective products in a batch is less than a certain number.

In other cases, the quality system has to produce a product with a number of specified parameter values and characteristics. Because some aspects of the process in any production system are inevitably subject to random variations, it necessarily follows that some tolerance has to be allowed in the specified parameters and characteristics of the product. Such tolerances can only be expressed in statistical terms, e.g. a 99.7 percent probability that a particular product parameter will be within say ± 1 percent of the target value. However, the statistical process means that if there is a 99.7 percent probability that a parameter is within ± 1 percent of the target value, there is a 0.3 percent probability that it is more than ± 1 percent away from it. In large-scale production systems, this means that a small percentage of the product will always be outside specification. This is often allowed for in the quality agreement made between the manufacturer and customer, with clauses such as: 'not more than 0.3 percent of the product will vary by more than ± 1 percent from the specified parameter value'. However, in some cases, no out-of-specification products must be delivered to the customer at all, and this demands even more rigorous inspection procedures.

12.1.2 Extension of Statistical Process Control to Product Testing

Where product quality is defined in terms of the acceptable degree of variation of product parameters from nominal values, one approach to product testing is to extend statistical process control and apply it to measurements of the important parameters in the product. In this approach, the measurements for each product parameter would normally be plotted on an XBAR chart, with the control limits for the chart calculated according to the maximum allowable deviations of the parameter.

This technique works well and almost guarantees finding and eliminating all defective units where it is feasible to measure every single product. Unfortunately, this is not feasible in many situations. In some cases, the production rate is just too high to make physical measurements on every product. In other cases, the cost of testing every product is uneconomic.

12.1.3 Reducing Inspection and Testing Costs by Sampling

The cost of final product testing depends very much on the nature of the product. In some cases, a product can be adequately tested by making physical measurements on it, e.g. measuring its mass, dimensions etc. In other cases, more detailed analysis of the product is necessary, which often requires testing it in a destructive way. Inspection costs are also influenced by whether testing is manual or automated.

A common approach to reducing inspection costs is to test just a sample of the product picked at random rather than the whole output. Whatever the cost of testing each unit of product, the total inspection cost clearly increases as the percentage of the product tested increases. Thus, there is a need to establish the minimum sample size that will ensure that testing is representative and effective but is achieved at the lowest cost possible.

12.2 PRINCIPLES OF PRODUCT SAMPLING AND TESTING

Most product sampling algorithms are designed for production systems where the product is produced in *batches* or *lots* of a certain size. A statistically significant number of samples are taken at random from each lot and tested. The usual course of action then is either to accept or reject the whole lot according to the results of the tests.

We will assume for the moment that the sample of products taken is entirely representative of the whole batch (although in practice this is not realizable as discussed later in this section). Then, if the acceptance criterion is based on one particular characteristic of the product, e.g. the wholeness of castings, the percentage of defective products in the whole batch will be the same as the percentage in the representative sample examined. Thus an acceptance/rejection decision based on the sample is very easy. If, on the other hand, the acceptance criterion is based on the percentage of the full batch that will have parameter values within certain limits, then it is necessary to use the analysis of random variations developed in Chapter 5.

The Gaussian Distribution curve for measurement errors in Figure 5.2 can be extended to analyse tolerances in manufactured products. Figure 12.1 shows the typical frequency distribution of a process parameter. Here, $F(x)$ is the probability that the product parameter will have some particular value x. The most likely value of x, x_m, corresponds to the peak of the curve. If the variations in the parameter values are Gaussian, the curve will be symmetrical about the line $x = x_m$, and x_m will represent the mean value of the parameter.

Equations similar to (5.4) and (5.5) can be written:

$$P(x_1 \leqslant x \leqslant x_2) = \int_{x_1}^{x_2} f(x)\,dx \tag{12.1}$$

$$P(x \leqslant x_o) = \int_{-\infty}^{x_o} f(x)\,dx \tag{12.2}$$

Also, by modifying equation (5.10), we obtain:

$$P(x_1 \leqslant x \leqslant x_2) = \int_{x_1}^{x_2} \frac{1}{\sqrt{\sigma(2\pi)}} \exp\left[-(x-\mu)^2/2\sigma^2\right] dx \tag{12.3}$$

where μ is the mean of the parameter values ($= x_m$).

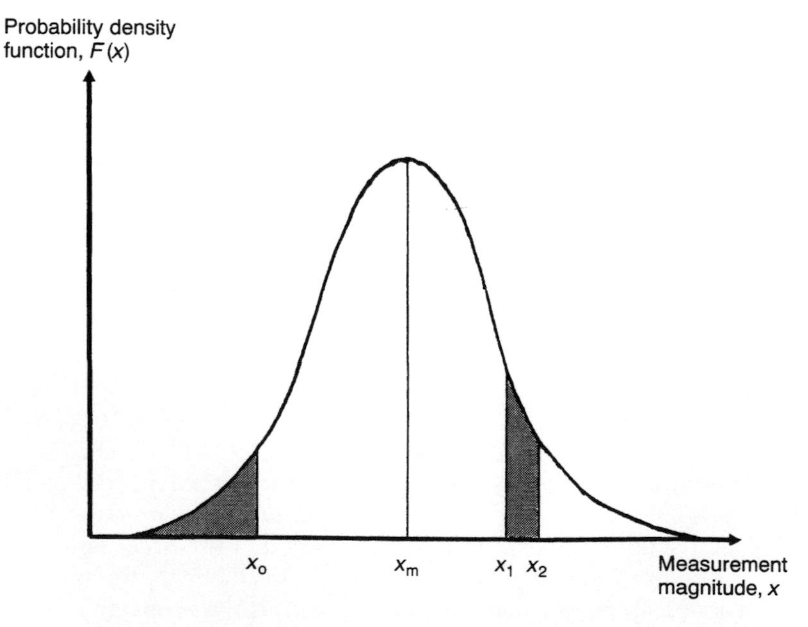

Figure 12.1. Frequency distribution of manufacturing tolerances

Having found the probability of any parameter value chosen at random lying within the range x_1 to x_2, the number of parameter values N lying within the range x_1 to x_2 can be calculated as follows:

$$N = \int_{x_1}^{x_2} \frac{n}{\sqrt{\sigma(2\pi)}} \exp[-(x-\mu)^2/2\sigma^2]\, dx \tag{12.4}$$

where n is the total number of products in the sample measured.

If the substitution $z = (x - \mu)/\sigma$ is made (see equation 5.9), equation (12.4) simplifies to:

$$N = \int_{z_1}^{z_2} \frac{n}{\sqrt{\sigma(2\pi)}} \exp[-z^2/2]\, dz \tag{12.5}$$

This is now in a form which can be evaluated using an error function table (see Table 5.2).

Numerical example

Bricks are produced in batches of 10 000. They have a mean length of 200 mm with a standard deviation of 20 mm. Calculate the following:

(a) the number of bricks with a length between 198 mm and 202 mm.

(b) the number of bricks with a length greater than 170 mm.

Solution

(a) The proportion of bricks where 198 < length < 202 is given by $P[x<202] - P[x<198]$. Calculating z from $z = (x-\mu)/\sigma$ (see equation 5.9):

For $x = 202$; $z = 0.1$ and for $x = 198$; $z = -0.1$

From error functions values in Table 5.2, $P[z<0.1] = 0.5398$. Using equation (5.11),

$$P[z<-0.1] = 1 - P[z<0.1] = 1 - 0.5398 = 0.4602$$

(a) Thus,

$$P[x<202] - P[x<198] = P[z<0.1] - P[z<-0.1]$$
$$= 0.5398 - 0.4602 = 0.0796$$

Thus $0.0796 \times 10^5 = 7960$ bricks out of the batch have a length in the range from 198 mm to 202 mm.

(b) The number of bricks with length > 170 is given by:

$$P[x > 170] = 1 - P[x < 170] = 1 - P[z < -1.5] = P[z < +1.5] = 0.9332$$

Thus, 93.32 percent i.e. 9332 bricks out of the batch have a length > 170.

The above analysis, using the measured mean and standard deviation of some samples of a product to calculate the percentage of a product having parameters within certain limits, has one quite serious limitation. This arises from the fact that the mean characteristics of any product sample will always differ to some extent from the mean characteristics of the whole product batch. This problem can be minimized by making the sample size large enough to ensure that this difference between the mean characteristics of the sample and the whole batch is small. However, there always remains some possibility that the sampling and testing procedure will misdiagnose a batch as being acceptable or otherwise. This risk is quantified by two parameters:

(a) a risk known as the *α risk* that sampling will incorrectly predict that a batch has too many defective products in it, causing it to be rejected unnecessarily.

(b) a risk known as the *β risk* that sampling will not detect a batch that should be rejected.

These two risks (α and β) in a sampling plan are quantified by its *operating characteristic*. An ideal operating characteristic is shown in Figure 12.2. This relates to a product where not more than 1 percent of products in a batch should be defective. This ideal plan accepts all batches where the number of defective products is less than or equal to 1 percent and rejects all other lots.

In practice, no sampling plan is ever able to achieve this ideal characteristic. Figure 12.3 shows a typical practical operating characteristic with the α and β risks marked (an α risk of a batch with a small percentage of defective products p_1 being rejected and a β risk of a batch with a larger percentage of defective products p_2 being accepted). Sampling plan design is therefore concerned with attempting to make the operating characteristic as close to the ideal curve as possible. Typical practical operating characteristics for different sampling plans are those marked as A, B and C in Figure 12.4. As the number of samples increases, the curve becomes steeper as in curve C and closer to the ideal characteristic.

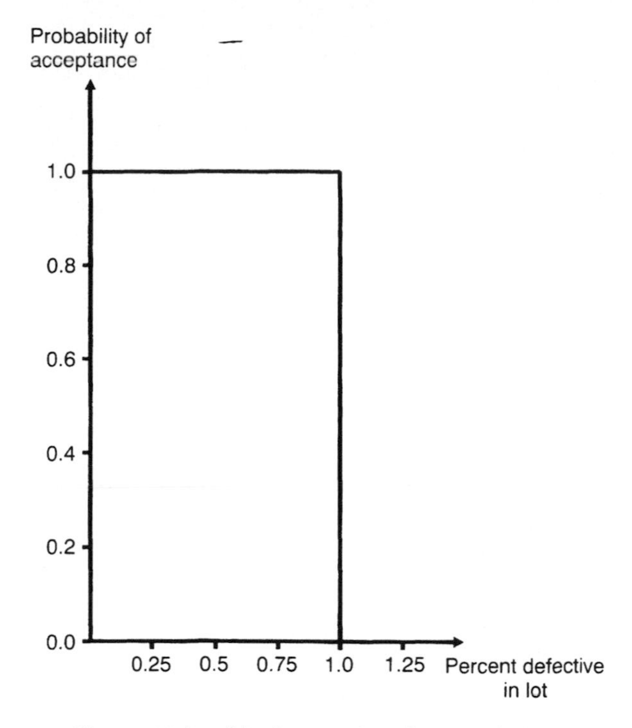

Figure 12.2. Ideal operating characteristic

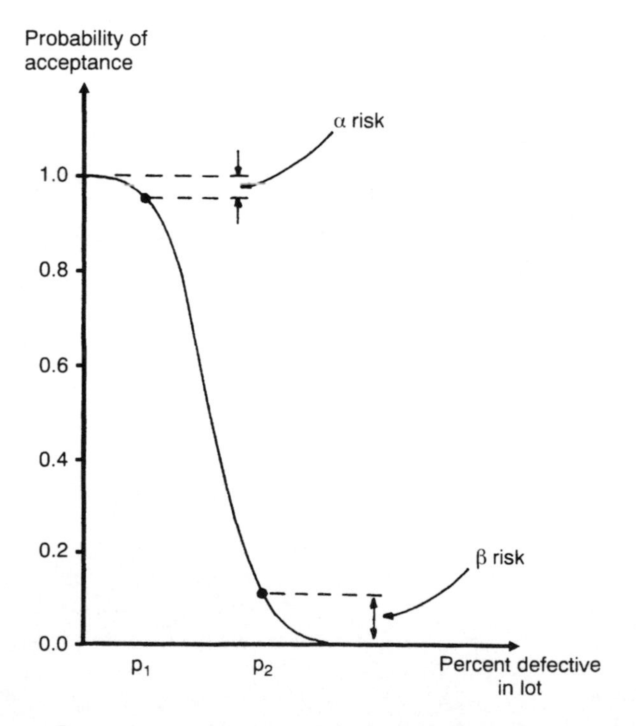

Figure 12.3. Typical practical operating characteristic with α and β risks shown

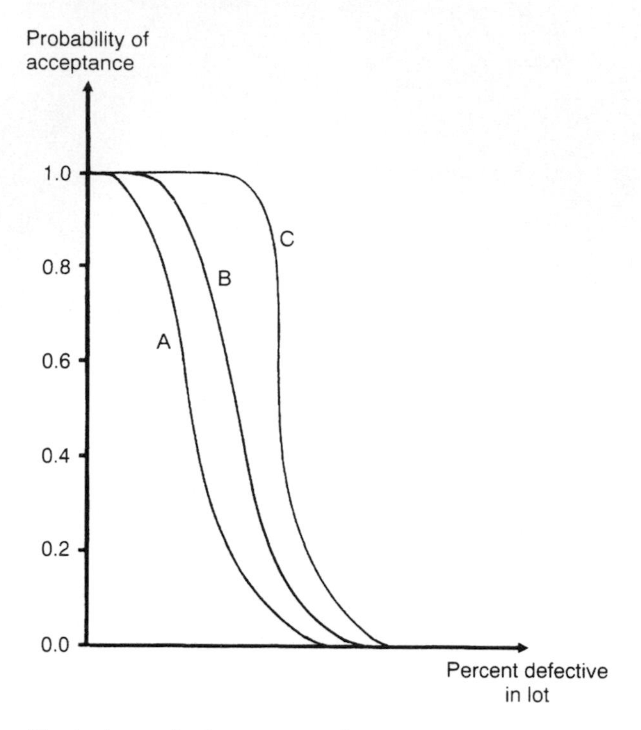

Figure 12.4. Typical practical operating characteristics (marked A, B and C)

The levels of the α and β sampling risks are affected by the sample size and the acceptance quality level, and increasing either of these has the effect of reducing the α and β risks. The lot size can also affect the α and β risks if it is small, but the effect becomes insignificant once the lot size becomes more than ten times larger than the sample size. However, when the lot size is large, the financial penalty involved in falsely rejecting 'good' lots is also large and therefore the sampling plan is usually designed more carefully for large lot sizes to reduce the α risk by moving the operating characteristic closer to the ideal curve.

12.2.1 Single, Double and Multiple Sampling

The scheme discussed above is a *single sampling plan*, where a single lot of n items are taken at random from a batch of products and tested. However, as an alternative to single sampling, double or multiple sampling can be carried out instead. These enable the average number of samples tested per lot to be reduced without affecting the magnitude of the α and β risks.

A *double sampling plan* involves first of all taking p items from a lot and testing them (where p is less than n above). If the number of defectives is large or small with respect to what is an acceptable number of defective products, then the batch is immediately rejected or accepted. If an immediate decision is not possible because the number of defectives is close to the limit of what is acceptable, then a second sample consisting of a further p items is tested to confirm whether the lot should be accepted or rejected. Most of the time, testing a second lot is not necessary. Thus, double sampling reduces the average number of samples taken and tested in each lot, but it is complicated to administer.

A *multiple sampling plan* extends the concept of double sampling by testing an unspecified number of successive lots of q samples (where q is less than p above), until there is a clear indication of whether the batch from which the successive samples are taken should be accepted or rejected. This further reduces the average number of product tests per batch that have to be carried out, but the plan is even more complicated to administer.

12.2.2 Testing Plans

A simple extension of statistical process control to product testing is unsatisfactory because the risk is too high that it will fail to detect defective products in the proportion not tested. More sophisticated product sampling and testing plans are therefore necessary which minimize the risk of failing to detect quality problems.

The two algorithms that are used most commonly for product sampling and testing are known as the *variables plan* and the *attributes plan*. A common feature of both of these is an *acceptance quality level* which the samples tested are compared against to determine whether lots should be accepted or rejected, but this is defined differently according to which sampling and testing plan is being used. This acceptance quality level is always specified in the quality target for the manufacturing process.

In the *variables plan*, certain parameters are measured for each product in the sample taken from a production batch and the average value for each of these parameters is calculated across all products in the sample. This information is used in one of two alternative ways in making the decision about whether to accept or reject that particular production batch from which the samples are taken.

In one form of the variables plan, the acceptance quality level is defined in terms of how close these mean parameter values are to some target value. The production batch in question will be accepted if none of these mean parameter values fall outside the acceptable range defined, or rejected otherwise.

In the second form of variables plan, the standard deviation of the parameter values in the sample is calculated and used to estimate how many products in the whole batch will have parameters values outside an acceptable range. In this case, the acceptance quality level for the batch has two components, one being the maximum tolerance allowable on specified product parameters and the second being the number of reject products allowable in the batch.

In the alternative *attributes plan*, every product in the sample is compared against some target quality level and classified as being either acceptable or not. The acceptance quality level in this case is simply the maximum number of unacceptable products allowed in the batch. This specified number might even be zero in production situations where no defects at all are permitted.

12.3 THE VARIABLES SAMPLING AND TESTING PLAN

A variables plan can only be used if the quality characteristics of the product vary in a continuous way and the variations follow a Gaussian distribution. However, provided that these conditions are satisfied, a variables plan is preferable to an attributes plan for several reasons.

(a) A smaller number of samples, and correspondingly reduced amount of testing, is necessary to verify a specified quality level.

(b) The necessary sample size to verify some specified quality level in a batch of a given size is easier to calculate.

As noted in Section 12.2.2, two principal types of variables plan exist. These differ according to the acceptance criterion for a batch, which is based on either the mean parameter values or on the percentage of defective products in a batch.

12.3.1 Batch Acceptance/Rejection Criterion Based on Mean Parameter Values

This is the simplest kind of variable sampling plan. Depending on the product requirements, the conditions for acceptance of a batch can be expressed in one of three ways, 1. the mean parameter value is within a specified range, 2. the mean parameter value is not more than a given amount less than some target value and 3. the mean parameter value is not more than a given amount greater than some target value. However, this is not as simple as it seems because the sampling risks also have to be taken into account.

Single Acceptance Limit

Where cases 2 or 3 above apply, i.e. the condition for batch acceptance is that the mean parameter values are not more than a given amount less/greater than some target values, this is known as a *single acceptance limit variables plan*. The first step in this kind of sampling plan involves calculating the sample size and this requires the following information:

(a) The value of the parameter, q_1, for which the probability of acceptance of a batch should be high.

(b) The α sampling risk (the probability that a batch where the mean parameter value is q_1 will be rejected).

(c) The value of the quality parameter, q_2, for which the probability of acceptance of a batch should be low.

(d) The β sampling risk (the probability that a batch where the mean parameter value is q_2 will be accepted)

The necessary sample size n can then be calculated as follows [2]:

$$n = \left[\frac{(z_2 - z_1)\sigma}{q_2 - q_1}\right]^2 \tag{12.6}$$

where σ is the usual standard deviation of the product parameter q, z_1 is the standard normal deviate (defined in Chapter 5, Section 5.2.2, equation 5.9) corresponding to $F(z_1) = 1 - \alpha$, and z_2 is the standard normal deviate corresponding to $F(z_2) = \beta$.

The acceptance limit q_a can be calculated as follows [2]:

$$q_a = \frac{z_2 q_1 - z_1 q_2}{z_2 - z_1} \tag{12.7}$$

The calculation of n and q_a will be understood better if a numerical example is considered. Suppose that a coil of copper wire is being cut into lengths of 30 mm in batches of 10 000 which will subsequently be sold to a manufacturer of copper nails. The important quality parameter is the length of each piece. The sampling and testing plan is required to have a 98 percent probability of accepting batches where the average length is 30.25 mm or more and a 95 percent probability of rejecting batches where the average length is 29.75 mm or less. Over a period of time, the standard deviation of the length has been established as 1.0 mm. The quantities necessary to evaluate the expression for n in equation (12.6) can be calculated as follows.

Values given: $q_1 = 30.25$; $q_2 = 29.75$; $\sigma = 1.0$

For an acceptance probability of 98 percent, the probability of rejecting a 'good' lot is therefore 2 percent, i.e. $\alpha = 0.02$. Thus $(1 - \alpha) = 0.98$.

For a rejection probability of 95 percent, the probability of accepting a 'bad' lot is 5 percent, i.e. $\beta = 0.05$.

Using error functions given in Table 5.2:

For $F(z_1) = 1 - \alpha = 0.98$, $z_1 = 2.054$.

For $F(z_2) = \beta = 0.05$, $z_2 = -1.645$.

(NB to obtain the value of z_2 from Table 5.2, use of the expression $F(-z) = 1 - F(z)$ in equation (5.11) is necessary).

Now putting the values into equation (12.6):

$$n = \left[\frac{(-1.645 - 2.054)\ 1}{29.75 - 30.25}\right]^2 = 54.73$$

Rounding the value up to the nearest integer, $n = 55$.

The acceptance limit can also be calculated, using equation (12.7), to give:

$$q_a = \frac{-1.645(30.25) - 2.054(29.75)}{-1.645 - 2.054} = 29.97 \text{ mm}$$

The required testing algorithm is therefore to take a sample of 55 wires from the batch and calculate the mean length. If this is 29.97 mm or greater, the batch should be accepted. Otherwise, it should be rejected.

Double Acceptance Limit

Where case 1 above applies (see start of Section 12.3.1), i.e. where the condition for batch acceptance is that the mean parameter value is within a specified range, this is known as a *double acceptance limits variables plan*. The method for calculating the minimum sample size is different to the single acceptance limit case and requires the following parameters to be specified.

(a) The upper value of the quality parameter q_U for which the probability of acceptance of a batch should be low.

(b) The lower value of the quality parameter q_L for which the probability of acceptance of a batch should be low.

(c) The β sampling risk (the probability that a batch will be accepted where the mean parameter value is greater than or equal to q_U or less than or equal to q_L)

(d) The α sampling risk (the probability that a batch where the mean parameter value is midway between q_U and q_L will be rejected).

The sample size and acceptance limits can then be calculated as follows:

$$q_m = (q_L + q_U)/2 \qquad (12.8)$$

Let
z_1 be the standard normal deviate corresponding to $F(z_1) = 1 - (\alpha/2)$,
z_2 be the standard normal deviate corresponding to $F(z_2) = 1 - (\beta/2)$,
σ be the standard deviation of the parameter q,
n be the required number of samples,
q_{max} be the upper acceptance limit,
q_{min} be the lower acceptance limit.

Four simultaneous equations can then be formed involving the unknown quantities n, q_{max} and q_{min} [2].

$$z_1 = (q_{max} - q_m)\sqrt{(n)}/\sigma \tag{12.9}$$

$$z_1 = (q_m - q_{min})\sqrt{(n)}/\sigma \tag{12.10}$$

$$z_2 = (q_{min} - q_L)\sqrt{(n)}/\sigma \tag{12.11}$$

$$z_2 = (q_U - q_{max})\sqrt{(n)}/\sigma \tag{12.12}$$

Solution of these then allows the calculation of n, q_{max} and q_{min}.

It should be noted [2] that the above equations and procedure for calculating n are only valid if

$$(q_U - q_L) \geqslant 6.48\sigma/\sqrt{n} \tag{12.13}$$

Again, considering a numerical example will aid the understanding of this. Referring back to the example concerning the lengths of copper wire, but this time defining double acceptance limits, suppose that wires with a mean length in the range from 29.5 mm to 30.5 mm are acceptable and outside this range they are unacceptable, σ is still equal to 1 mm, and the sampling risk parameters are specified as $\alpha = 0.02$ and $\beta = 0.05$.

Using Table 5.2,

$$\text{For } F(z_1) = 1 - (\alpha/2) = 0.99, \quad z_1 = 2.327$$

$$\text{For } F(z_2) = 1 - (\beta/2) = 0.975, \quad z_2 = 1.960$$

$$q_U = 30.5; \quad q_L = 29.5 \text{ (given)}$$

Hence, from equation (12.8), $q_m = 30.0$. Inserting values into equations (12.9) to (12.12):

$$2.327 = (q_{max} - 30.0)\sqrt{n} \tag{12.14}$$

$$2.327 = (30.0 - q_{min})\sqrt{n} \tag{12.15}$$

$$1.960 = (q_{min} - 29.5)\sqrt{n} \tag{12.16}$$

$$1.960 = (30.5 - q_{max})\sqrt{n} \tag{12.17}$$

Equations (12.14) to (12.17) represent four simultaneous equations with only three unknowns n, q_{max} and q_{min}. Thus only three out of the four equations are needed, and the values of the unknown parameters can be extracted in various ways. For example, the two equations (12.15) and (12.16) involve just the two unknown parameters n and q_{min}.

From (12.15), $2.327/\sqrt{n} = 30.0 - q_{min}$ (12.18)
From (12.16), $1.960/\sqrt{n} = q_{min} - 29.5$ (12.19)
Adding (12.18) and (12.19), $4.287/\sqrt{n} = 0.5$
Hence $n = 73.51$ or, rounding up, $n = 74$.
Then, from (12.19), $q_{min} = 1.960/\sqrt{n} + 29.5 = 29.73$
From (12.14), $q_{max} = 2.327/\sqrt{n} + 30.0 = 30.27$

The required testing algorithm for this double acceptance limits case is therefore to take a sample of 74 wires from the batch and calculate the mean length. If this is greater than 29.73 mm or less than 30.27 mm, the batch should be accepted. Otherwise, it should be rejected.

Having calculated n etc., a final step of applying equation (12.13) is necessary to check the validity of the procedure used:

$$q_U - q_L = 1.0$$

$$6.48\sigma/\sqrt{n} = 0.76$$

Thus, $(q_U - q_L) > 6.48\sigma/\sqrt{n}$ and equation (12.13) is therefore satisfied, making the procedure used for calculating n etc. valid.

12.3.2 Batch Acceptance/Rejection Criterion Based on Percentage of Batch Defective

As the percentage of defective products in a batch is directly related to the standard normal deviate, the acceptance criterion can in fact be specified just in terms of a minimum value of the standard normal deviate of a sample rather than as a minimum value of the percentage of non-defective products. This avoids having to use error function tables to actually calculate the percentage of defective products from the standard normal deviate.

The sampling plan can therefore be expressed as:

1. Take n samples of the product and calculate the mean parameter value x_{mean}.

2. Calculate the standard normal deviate of the sample as:

$$z_{sample} = (x_{mean} - x_{min})/\sigma \qquad (12.20)$$

where x_{min} is the minimum value allowable in the product parameter and σ is the standard deviation of the product parameter.

3. Accept the batch if $z_{sample} \geqslant z_{min}$, where z_{min} is the minimum value of the standard normal deviate allowable. Otherwise, reject the batch.

Design of the sampling plan in terms of calculating n and z_{min} for this case is somewhat different and full details can be found in [2]. The equations for calculating n, the required sample size, and z_{min}, the minimum value of the standard normal deviate for which a product batch should be accepted, are as follows:

$$n = \left[\frac{z_\alpha + z_\beta}{z_1 - z_2}\right]^2 \tag{12.21}$$

where

α is the sampling risk that a batch will be rejected when the percentage nonconforming is p_1, and p_1 is the percentage nonconforming value for which the probability of acceptance of a batch should be high,

β is the sampling risk that a batch will be accepted when the percentage non-conforming is p_2, and p_2 is the percentage nonconforming value for which the probability of acceptance of a batch should be low,

z_α is the standard normal deviate for which the area under the frequency distribution curve is $(1 - \alpha)$, i.e. $z_\alpha = F(1 - \alpha)$,

z_β is the standard normal deviate for which the area under the frequency distribution curve is $(1 - \beta)$, i.e. $z_\beta = F(1 - \beta)$,

z_1 is the standard normal deviate for which the area under the frequency distribution curve greater than z_1 is p_1, i.e.

$$z_1 = F(z_1) = 1 - p_1,$$

z_2 is the standard normal deviate for which the area under the frequency distribution curve greater than z_2 is p_2, i.e.

$$z_2 = F(z_2) = 1 - p_2$$

Now, the minimum value of z can be calculated from

$$z_{min} = z_1 - z_\alpha/\sqrt{n} \tag{12.22}$$

A numerical example will now be considered. Suppose that loaves of bread are produced in a bakery in batches of 1000. The sampling and testing plan is required to accept batches with a probability of 98 percent if not more than 0.5 percent (1 in 200) of the loaves are below an acceptable mass, and reject batches with a probability of 95 percent if 5 percent or more of the loaves are below the minimum acceptable mass. The standard deviation of the mass is given as 8 g.

From the given data, the following parameter values can be written down:

$$\alpha = 0.02; \quad \beta = 0.05; \quad p_1 = 0.5\% = 0.005; \quad p_2 = 5\% = 0.05$$

The standard normal deviates can now be calculated, using Table 5.2:

$$F(z_1) = 1 - p_1 = 0.995; \qquad \text{Hence, } z_1 = 2.575$$

$$F(z_2) = 1 - p_2 = 0.95; \qquad \text{Hence, } z_2 = 1.645$$

$$F(z_\alpha) = 1 - \alpha = 0.98; \qquad \text{Hence, } z_\alpha = 2.054$$

$$F(z_\beta) = 1 - \beta = 0.95; \qquad \text{Hence, } z_\beta = 1.645$$

Now substituting values into equation (12.21):

$$n = \left[\frac{1.645 + 2.054}{2.575 - 1.645}\right]^2 = \left[\frac{3.699}{0.93}\right]^2 = [3.977]^2 = 15.8$$

Rounding up, $n = 16$.

Then calculating $z_{\min}$ from (12.22)

$$z_{\min} = z_1 - \frac{z_\alpha}{\sqrt{n}} = 2.575 - \frac{2.054}{3.977} = 2.06$$

Suppose that a sample of 16 loaves is taken and the average mass is measured as 420 g. Should the batch from which the samples were taken be accepted or not if the minimum acceptable mass for a loaf is 400 g?

Using equation (12.20)

$$z = (420 - 400)/8 = 2.5$$

Therefore, $z > z_{\min}$ (because $2.5 > 2.06$), and the batch should be accepted.

12.4 THE ATTRIBUTES SAMPLING AND TESTING PLAN

The Attributes plan classifies the acceptability of product batches purely on the goodness or otherwise of one product characteristic, and accepts or rejects a batch of products according to the proportion in the batch that are defective. One advantage of this over the variables plan is that the testing criterion is applied only once. In the case of a variables plan, the testing algorithm usually has to be applied a number of times, once for each separate measured parameter.

The operating characteristic curve in attribute sampling is governed by the binomial distribution. The equations for calculating the sample size n for a given maximum number of defective products in a batch are non-linear, and thus solution for n is much harder than it is for the variable sampling plan. Solution is obtained either by a trial and error approach or by using a *nomogram*, which consists of multiple plots of n and c for different cases. Look-up tables can also be used. The detailed explanation of these is outside the scope of this book and the interested reader is therefore referred elsewhere (e.g. to [2]).

12.5 SUMMARY

This chapter should perhaps close with a warning that it has been possible to provide only an introduction to the subject of product sampling and testing within the space available. Some simple cases have been covered but, in practice, the analysis and design involved in testing plans is often considerably more involved. For example, nothing has been said in detail about double sampling and multiple sampling schemes. Also, in the examples given for calculating the sample size, it is assumed that the standard deviation is known. However, if the standard deviation is not known, special formulae are needed to calculate the sample size (these as expected show that a greater sample size is needed when the standard deviation is unknown).

More detailed information on sampling and testing can be found in references [1] and [2]. It should also be mentioned that formal standards exist for sampling schemes, e.g. ANSI/ASQC Z1.4 for the attributes plan and ANSI/ASQC Z1.9 for the variables plan. These propose rules for the selection of sampling plans and possible switching between them to either increase or reduce the proportion of products in a batch inspected if the quality level deteriorates or improves. The schemes proposed by such standards often provide for alternative sampling and testing plans depending on the level of quality expected, and, where an accredited quality system is in operation, the least rigorous plan out of the options specified can be used.

REFERENCES

1. Juran, J. M. and Gryna, F. M., *Quality planning and analysis*, Tata McGraw-Hill, New Delhi (1982).
2. Banks, J., *Principles of quality control*, John Wiley & Sons, New York (1989).

CHAPTER 13

Temperature Calibration

13.1 REVIEW OF TEMPERATURE MEASURING INSTRUMENTS

Instruments to measure temperature can be divided into six separate classes according to the physical principle on which they operate. These principles are:

1. Thermal expansion
2. The thermoelectric effect
3. Resistance change
4. Resonant frequency change
5. Velocity of sound
6. Radiative heat emission

Thermal expansion methods make use of the fact that the dimensions of all substances, whether solids, liquids or gases, change with temperature. Instruments operating on this physical principle include the liquid-in-glass thermometer, the bimetallic thermometer and the pressure thermometer.

Thermoelectric effect instruments rely on the physical principle that, when any two different metals are connected together, an e.m.f., which is a function of the temperature, is generated at the junction between the metals. The general form of this relationship is:

$$e = a_1 T + a_2 T^2 + a_3 T^3 \ldots + a_n T^n$$

This is clearly non-linear, which is inconvenient for measurement applications. Fortunately, for certain pairs of materials, the terms involving squared and higher powers of T ($a_2 T^2$, $a_3 T^3$ etc.) are approximately zero and the e.m.f.–temperature relationship is approximately linear according to:

$$e \approx a_1 T$$

Wires of such pairs of materials are connected together at one end, and in this form are known as thermocouples. Thermocouples are a very important class of device as they provide the most commonly used method of measuring temperatures in industry.

Varying resistance devices rely on the physical principle of the variation of resistance with temperature. The instruments working on this principle are known as either resistance thermometers or thermistors according to whether the material used for their construction is a metal or a semiconductor material.

The principle of resonant frequency change with temperature is used in the quartz thermometer. The resonant frequency of materials such as quartz is a function of temperature, which enables temperature changes to be translated into frequency changes.

Another parameter that changes with temperature is the velocity of sound in a gas. This principle is used in acoustic thermometers.

The final physical principle that can be used in measuring temperature is radiation emission from a body. All bodies emit electromagnetic radiation as a function of their temperature above absolute zero. Measurement of the radiation emission therefore allows the temperature of the body to be calculated. Instruments using this principle are known as radiation thermometers.

Liquid-in-glass Thermometers

The liquid-in-glass thermometer has a fluid consisting of either mercury or coloured alcohol contained within a bulb and capillary tube, as shown in Figure 13.1. As the temperature rises, the fluid expands along the capillary tube and the meniscus level is read against a calibrated scale etched on the tube. Typically, the measurement inaccuracy of such instruments is quoted as ±1 percent of full scale. However, an inaccuracy of only ±0.15 percent can be obtained in the best industrial instruments. Industrial versions of the liquid-in-glass thermometer are normally used to measure temperatures in the range between −200 °C and +1000 °C, although instruments are available to special order which can measure temperatures up to 1500 °C.

The major source of measurement error arises from the difficulty of correctly estimating the position of the curved meniscus of the fluid against the scale. In the longer term, additional errors are introduced owing to volumetric changes in the glass. Such changes occur because of creep-like processes in the glass, but occur only over a time-scale of years. Annual calibration checks are therefore advisable.

Bimetallic Thermometer

The bimetallic principle is probably more commonly known in connection with its use in thermostats. It is based on the fact that if two strips of

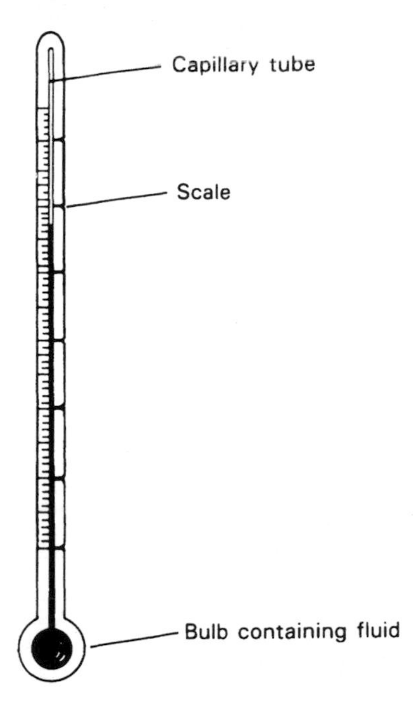

Figure 13.1. The liquid-in-glass thermometer

different metals are bonded together, any temperature change will cause the strip to bend, as this is the only way in which the differing rates of change of length of each metal in the bonded strip can be accommodated. In the bimetallic thermostat, this is used as a switch in control applications.

If the magnitude of bending is measured, the bimetallic device becomes a thermometer. For such purposes, the strip is often arranged in a spiral configuration, as shown in Figure 13.2, as this gives a relatively large displacement of the free end for any given temperature change. Strips in a helical shape are an alternative for this purpose. The measurement sensitivity is increased further by choosing the pair of materials carefully so that the degree of bending is maximized, with Invar (a nickel-steel alloy) and brass being commonly used.

For visual indication purposes, the end of the strip can be made to turn a pointer mounted in low friction bearings which moves against a calibrated scale. Another suitable form of translational displacement transducer is the LVDT (linear variable differential transformer).

Bimetallic thermometers are used to measure temperatures between −75 °C and +1500 °C. The inaccuracy of the best instruments can be as little as ±0.5 percent but such devices are quite expensive. Many instrument applications do not require this degree of accuracy in temperature measurements, and in such cases much cheaper bimetallic thermometers with substantially inferior accuracy specifications are used.

All such devices are liable to suffer changes in characteristics owing to contamination of the metal components exposed to the operating

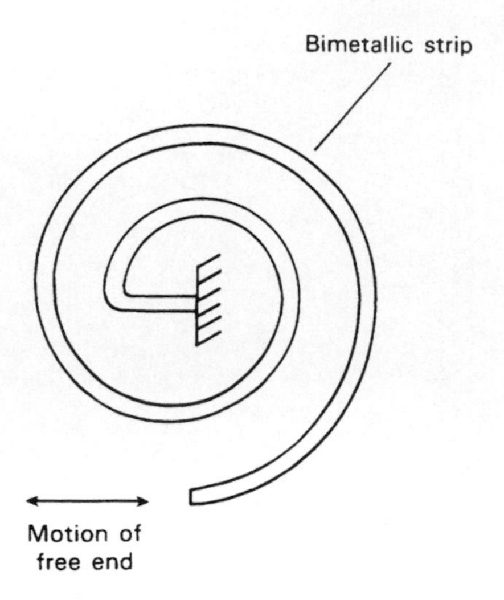

Figure 13.2. The bimetallic thermometer

environment. Further changes are to be expected arising from mechanical damage during use, particularly if they are mishandled or dropped. As the magnitude of these effects varies with their application, the required calibration interval must be determined by practical experimentation.

Pressure Thermometers

The pressure thermometer measures the variation in pressure of a gas constrained inside a bulb of fixed volume as the temperature changes. As such, it does not strictly belong to the thermal expansion class of instruments but is included because of the relationship between volume and pressure according to Boyle's gas law.

The change in pressure of the gas is measured by a suitable pressure transducer such as a Bourdon tube. This transducer is located remotely from the bulb and connected to it by a capillary tube as shown in Figure 13.3. The need to protect the pressure-measuring instrument from the environment in which the temperature is being measured can require the use of capillary tubes up to 5 m long, and the temperature gradient and hence pressure gradient along the tube act as an environmental input which can introduce a significant measurement error. Correction for this by introducing an opposing environmental input, as discussed in Chapter five, can be carried out according to the scheme shown in Figure 13.4. This includes a second, dummy capillary tube whose temperature gradient is measured by a second Bourdon tube. The outputs of the two Bourdon tubes are connected together in such a manner that the output from the second tube is subtracted from the output of the first, thus eliminating the error due to the temperature gradient along the tube.

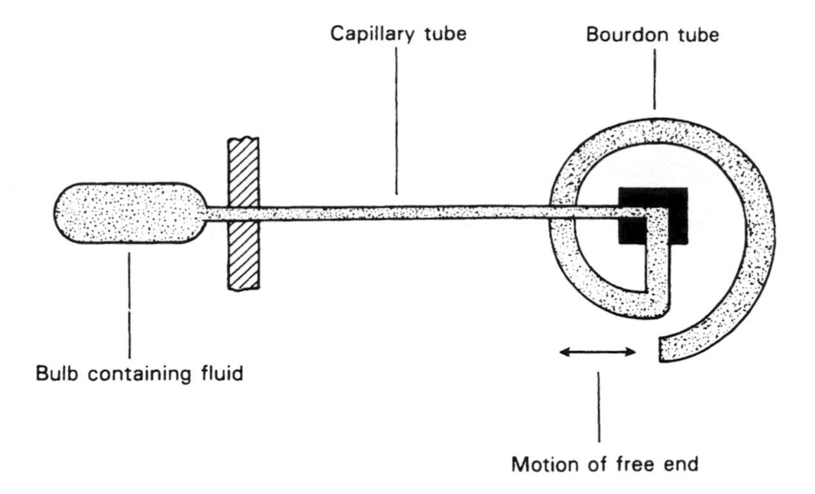

Figure 13.3. The pressure thermometer

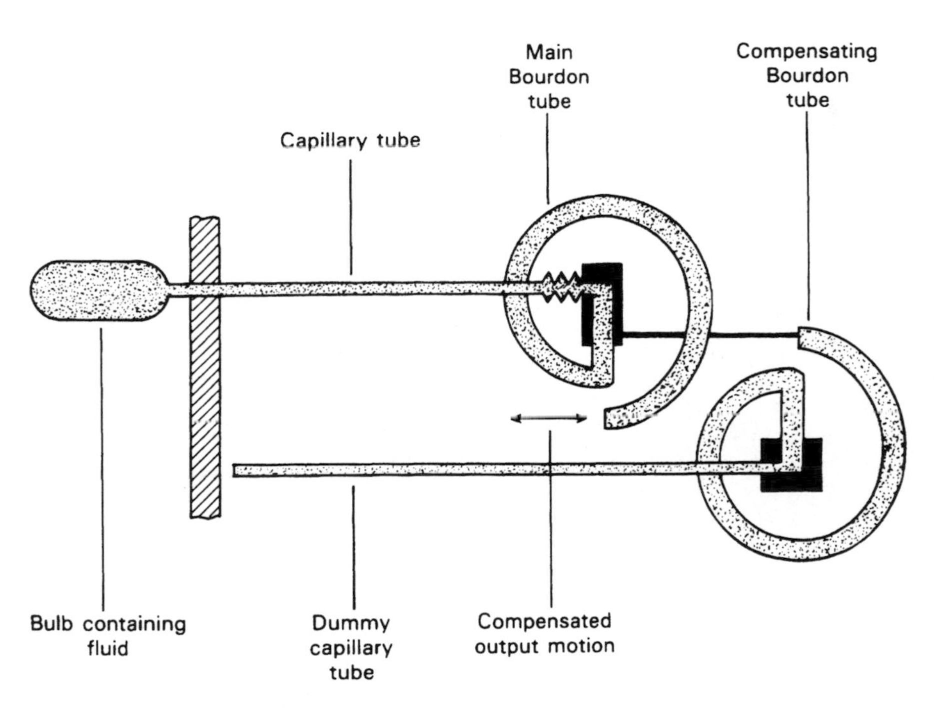

Figure 13.4. Correction for temperature gradient along capillary tube in pressure thermometer

Pressure thermometers are used to measure temperatures in the range between −250 °C and +2000 °C and their typical inaccuracy is ±0.5 percent of full-scale reading. Errors occur in the short term owing to mechanical damage and in the longer term owing to small volumetric changes in the glass components. The rate of increase in these errors is mainly use-related and therefore the required calibration interval must be determined by practical experimentation.

Thermocouples

Thermocouples [1] are manufactured from various combinations of the base metals copper and iron, the base metal alloys of alumel (Ni/Mn/Al/Si), chromel (Ni/Cr), constantan (Cu/Ni), nicrosil (Ni/Cr/Si) and nisil (Ni/Si/Mn), the noble metals platinum and tungsten, and the noble metal alloys of platinum/rhodium and tungsten/rhenium. The particular combinations which are used as thermocouples are known by internationally recognized type letters, for instance type K being chromel–alumel. The e.m.f.–temperature characteristics for some of these standard thermocouples are shown in Figure 13.5. Certain of these exhibit reasonable linearity over short temperature ranges and their characteristic can therefore be approximated by a series of straight-line relationships for use in intelligent instruments containing thermocouples. In general, however, the relationships do not

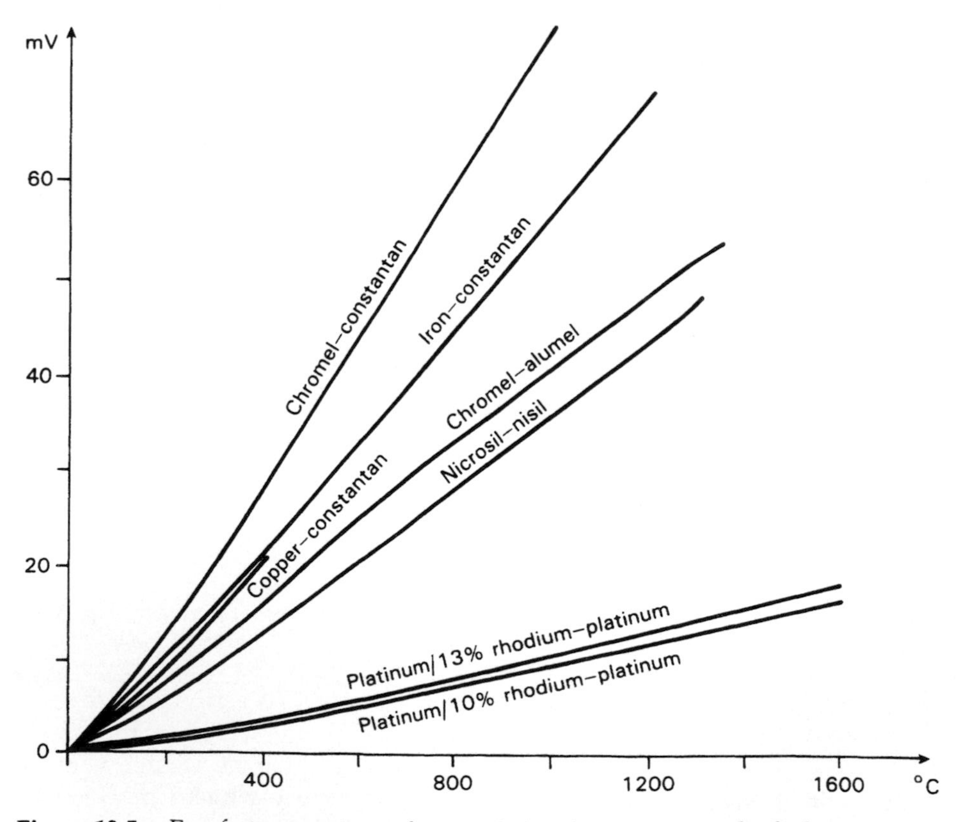

Figure 13.5. E.m.f. temperature characteristics for some standard thermocouple materials

approximate sufficiently to straight-line ones, and the temperature indicated by a given e.m.f. output measurement has to be calculated from published thermocouple tables.

Thermocouple tables are calculated assuming that the reference junction of the thermocouple is at 0 °C. This condition can be achieved by immersing the reference junction in an ice bath. However, it is not easy in many applications to maintain such a low reference temperature. In such circumstances, the reference junction can be placed in an environment maintained at some higher temperature by an electrical heating element. Correction then has to be made to the temperature indicated by the thermocouple tables to allow for the different reference junction temperature.

In order to make a thermocouple conform to some precisely defined e.m.f.–temperature characteristic described by standard tables, it is necessary that all metals used be refined to a high degree of pureness and all alloys be manufactured to an exact specification. This makes the materials used expensive, and consequently thermocouples are typically only a few centimetres long. It is clearly impractical to connect a voltage-measuring instrument to measure the thermocouple output in such close proximity to the environment whose temperature is being measured, and therefore extension leads up to several metres long are normally connected between the thermocouple and the measuring instrument. It is important that the voltage measured at the open ends of the extension wires arises only from the hot junction of the thermocouple immersed in the medium whose temperature is being measured. Therefore, no e.m.f. must be generated at the junction between the thermocouple and its extension leads and, consequently, the thermocouple and the extension wires must have very similar thermoelectric properties. This condition is usually met by choosing extension leads of the same basic materials as the thermocouple but manufactured to a lower specification. However, this solution is still prohibitively expensive in the case of noble-metal thermocouples, and base-metal extension leads which have a similar thermoelectric behaviour to the noble-metal thermocouple are generally used. In this form, the extension leads are known as compensating leads. A typical example is the use of nickel/copper–copper extension leads connected to a platinum/rhodium–platinum thermocouple.

The five standard base-metal thermocouples are chromel–constantan (type E), iron–constantan (type J), chromel–alumel (type K), nicrosil–nisil (type N) and copper–constantan (type T). These are all relatively cheap to manufacture but they become inaccurate with age and have a short life.

Chromel–constantan devices give the highest measurement sensitivity of 80 μV/°C, with an inaccuracy of ±0.75 percent and a useful measuring range of −200 °C up to 900 °C. Unfortunately, whilst they can operate satisfactorily in oxidizing environments, their performance and life are seriously affected by reducing atmospheres. Iron–constantan thermocouples have a sensitivity of 60 μV/°C and are the preferred type for general-purpose measurements in the temperature range of −150 °C to +1000 °C where the typical measurement inaccuracy is ±1 percent. Their performance is little affected by either oxidizing or reducing atmospheres. Copper–constantan devices have a similar measurement sensitivity of 60 μV/°C and find their main application in measuring subzero temperatures down to −200 °C, with an inaccuracy of ±0.5 percent. They can also be used in both oxidizing and reducing atmospheres to measure temperatures up to 350 °C. Chromel–alumel thermocouples have a measurement sensitivity of only 45 μV/°C , although their characteristic is particularly linear over the temperature range between 700 °C and 1200 °C and this is therefore their main application. Like chromel–constantan devices, they are suitable for oxidizing atmospheres but not for reducing ones. The measurement inaccuracy for them is ±0.75 percent. Nicrosil–Nisil thermocouples are a recent development which resulted from attempts to improve the performance and stability of chromel–alumel thermocouples. Their thermoelectric characteristic has a very similar shape to type K devices with equally good linearity over the high temperature measurement range and a measurement sensitivity of 40 μV/°C. The operating environment limitations are the same as for chromel–alumel devices but their long-term stability and life are at least three times better.

Noble-metal thermocouples are always expensive but enjoy high stability and long life in conditions of high temperature and oxidizing environments. They are chemically inert except in reducing atmospheres. Thermocouples made from platinum and a platinum/rhodium alloy have a low inaccuracy of ±0.2 percent and can measure temperatures up to 1500 °C, but their measurement sensitivity is only 10 μV/°C. Alternative devices made from tungsten and a tungsten/rhenium alloy have a better sensitivity of 20 μV/°C. and can measure temperatures up to 2300 °C.

Thermocouples are delicate instruments which must be treated carefully if their specified operating characteristics are to be reproduced. One major source of error is induced strain in the hot junction which reduces the e.m.f. output, and precautions are normally taken to minimize this by mounting the thermocouple horizontally rather than vertically. In some operating environments, no further protection is necessary to obtain satisfactory performance from the instrument. However, thermocouples are prone to contamination by various metals and protection is often necessary to minimize this. Such contamination alters the thermoelectric behaviour of the device, so that its characteristic varies from that published in standard tables. It also becomes brittle and its life is therefore shortened.

Protection takes the form of enclosing the thermocouple in a sheath. Some common sheath materials and their maximum operating temperatures are shown in Table 13.1. Whilst the thermocouple is an instrument with a naturally first-order type of step response characteristic, the time constant is usually so small as to be negligible when the thermocouple is used unprotected. When enclosed in a sheath, however, the time constant of the combination of thermocouple and sheath is significant. The size of the thermocouple and hence the diameter required for the sheath have a large effect on the importance of this. The time constant of a thermocouple in a 1 mm diameter sheath is only 0.15 s and this has little practical effect in

Table 13.1. Common sheath materials for thermocouples

Material	Maximum operating temperature (°C)
Mild steel	900
Nickel–chromium	900
Fused silica	1000
Special steel	1100
Mullite	1700
Recrystallized alumina	1850
Beryllia	2300
Magnesia	2400
Zirconia	2400
Thoria	2600

Note: The maximum operating temperature quoted assume oxidizing or neutral atmospheres. For operation in reducing atmospheres, the maximum allowable temperature is usually reduced.

most measurement situations, whereas a larger sheath of 6 mm diameter gives a time constant of 3.9 s which cannot be ignored so easily.

Thermocouples are manufactured by connecting together two wires of different materials, where each material is produced so as to conform precisely with some defined composition specification. This ensures that its thermoelectric behaviour accurately follows that for which standard thermocouple tables apply. The connection between the two wires is effected by welding, soldering or in some cases just by twisting the wire ends together. Welding is the most common technique used generally, with silver–soldering being reserved for copper–constantan instruments.

The diameter of wire used to construct thermocouples is usually in the range between 0.4 mm and 2 mm. The larger diameters are used where ruggedness and long life are required, although these advantages are gained at the expense of increasing the measurement time constant. In the case of noble-metal thermocouples, the use of large-diameter wire incurs a substantial cost penalty. Some special applications have a requirement for a very fast response time in the measurement of temperature, and in such cases wire diameters as small as 0.1 μm (0.1 microns) can be used.

The mode of construction of thermocouples means that their characteristics can be incorrect even when they are new, owing to faults in either the homogeneity of the thermocouple materials or in the construction of the device. Therefore, calibration checks should be carried out on all new thermocouples before they are put into use. Thereafter, the rate of change of thermoelectric characteristics with time is entirely dependent upon the operating environment and the degree of exposure to it. Particularly relevant factors in the environment are the type and concentration of trace metal elements and the temperature (the rate of contamination of thermocouple materials with trace elements of metals is a function of temperature). A suitable calibration frequency can therefore only be defined by practical experimentation, and this must be reviewed whenever the operating environment and conditions of use change.

The Thermopile

The thermopile is the name given to a temperature-measuring device which consists of several thermocouples connected together in series, such that all the reference junctions are at the same cold temperature and all the hot junctions are exposed to the temperature being measured, as shown in Figure 13.6. The effect of connecting n thermocouples together in series is to increase the measurement sensitivity by a factor of n. A typical thermopile manufactured by connecting together 25 chromel–constantan thermocouples gives a measurement resolution of 0.001 °C. As the thermopile is essentially a multiple thermocouple, all the previous discussion about thermocouples applies equally well to thermopiles and the calibration requirements are also identical.

Resistance Thermometers

Resistance thermometers rely on the principle that the resistance of a metal varies with temperature according to the relationship:

$$R = R_0(1 + a_1T + a_2T^2 + a_3T^3 + \ldots + a_nT^n) \qquad (13.1)$$

This equation is non-linear and so is inconvenient for measurement purposes. The equation becomes linear if all the terms in a_2T^2 and higher powers of T are negligible. This is approximately true for some metals over

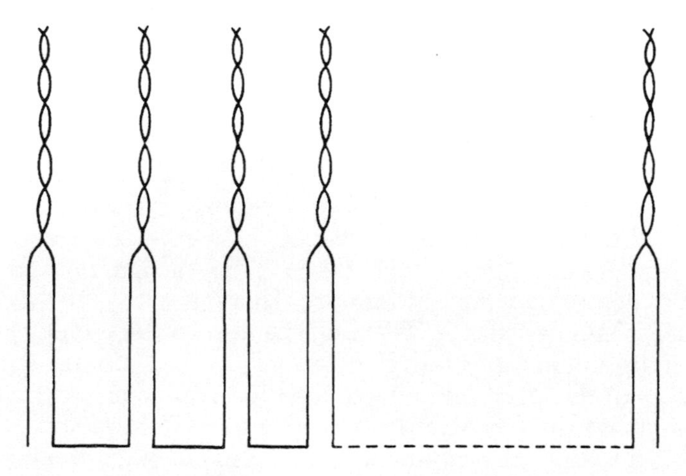

Figure 13.6. The thermopile

a limited temperature range and, in such cases, the resistance and temperature are related according to:

$$R \approx R_0(1 + a_1 T) \tag{13.2}$$

Platinum is one such metal where the resistance–temperature relationship is linear within ±0.4 percent over the temperature range between −200 °C and +40 °C. Even at +1000 °C, the quoted maximum inaccuracy figure is only ±1.2 percent. The characteristics of platinum and other metals are summarized in Figure 13.7.

Its good linearity and chemical inertness makes platinum the first choice for resistance thermometers in many applications. Platinum is very expensive, however, and consequently the cheaper but less accurate alternatives nickel and copper are sometimes used. These two metals are very susceptible to oxidation and corrosion and so the range of applications where they can be used is strictly limited even if their reduced accuracy is acceptable. Another metal, tungsten, is also used in some circumstances, particularly for high-temperature measurements. The working range of each of these four types of resistance thermometer are as shown below:

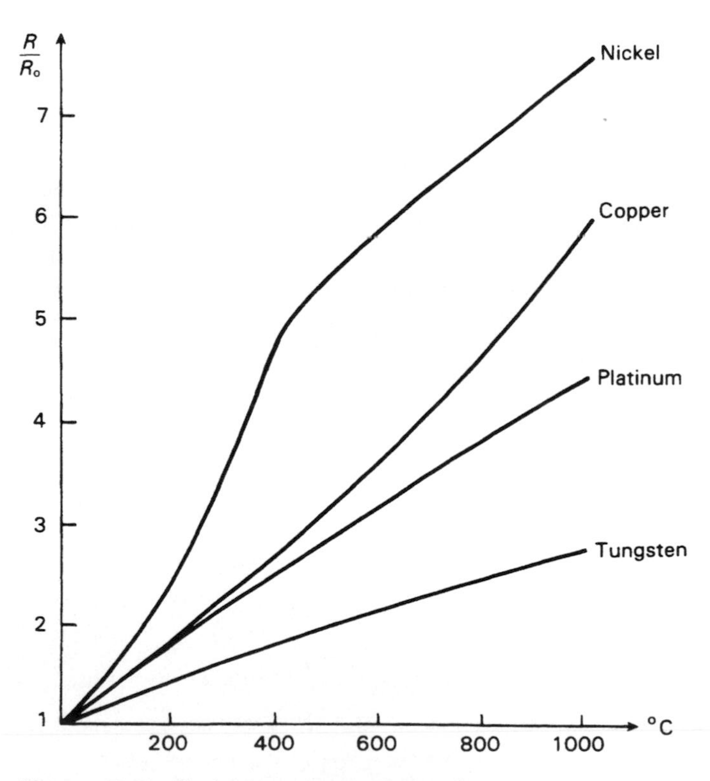

Figure 13.7. Resistance–temperature characteristics of metals

Platinum: −270 °C to 1000 °C (though use above 650 °C is uncommon)

Copper: −200 °C to 260 °C

Nickel: −200 °C to 430 °C

Tungsten: −270 °C to 1100 °C

In the case of non-corrosive and non-conducting environments, resistance thermometers are used without protection. In all other applications, they are protected inside a sheath. As in the case of thermocouples, such protection affects the speed of response of the system to rapid changes in temperature. A typical time constant for a sheathed platinum resistance thermometer is 0.4 s.

The resistance thermometer exists physically as a coil of metal resistance wire. The different instruments available have resistance elements ranging from 10 Ω right up to 25 kΩ. The devices with high resistance have several operational advantages. The first of these is that any connection resistances within the circuit become negligible in their effect. The second advantage is that the relatively high-voltage output produced by these instruments makes any induced e.m.f.'s produced by thermoelectric behaviour at the junction with connection leads negligible in magnitude.

The normal method of measuring resistance is to use a d.c. bridge. The excitation voltage of the bridge has to be chosen very carefully because, although a high value is desirable for achieving high measurement sensitivity, the self-heating effect of high currents flowing in the temperature transducer creates an error by increasing the temperature of the device and so changing the resistance value.

The frequency at which a resistance thermometer should be calibrated depends upon the material it is made from and upon the operating environment. Practical experimentation is therefore needed to determine the necessary frequency and this must be reviewed if the operating conditions change.

Thermistors

Thermistors are manufactured from beads of semiconductor material prepared from oxides of the iron group of metals such as chromium, cobalt, iron, manganese, and nickel. The resistance of such materials varies with temperature according to the following expression:

$$R = R_0 \cdot e^{\beta \cdot (1/T - 1/T_0)} \tag{13.3}$$

This relationship, as illustrated in Figure 13.8, exhibits a large negative temperature coefficient (i.e. the resistance decreases as the temperature increases), and so is fundamentally different from the relationship for the

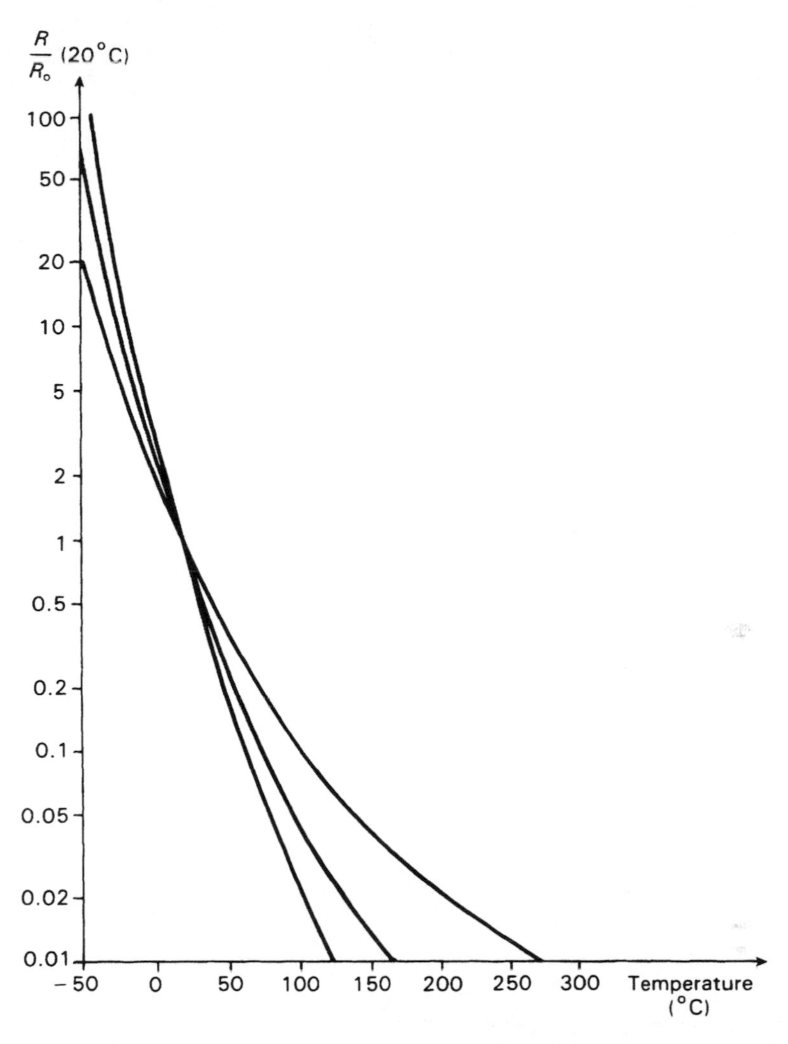

Figure 13.8. Resistance–temperature characteristics of thermistor materials

resistance thermometer, which shows a positive temperature coefficient. The form of equation (13.3) is such that it is not possible to make a linear approximation to the curve over even a small temperature range, and hence the thermistor is very definitely a non-linear instrument.

The major advantages of thermistors are their relatively low cost and their small size. This size advantage means that the time constant of thermistors operated in sheaths is small. However, the size reduction also decreases their heat dissipation capability, and so makes the self-heating effect greater. In consequence, thermistors have to be operated at generally lower current levels than resistance thermometers and so the measurement sensitivity afforded is less.

Similarly to the resistance thermometer, the resistance of a thermistor is measured by a d.c. bridge, and similar comments apply regarding the required calibration frequency.

Quartz Thermometers

The quartz thermometer is a fairly expensive device which has been developed recently and makes use of the principle that the resonant frequency of a material such as quartz varies with temperature. The temperature-sensing element consists of a quartz crystal enclosed within a probe (sheath). The crystal is connected electrically so as to form the resonant element within an electronic oscillator. Measurement of the oscillator frequency therefore allows the measured temperature to be calculated.

The instrument has a very linear output characteristic over the temperature range between −40 °C and +230 °C, with a measurement resolution of 0.1 °C. The characteristics of the instrument are generally very stable over long periods of time and therefore only infrequent calibration will be necessary. However, the exact calibration frequency required should be determined experimentally in the usual way.

Acoustic Thermometers

The principle of acoustic thermometry was invented as long ago as 1873, but until very recently it has only been used for measuring cryogenic (very low) temperatures. Acoustic thermometers use the fact that the velocity of sound through a gas varies with temperature according to the equation

$$v = \sqrt{(\alpha RT/M)} \qquad (13.4)$$

where v is the sound velocity, T is the gas temperature, M is the molecular weight of the gas and both R and α are constants. The various versions of acoustic thermometer which are available differ according to the technique used for generating sound and measuring its velocity in the gas. Further information can be found in [2].

Radiation Thermometers

The term 'radiation thermometer' describes a whole class of devices which measure the radiation emission from a body, and includes the optical pyrometer and various forms of radiation pyrometer. Between them, the various devices cover the temperature range from −20 °C up to +1800 °C. The total rate of radiation emission per second is given by:

$$E = K \cdot T^4 \qquad (13.5)$$

where T is the temperature of the body in kelvin.

The power spectral density of this emission varies with temperature in the manner shown in Figure 13.9. The major part of the frequency spectrum lies within the band of wavelengths between 0.3 μm and 40 μm, which corresponds to the visible (0.3–0.72 μm) and infra-red (0.72–1000 μm) ranges. Choice of the best method of measuring the emitted radiation depends on the temperature of the body. At low temperatures, the peak of the power spectral density function lies in the infra-red region, whereas at higher temperatures it moves towards the visible part of the spectrum. This phenomenon is observed as the red glow which a body begins to emit as its temperature is increased beyond 600 °C.

Radiation thermometers have one major advantage in that they do not require to be in contact with the hot body in order to measure its temperature. This makes them especially suitable for measuring high temperatures

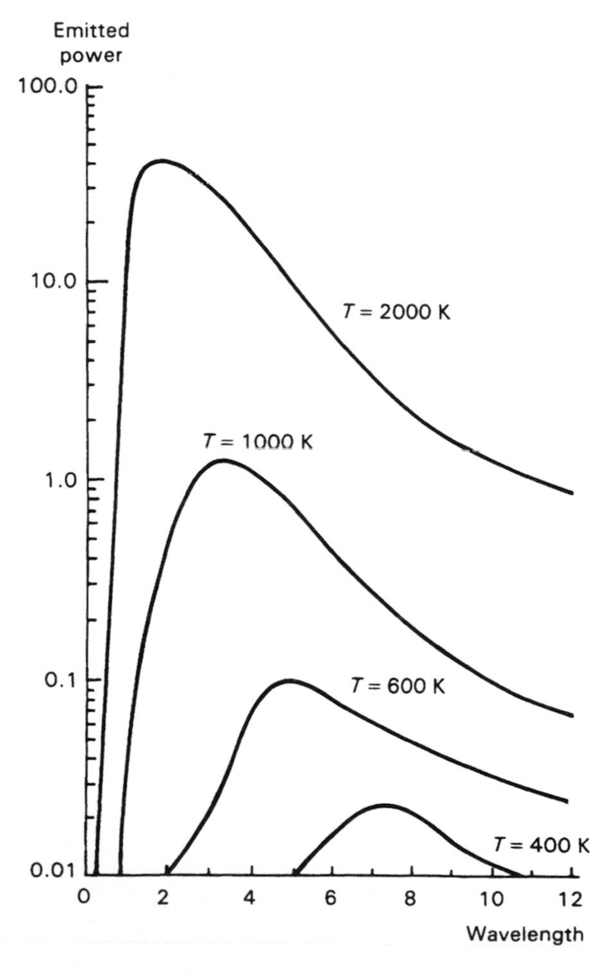

Figure 13.9. Power spectral density of radiated energy emission at various temperatures

which are beyond the capabilities of contact instruments such as thermocouples, resistance thermometers and thermistors. They are also capable of measuring moving bodies, for instance, the temperature of steel bars in a rolling mill. Their use is not as straightforward as the discussion so far might have suggested, however, because the radiation from a body varies with the composition and surface condition of the body as well as with temperature. This dependence on surface condition is quantified by a parameter known as the *emissivity* of the body. The use of radiation thermometers is further complicated by absorption and scattering of the energy between the emitting body and the radiation detector. Energy is scattered by atmospheric dust and water droplets and absorbed by carbon dioxide, ozone and water-vapour molecules. Therefore, all radiation thermometers have to be carefully calibrated for each separate application.

Optical Pyrometer

The optical pyrometer, illustrated in Figure 13.10, is designed to measure temperatures where the peak radiation emission is in the red part of the visible spectrum, i.e. where the measured body glows a certain shade of red according to the temperature. This limits the instrument to measuring temperatures above 600 °C. The instrument contains a heated tungsten filament within its optical system. The current in the filament is increased until its colour is the same as the hot body: under these conditions the filament apparently disappears when viewed against the background of the hot body. Temperature measurement is therefore obtained in terms of the current flowing in the filament. As the brightness of different materials at any particular temperature varies according to the emissivity of the material, the calibration of the optical pyrometer must be adjusted

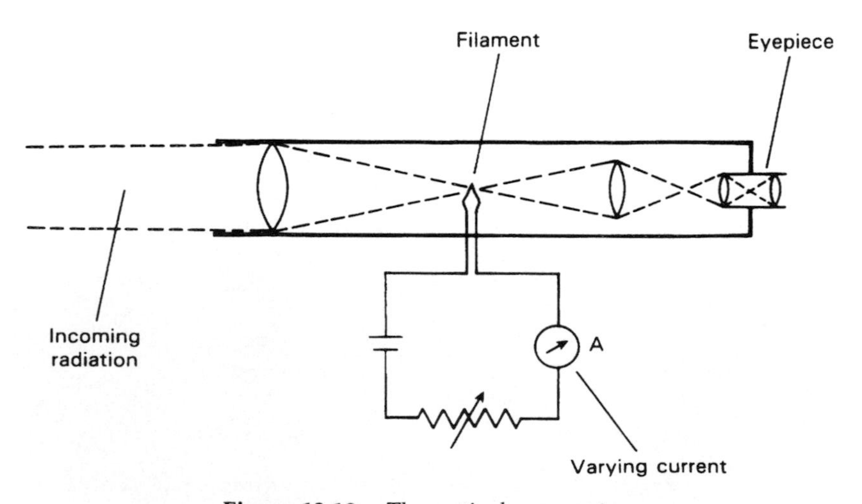

Figure 13.10. The optical pyrometer

according to the emissivity of the target. Manufacturers provide tables of standard material emissivities to assist with this.

The inherent measurement inaccuracy of an optical pyrometer is ±5 °C. However, in addition to this error, there can be a further operator-induced error of ±10 °C arising out of the difficulty in judging the moment when the filament 'just' disappears. Measurement accuracy can be improved somewhat by employing an optical filter within the instrument which passes a narrow band of frequencies of wavelength around 0.65 μm corresponding to the red part of the visible spectrum.

The instrument cannot be used in automatic temperature control schemes because the eye of the human operator is an essential part of the measurement system. The reading is also affected by fumes in the sight path. Because of these difficulties and its low accuracy, hand-held radiation pyrometers are rapidly overtaking the optical pyrometer in popularity, although the instrument is still widely used in industry for measuring temperatures in furnaces and similar applications at present.

Radiation Pyrometers

All the alternative forms of radiation pyrometer described below have an optical system which is similar to that in the optical pyrometer and focuses the energy emitted from the measured body. They differ, however, by omitting the filament and eyepiece and using instead an energy detector in the same focal plane as the eyepiece was, as shown in Figure 13.11. The radiation detector is either a thermal detector, which measures the temperature rise in a black body at the focal point of the optical system, or a photon detector. Thermal detectors respond equally to all wavelengths in the frequency spectrum whereas photon detectors respond selectively to a particular band within the full spectrum.

Thermopiles, resistance thermometers and thermistors are all used as thermal detectors in different versions of these instruments. These typically have time constants of several milliseconds, because of the time taken for the black body to heat up and the temperature measuring instrument to respond to the temperature change.

Photon detectors are usually of the photoconductive or photovoltaic type.

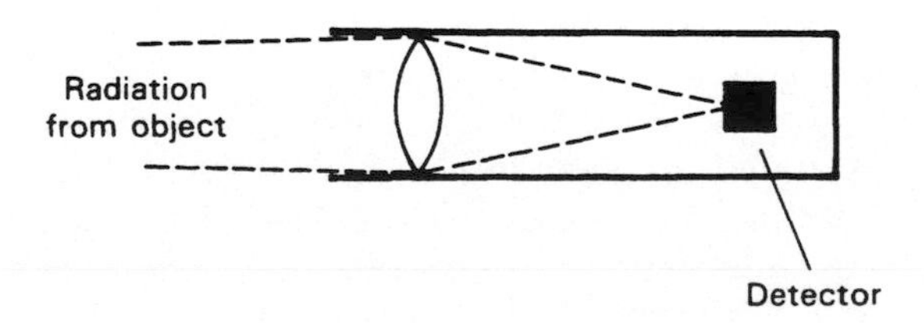

Figure 13.11. Structure of a radiation thermometer

Both of these respond very much faster to temperature changes than thermal detectors because they involve atomic processes, and typical measurement time constants are a few microseconds.

Various forms of electrical output are available from the radiation detector: these are functions of the energy incident on the detector and are therefore functions of the temperature of the measured body. Whilst this therefore makes such instruments of use in automatic control systems, their accuracy is often inferior to optical pyrometers. This reduced accuracy arises firstly because a radiation pyrometer is sensitive to a wider band of frequencies than an optical instrument and the relationship between emitted energy and temperature is less well defined. Secondly, the magnitude of energy emission at low temperatures gets very small, according to equation (13.5), increasing the difficulty of accurate measurement.

The various forms of radiation pyrometer differ mainly in the technique used to measure the emitted radiation. They also differ in the range of energy wavelengths that they are designed to measure and hence in the temperature range measured. One further difference is the material used to construct the energy-focusing lens. Outside the visible part of the spectrum, glass becomes almost opaque to infra-red wavelengths, and other lens materials such as arsenic trisulphide are used.

Intelligent Temperature Measuring Instruments

Intelligent temperature transmitters have recently been introduced into the catalogues of instrument manufacturers, and they bring about the usual benefits associated with intelligent instruments. Such transmitters are designed for use with transducers which have either a d.c. voltage output in the mV range or an output in the form of a resistance change. They are therefore suitable for use in conjunction with thermocouples, thermopiles, resistance thermometers, thermistors and broad-band radiation pyrometers. All of the transmitters presently available have non-volatile memories where all constants used in correcting output values for environmental inputs, etc., are stored, thus enabling the instrument to survive power failures without losing such information.

The transmitters now available include adjustable damping, noise rejection, self-adjustment for zero and sensitivity drifts and expanded measurement range. These features allow an inaccuracy level of only ± 0.05 percent of full scale to be specified.

Intelligent instruments cost significantly more than their non-intelligent counterparts, and justification purely on the grounds of their superior accuracy is hard to make. However, their expanded measurement range means immediate savings are made in terms of the reduction in the number of spare instruments needed to cover a number of measurement ranges. Their capability for self-diagnosis and self-adjustment means that they require attention much less frequently, giving additional savings in maintenance costs.

An intelligent temperature-measuring instrument should be properly regarded as a two-part measurement system consisting of a temperature transducer and a signal processing element. Although the signal processing element is largely self-calibrating, the appropriate calibration routines described earlier have to be applied to the transducer.

13.2 INTRODUCTION TO CALIBRATION

The primary standard of temperature is fundamentally different in character from the primary standards of mass, length and time. If two bodies of lengths l_1 and l_2 are connected together end to end, the result is a body of length $l_1 + l_2$. A similar relationship exists between separate masses and separate times. However, if two bodies at the same temperature are connected together, the joined body has the same temperature as each of the original bodies.

This is a root cause of the fundamental difficulties which exist in establishing an absolute standard for temperature in the form of a relationship between it and other measurable quantities for which a primary standard unit exists.

One obvious way of defining temperatures in absolute units would seem to be to measure the heat flowing between two bodies at different temperatures. An attempt to do this was made in 1848 by Kelvin. He proposed a thermodynamic temperature scale, based on the Carnot cycle, which defines the temperature ratio between two bodies A and B. If two heat reservoirs of infinite capacity, A and B, at temperatures T_A and T_B are put into contact, the heat flow from A, Q_A, and the heat flow to B, Q_B, are related by:

$$\frac{T_A}{T_B} = \frac{Q_A}{Q_B}$$

Unfortunately, this is not realizable physically because it is based on an ideal Carnot cycle which cannot be achieved practically.

An identical temperature scale can be defined by a constant volume or constant pressure gas thermometer using an ideal gas, and we have for example:

$$\frac{T_1}{T_2} = \frac{P_1}{P_2} \quad \text{(from ideal gas law: } PV = RT\text{)}$$

Again, however, this is of little practical use since real gases only approach the behaviour of ideal gases as their pressure is reduced towards zero. The procedure that must be followed to make use of this relationship is to measure the pressure ratio at several low pressures and extrapolate the result to zero pressure. This would be a very time-consuming procedure for normal calibration procedures and in any case the accuracy yielded by a gas thermometer is inadequate.

In the absence of a convenient relationship that relates the temperature of a body to another measurable quantity and can be expressed in primary standard units, it is necessary to establish fixed, reproducible reference points for temperature in the form of freezing and boiling points of substances where the transition between solid, liquid and gaseous states is sharply defined. The International Practical Temperature Scale (IPTS) uses this philosophy and defines six *primary fixed points* for reference temperatures in terms of:

the triple point of equilibrium hydrogen:	−259.34 °C
the boiling point of oxygen:	−182.962 °C
the boiling point of water:	100.0 °C
the freezing point of zinc:	419.58 °C
the freezing point of silver:	961.93 °C
the freezing point of gold:	1064.43 °C

(all at standard atmospheric pressure)

The thermodynamic temperatures of these states are established for calibration purposes by gas thermometers. For calibrating intermediate temperatures, interpolation between the fixed points is carried out by the following reference instruments:

(a) a platinum resistance thermometer in the temperature range of −182.962 °C to +630.74 °C and

(b) a platinum–platinum/10 percent rhodium thermocouple in the temperature range of +630.74 °C to +1064.43 °C (630.74 °C is the melting temperature of the secondary fixed point metal antimony)

Above the gold melting point, standard temperatures are defined by a narrow-band radiation thermometer using the formula

$$\frac{J_t}{J_G} = \frac{e^{1.438/1336.15W} - 1}{e^{1.438/Wt} - 1}$$

where J_t is the spectral radiance of a black body at temperature t, J_G is the spectral radiance of the black body at the boiling point of gold, and W is the wavelength of the pyrometer.

The freezing points of certain other metals are also used as *secondary fixed points* to provide additional reference points during calibration procedures.

These are of particular use for calibrating instruments measuring high temperatures. Some examples are:

Freezing point of Tin:	231.968 °C
Freezing point of Lead:	327.502 °C
Freezing point of Zinc:	419.58 °C
Freezing point of Antimony:	630.74 °C
Freezing point of Aluminium:	660.37 °C
Freezing point of Copper:	1084.5 °C
Freezing point of Nickel:	1455 °C
Freezing point of Palladium:	1554 °C
Freezing point of Platinum:	1772 °C
Freezing point of Rhodium:	1963 °C
Freezing point of Iridium:	2447 °C
Freezing point of Tungsten:	3387 °C

The accuracy of temperature calibration procedures is fundamentally dependent on how accurately points on the IPTS can be reproduced. The present limits are:

1 K	0.3%	800 K	0.001%
10 K	0.1%	1500 K	0.02%
100 K	0.005%	4000 K	0.2%
273.15 K	0.0001%	10 000 K	6.7%

13.3 CALIBRATION INSTRUMENTS AND EQUIPMENT

Almost any instrument from the list presented in Section 13.1 might be used for calibration duties in certain circumstances. Where involved in such duties, of course, the instrument used would be one of high accuracy which was reserved solely for calibration duties. The list of calibration instruments therefore includes mercury-in-glass thermometers, base-metal thermocouples (type K), noble-metal thermocouples (types B, R and S), platinum resistance thermometers and radiation pyrometers.

Although no special instruments have been developed for temperature calibration, certain purpose-designed equipment is required to provide the controlled temperatures necessary for the calibration process. Such equipment is commercially available from several sources.

For calibration of all temperature transducers other than radiation thermometers, a furnace consisting of an electrically heated ceramic tube is commonly used. The temperature of such a furnace can typically be controlled within limits of ±2 °C over the range from 20 °C to 1600 °C. Up to 630 °C, the platinum resistance thermometer is commonly used as a reference standard. Above that temperature, a type R (platinum/13 percent rhodium–platinum) thermocouple is usually employed. Type K (chromel–alumel) thermocouples are also used as an alternative reference standard for temperature calibration up to 1000 °C.

Below 20 °C, a stirred water bath is used to provide a constant reference temperature, and the same equipment can be used for temperatures up to 100 °C. Similar stirred liquid baths containing oil or salts (potassium/sodium nitrate mixtures) can be used to provide reference temperatures up to 600 °C.

All of the above equipment provides an environment in which calibration is performed by comparing one instrument with another. Absolute reference standards of temperature can only be obtained by the freezing point method in which an ingot of pure metal (better than 99.99 percent pure), protected against oxidation inside a graphite crucible with a close-fitting lid, is heated beyond its melting point and allowed to cool. If its temperature is monitored, an arrest period is observed in its cooling curve at the freezing point of the metal. Electric resistance furnaces are available to carry out this procedure. Up to 1100 °C, a measurement uncertainty of less than ±0.5 °C is achievable. It should be noted that this technique only provides standard reference temperatures at the fixed melting points of the metals used. However, this is still a very useful check on the accuracy of reference standard instruments, and, once their accuracy has been verified at such points within their measuring range, much greater confidence is created in their accuracy at intermediate temperatures.

For the calibration of radiation thermometers, a radiation source which approximates as closely as possible to the behaviour of a black body is required. The actual value of the emissivity of the source must be measured by a surface pyrometer. Some form of optical bench is also required so that instruments being calibrated can be held firmly and aligned accurately.

The simplest form of radiation source is a hot plate heated by an electrical element. The temperature of such devices can be controlled within limits of ±1 °C over the range from 0 °C to 650 °C and the typical emissivity of the plate surface is 0.85. Type R noble-metal thermocouples embedded in the plate are normally used as the reference instrument.

A black-body cavity provides a heat source with a much better emissivity. This can be constructed in various alternative forms according to the temperature range of the radiation thermometers to be calibrated, though a common feature is a blackened conical cavity with a cone angle of about 15°.

For calibrating low-temperature radiation pyrometers (measuring temperatures in the range of 20 °C to 200 °C), the black-body cavity is maintained at a constant temperature (±0.5 °C) by immersing it in a liquid bath.

The typical emissivity of a cavity heated in this way is 0.995. Water is suitable for the bath in the temperature range of 20 °C–90 °C and a silicone fluid is suitable for the range of 80 °C–200 °C. Within these temperature ranges, a mercury-in-glass thermometer is commonly used as the standard reference calibration instrument, although a platinum resistance thermometer is used when better accuracy is required.

Another form of black-body cavity is one lined with a refractory material and heated by an electrical element. This gives a typical emissivity of 0.998 and is used for calibrating radiation pyrometers at higher temperatures. Within the range of 200 °C–1200 °C, temperatures can be controlled within limits of ±0.5 °C and a type R thermocouple is generally used as the reference instrument. At the higher range of 600 °C–1600 °C, temperatures can be controlled within limits of ±1 °C and a type B thermocouple (30 percent rhodium–platinum/6 percent rhodium–platinum) is normally used as the reference instrument. As an alternative to thermocouples, radiation thermometers can also be used as a standard within ±0.5 °C over the temperature range from 400 °C to 1250 °C.

To provide reference temperatures above 1600 °C, a carbon-cavity furnace is used. This consists of a graphite tube with a conical radiation cavity at its end. Temperatures up to 2600 °C can be maintained with an accuracy of ±5 °C. Narrow-band radiation thermometers are used as the reference standard instrument.

Again, the above equipment merely provides an environment in which radiation thermometers can be calibrated against some other reference standard instrument. To obtain an absolute reference standard of temperature, a fixed-point, black-body furnace is used. This has a radiation cavity consisting of a conical-ended cylinder which contains a crucible of 99.999 percent pure metal. If the temperature of the metal is monitored as it is heated up at a constant rate, an arrest period is observed at the melting point of the metal where the temperature ceases to rise for a short time. Thus the melting point, and hence the temperature corresponding to the output reading of the monitoring instrument at that instant, are defined exactly. The measurement uncertainty is of the order of ±0.3 °C. The list of metals, and their melting points, used to provide such primary and secondary reference temperatures can be found in Section 13.2.

In the calibration of radiation thermometers, knowledge of the emissivity of the hot plate or black-body furnace used as the radiation source is essential. This is measured by special types of surface pyrometer. Such instruments contain a hemispherical, gold-plated surface which is supported on a telescopic arm that allows it to be put into contact with the hot surface. The radiation emitted from a small hole in the hemisphere is independent of the surface emissivity of the measured body and is equal to that which would be emitted by the body if its emissivity value was 100. This radiation is measured by a thermopile with its cold junction at a controlled temperature. A black hemisphere is also provided with the instrument, which can be inserted to cover the gold surface. This allows the instrument to measure the normal radiation emission from the hot body and so allows the surface emissivity to be calculated by comparing the two radiation measurements.

Within this list of special equipment, mention must also be made of standard tungsten strip lamps which are used for providing constant known temperatures in the calibration of optical pyrometers. The various versions of these provide a range of standard temperatures between 800 °C and 2300 °C to an accuracy of ±2 °C.

13.4 CALIBRATION FREQUENCY

The manner in which the required frequency for calibration checks is determined for the various temperature-measuring instruments available was discussed in the instrument review presented in Section 13.1. The simplest instruments from a calibration point of view are liquid-in-glass thermometers. The only parameter able to change within these is the volume of the glass used in their construction. This only changes very slowly with time, and hence only infrequent (e.g. annual) calibration checks are required.

The required frequency for calibration of all other instruments is either (a) dependent upon the type of operating environment and the degree of exposure to it or (b) use-related. In some cases, both of these factors are relevant.

Resistance thermometers and thermistors are examples of instruments where the drift in characteristics depends on the environment they are operated in and on the degree of protection they have from that environment. Devices such as gas thermometers and quartz thermometers suffer characteristics drift which is largely a function of how much they are used (or misused!), though in the case of quartz thermometers, any drift is likely to be small and only infrequent calibration checks will be required. Any instruments not mentioned so far suffer characteristics drift due to both environmental and use-related factors. The list of such instruments includes bimetallic thermometers, thermocouples, thermopiles and radiation thermometers. In the case of thermocouples and thermopiles, it must be remembered that error in the required characteristics is possible even when the instruments are new, as discussed in Section 13.1, and therefore their calibration must be checked before use.

As the factors responsible for characteristics drift vary from application to application, the required frequency of calibration checks can only be determined experimentally. The procedure for doing this is to start by checking the calibration of instruments used in new applications over very short intervals of time, and then to progressively lengthen the interval between calibration checks until a significant deterioration in instrument characteristics is observed. The required calibration interval is then defined as that time interval which is predicted to elapse before the characteristics of the instrument have drifted to the limits that are allowable in that particular measurement application. A suggested mechanism for this procedure was described more fully in Section 8.4.

Working and reference standard instruments and ancillary equipment must also be calibrated periodically. An interval of two years is usually recommended between such calibration checks although monthly checks are advised for the black-body-cavity furnaces used to provide standard reference temperatures in pyrometer calibration. Standard resistance thermometers and thermocouples may also need more frequent calibration checks if the conditions (especially of temperature) and frequency of use demand them.

13.5 CALIBRATION PROCEDURES

The standard way of calibrating temperature transducers is to put them into a temperature-controlled environment together with a standard instrument [2, 3], or to use a radiant heat source of controlled temperature with high emissivity in the case of radiation thermometers. In either case, the controlled temperature must be measured by a standard instrument whose calibration is traceable to reference standards. This is a suitable method for most instruments in the calibration chain but is not necessarily appropriate or even possible for the process instruments at the lower end of the chain.

In the case of many process instruments, their location and mode of fixing makes it difficult or sometimes impossible to remove them to a laboratory for calibration checks to be carried out. In this event, it is standard practice to calibrate them in their normal operational position, using a reference instrument that is able to withstand whatever hostile environment may be present. If this practice is followed, it is imperative that the working standard instrument is checked regularly to ensure that it has not been contaminated.

Such *in-situ* calibration may also be required where process instruments have characteristics which are sensitive to the environment in which they work, so that they are calibrated under their usual operating conditions and are therefore accurate in normal use. However, the preferred way of dealing with this situation is to calibrate them in a laboratory with ambient conditions (of pressure, humidity etc.) set up to mirror those of the normal operating environment. This alternative avoids having to subject reference calibration instruments to harsh chemical environments which are commonly associated with manufacturing processes.

For instruments at the lower end of the calibration chain, i.e. those measuring process variables, it is common practice to calibrate them against an instrument which is of the same type but of higher accuracy and reserved only for calibration duties. If a large number of different types of instrument have to be calibrated, however, this practice leads to the need to keep a large number of different calibration instruments. To avoid this, various reference instruments are available which can be used to calibrate all process instruments within a given temperature-measuring range.

Examples are the liquid-in-glass thermometer (0 °C to 200 °C), platinum resistance thermometer (−200 °C to 630 °C) and type B thermocouple (600 °C to 1600 °C). The optical pyrometer is also often used as a reference instrument at this level for the calibration of other types of radiation thermometer.

For calibrating instruments further up the calibration chain, particular care is needed with regard to both the instruments used and the conditions they are used under. It is difficult and expensive to meet these conditions and hence this function is subcontracted by most companies to specialist laboratories. The reference instruments used are the platinum resistance thermometer in the temperature range of −200 °C to +630 °C, the platinum–platinum/10 percent rhodium (type S) thermocouple in the temperature range of +630 °C to +1064 °C and a narrow-band radiation thermometer at higher temperatures. An exception is *optical pyrometers*, which are calibrated as explained in the final paragraph of this chapter. A particular note of caution must be made where platinum–rhodium thermocouples are used as a standard. These are very prone to contamination and, if they need to be handled at all, this should be done with very clean hands.

Before ending this chapter, it is appropriate to mention one or two further points which concern the calibration of thermocouples and radiation thermometers.

As mentioned earlier, apart from calibration checks in the normal way, the calibration even of new *thermocouples* must be checked before use. The procedure for this is to immerse both junctions of the thermocouple in an ice-bath and measure its output with a high accuracy digital voltmeter (±5 μV). Any output greater than 5 μV would indicate a fault in the thermocouple material and/or its construction. When carrying out any calibration operations on thermocouples, it is important to exclude any source of electrical or magnetic fields, as these will induce erroneous voltages in the sensor.

Special comments are also relevant regarding the calibration of a *radiation thermometer*. As well as the normal accuracy checks, its long-term stability must also be verified by testing its output over a period which is one hour longer than the manufacturer's specified 'warm-up' time. This shows up any components within the instrument that are suffering from temperature-induced characteristics drift. It is also necessary to calibrate radiation thermometers according to the emittance characteristic of the body whose temperature is being measured and according to the level of energy losses in the radiation path between the body and the measuring instrument. Such emissivity calibration must be carried out for every separate application that the instrument is used for, using a surface pyrometer.

Optical pyrometers are calibrated by sighting them on the filament of a tungsten strip lamp in which the current is accurately measured. This method of calibration can be used at temperatures up to 2500 °C. Alternatively, they can be calibrated against a standard radiation pyrometer.

REFERENCES

1. NAMAS Document B 5551: *Thermocouples, Reference tables and Traceability*, NAMAS Executive, National Physical Laboratory, Middlesex, UK (1975).
2. Moore, G., 'Acoustic thermometry', *Electronics and Power*, 30 675–677 (1984).
3. NAMAS Document NIS 7: *Traceability – thermometers, thermocouples and platinum resistance thermometers*, NAMAS Executive, National Physical Laboratory, Middlesex, UK (1984).

CHAPTER 14
Pressure Calibration

14.1 REVIEW OF PRESSURE-MEASURING INSTRUMENTS

Pressure-measuring instruments normally measure gauge pressure rather than absolute pressure because the measurement of absolute pressure is very difficult. The exception to this is one form of U-tube manometer which does measure absolute pressure, although with limited accuracy.

The instruments available for measuring pressures can be divided into four categories according to the range of pressure that they are designed to measure. These four pressure ranges are

(a) high pressures greater than 7000 bar (6910 atmospheres)

(b) mid-range pressures in the span between 1.013 bar and 7000 bar (1 to 6910 atmospheres)

(c) low pressures between 0.001 mbar and 1.013 bar (1 atmosphere)

(d) very low pressures less than 0.001 mbar

Apart from certain types of diaphragm-type instrument which use either a piezoelectric crystal or a capacitive transducer to measure the pressure-induced displacement, all the instruments discussed in this chapter are restricted to measuring only static pressures. Instruments and techniques for measuring dynamic pressures have not been included because the measurement of such is a very specialized area which is not of general interest.

Measurement of high pressures above 7000 bar is normally carried out electrically by monitoring the change of resistance of wires of special materials. Materials having resistance/pressure characteristics which are suitably linear and sensitive include gold–chrome alloys and manganin.

Mid-range pressures are the ones measured most commonly in industrial applications. Suitable instruments for this range are the U-tube manometer, the well-type manometer, the inclined manometer, the diaphragm, Bourdon tubes, bellows-type instruments and resonant wire devices.

Low pressures are measurable by special versions of the common mid-range pressure-measuring instruments. Suitable instruments include special manometers, Bourdon tubes, bellows-type instruments and diaphragms.

For the very-low-pressure range, special instruments have been developed. These include the thermocouple gauge, Pirani gauge, thermistor gauge and ionization gauge.

The mode of operation and characteristics of these various pressure-measuring devices are considered below.

U-tube Manometers

All U-tube manometers consist of a glass vessel shaped into a letter U and filled with a liquid. Two alternative forms of the instrument exist.

In one form, shown in Figure 14.1, one end of the U-tube is sealed and evacuated. This is the form mentioned earlier which measures *absolute pressure*. The absolute pressure is measured in terms of the difference between the mercury levels in the two halves of the tube. Thus

$$p_1 \text{ (absolute pressure)} = hd$$

where d is the specific gravity of the fluid. Quite apart from the difficulty in judging exactly where the meniscus levels in the mercury are, such an instrument cannot give a perfect measurement because of the impossibility of achieving a total vacuum at the sealed end of the tube. Although it is possible by modern techniques to design an instrument that does give a reasonably accurate measurement of absolute pressure, the problem is usually avoided in practice by measuring gauge pressure instead of absolute pressure.

The alternative form of U-tube manometer is the one shown in Figure 14.2, where both ends of the tube are open. If an unknown pressure is applied to one end and the other end is left open to the atmosphere, *gauge pressure* is measured. The unknown gauge pressure of the fluid (p_2) is

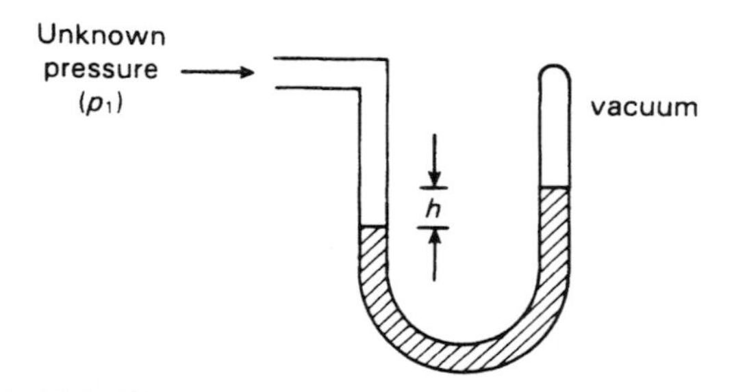

Figure 14.1. Sealed U-tube measuring absolute pressure

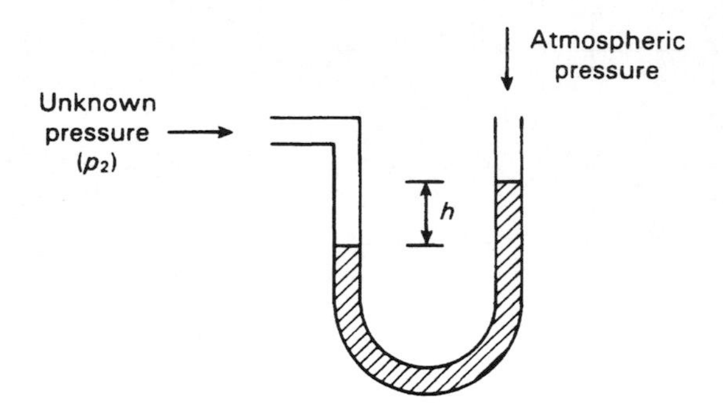

Figure 14.2. U-tube measuring gauge pressure.

related to the difference (h) in the levels of fluid in the two halves of the tube by the expression

$$p_2 \text{ (gauge pressure)} = hd$$

where d is the specific gravity of the fluid.

The form of U-tube manometer with two open ends can also be used to measure *differential pressure* as shown in Figure 14.3. The unknown pressures (p_3 and p_4) are applied to the ends of the tube and the differential pressure ($p_3 - p_4$) is then given by:

$$(p_3 - p_4) = hd$$

The type of liquid used in the instrument depends on the pressure and characteristics of the fluid being measured. Water is a convenient and certainly cheap choice, but evaporates easily and is difficult to see. It is nevertheless used extensively, with the major obstacles to its use being overcome by using coloured water and regularly topping up the tube to counteract evaporation.

Situations where water is definitely not used as a U-tube manometer fluid include the measurement of fluids that react with or dissolve in water, and

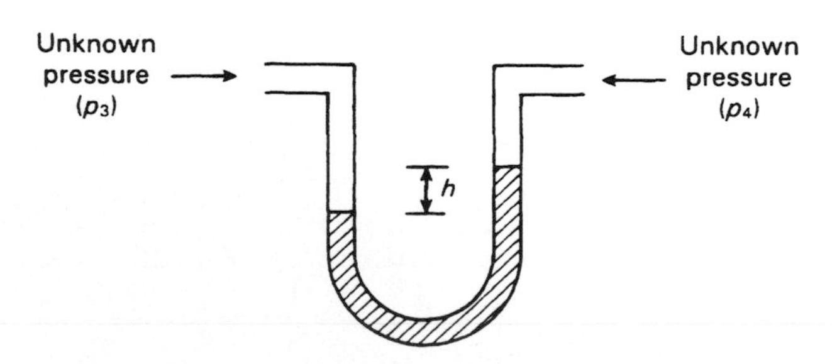

Figure 14.3. U-tube measuring differential pressure

also where higher-magnitude pressure measurements are required. In such circumstances, liquids such as aniline, carbon tetrachloride, bromoform, mercury or transformer oil are used.

U-tube manometers are normally used to measure mid-range fluid pressures up to 2 bar in magnitude. However, special versions exist which are designed to measure low pressures down to 0.1 mbar.

As far as calibration requirements are concerned, there are very few reasons for the measurement characteristics of U-tube manometers to drift. In the longer term, however, small errors can be introduced through volumetric changes in the glass. Therefore, annual calibration checks are advisable.

Well-type Manometer (Cistern Manometer)

The well-type manometer, shown in Figure 14.4, is effectively a U-tube manometer in which one half of the tube is made very large and takes the form of a well. The change in the level of the well as the measured pressure varies is negligible. Therefore, the liquid level in only one tube has to be measured, which makes the instrument much easier to use than the U-tube manometer. Gauge pressure is given by:

$$p_1 = hd$$

It might appear that the instrument would give a better measurement accuracy than the U-tube manometer because the need to subtract two liquid-level measurements in order to arrive at the pressure value is avoided. However, this benefit is swamped by errors that arise owing to the typical cross-sectional area variations in the glass used to make the tube. Such variations do not affect the accuracy of the U-tube manometer. Annual calibration checks are advised as for the U-tube manometer.

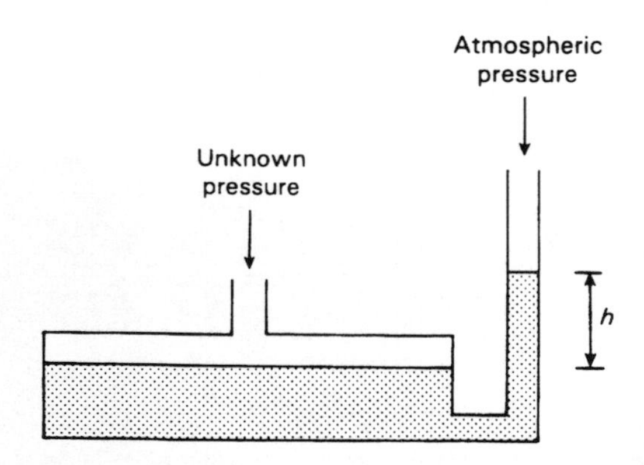

Figure 14.4. Well-type manometer

Inclined Manometer (Draft Gauge)

The inclined manometer, shown in Figure 14.5 is a variation on the well-type manometer in which one leg of the tube is inclined to increase measurement sensitivity. Comments similar to those above apply to its accuracy and calibration requirements.

Diaphragm

The diaphragm is one of the three common types of elastic-element pressure transducer, and is shown schematically in Figure 14.6. Applied pressure causes displacement of the diaphragm and this movement is measured by a displacement transducer. Both gauge pressure and differential pressure can be measured by different versions of diaphragm-type instrument. In the case of differential pressure, the two pressures are applied to either side of the diaphragm and the displacement of the diaphragm corresponds to the pressure difference. The typical magnitude of displacement in either version is 0.1 mm, which is well suited to a strain-gauge type of measuring transducer.

Four strain gauges are normally used in a bridge configuration, in which an excitation voltage is applied across two opposite points of the

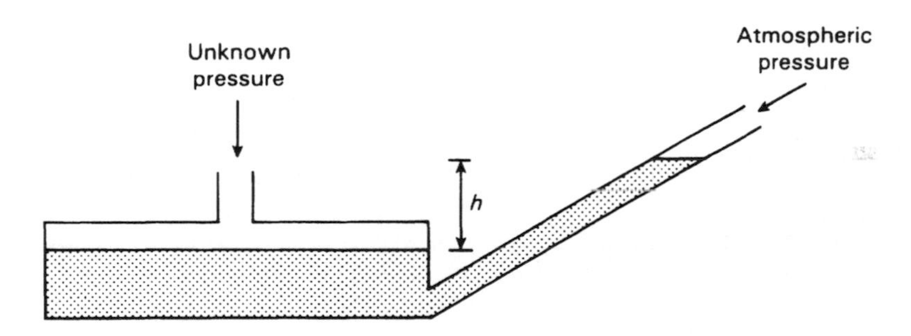

Figure 14.5. Inclined manometer

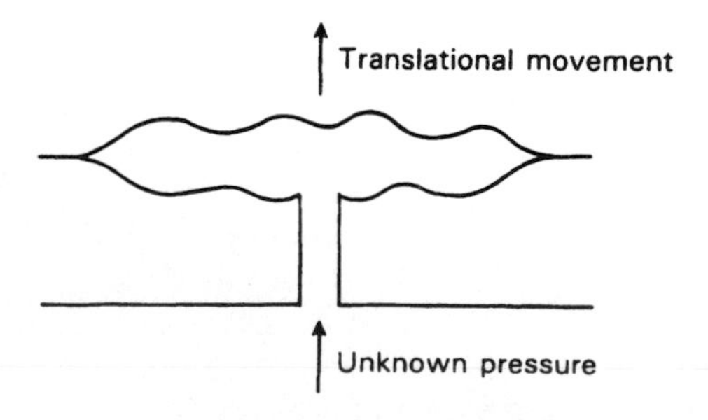

Figure 14.6. Diaphragm

bridge. The output voltage measured across the other two points of the bridge is then a function of the resistance change due to the strain in the diaphragm. This arrangement automatically provides compensation for environmental temperature changes. Older pressure transducers of this type used metallic strain gauges bonded to a diaphragm typically made of stainless steel. Apart from manufacturing difficulties arising from the problem of bonding the gauges, metallic strain gauges have a low gauge factor. This means that the strain-gauge bridge only has a low output which has to be amplified by an expensive direct-current amplifier. The development of semiconductor (piezoresistive) strain gauges provided a solution to the low output problem, as they have gauge factors up to one hundred times greater than metallic gauges. However the difficulty of bonding gauges to the diaphragm remained and a new problem emerged regarding the highly non-linear characteristic of the strain–output relationship.

The problem of strain gauge bonding was solved with the emergence of monolithic piezoresistive pressure transducers, and these are now the most commonly used type of diaphragm pressure transducer. The monolithic cell consists of a diaphragm made of a silicon sheet into which resistors are diffused during the manufacturing process. Besides avoiding the difficulty with bonding, such monolithic silicon measuring cells have the advantage of being very cheap to manufacture in large quantities and their sensitivity is three times better than that of unbonded strain gauges. Another advantage is that they can be made extremely small, down to 0.75 mm in diameter, and in this form find application in specialist areas such as on the tip of medical catheters. Although the inconvenience of a non-linear characteristic remains, this is normally overcome by processing the output signal with an active linearization circuit or incorporating the cell into a microprocessor-based intelligent measuring transducer. The latter usually provides analogue-to-digital conversion and interrupt facilities within a single chip and gives a digital output which is readily integrated into computer control schemes. Such instruments can also offer automatic temperature compensation, built in diagnostics, and simple calibration procedures. These features allow measurement inaccuracies as low as ± 0.1 percent of full-scale reading.

As an alternative to strain-gauge type displacement measurement, capacitive or inductive transducers are sometimes used. As a further option, developments in optical fibres have been exploited in the *fotonic sensor* (Figure 14.7), which is a diaphragm-type device in which the displacement is measured by optoelectronic means. In the device, light travels from a light source down an optical fibre, is reflected back from the diaphragm and travels back along a second fibre to a photodetector. There is a characteristic relationship between the light reflected and the distance from the fibre ends to the diaphragm, thus making the amount of reflected light dependent upon the measured pressure.

Normally, diaphragm-type instruments are used to measure mid-range gauge and differential pressures in the range up to 10 bar. However,

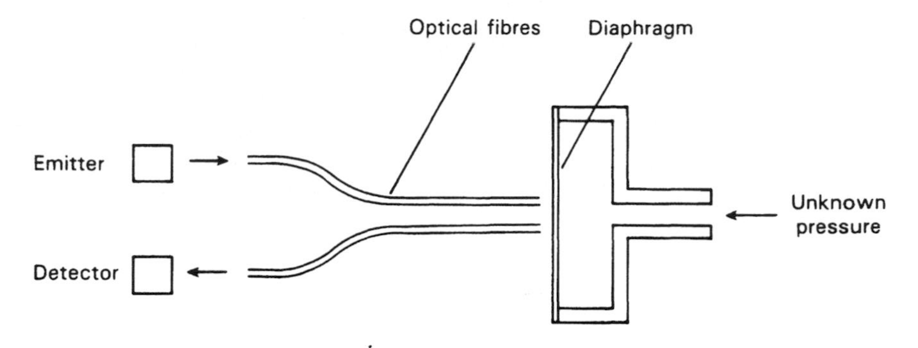

Figure 14.7. Fotonic sensor

special versions are available which can measure low pressures down to 0.001 mbar.

Whilst the discussion so far has been about measuring static pressures, diaphragm-type instruments can also be used to measure *dynamic pressures*. The principal displacement-measuring device used in such circumstances is the piezoelectric crystal. However, the strain gauge in its various forms can be used as an alternative, and there are particular advantages in its use in respect of its greater measurement sensitivity and also its ability to measure static pressures. A further alternative for measuring dynamic pressures is the diaphragm-based capacitive displacement transducer.

As diaphragms have mechanical components, their measurement characteristics can be affected by their operating environment and also by mishandling. As the magnitude of this effect is use-related, the necessary interval between recalibrations must be determined by practical experimentation.

Bellows

The bellows, illustrated in Figure 14.8, is another elastic-element type device. It operates on a very similar principle to the diaphragm, although its use is much less common. The sensitivity of a bellows is greater than that of a diaphragm, which is its principal attribute. Pressure changes within the

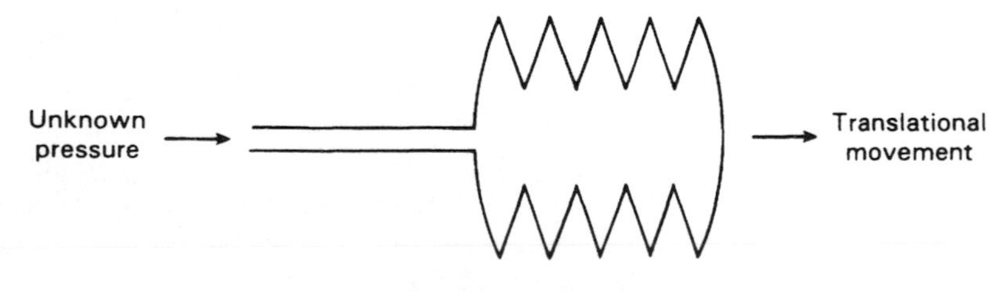

Figure 14.8. Bellows

bellows produce translational motion of the end of the bellows which can be measured by capacitive, inductive (LVDT) or potentiometric transducers according to the range of movement produced. A typical measuring range for a bellows-type instrument is 0–1 bar (gauge pressure). However, special versions are available which are designed to measure low pressures down to 0.1 mbar.

As already stated, bellows are very similar to diaphragms in their operational characteristics and their required calibration frequency must be determined in the same manner.

Bourdon Tube

The bourdon tube is the third type of elastic-element pressure transducer and is a very common industrial measuring instrument which is used for measuring the pressure of both gaseous and liquid fluids. It consists of a specially shaped piece of oval-section flexible tube which is fixed at one end and free to move at the other. When pressure is applied at the fixed end of the tube, the oval cross-section becomes more circular. As the cross-section of the tube tends towards a circular shape, a deflection of the closed, free end of the tube is caused. When used as a pressure-indicator, this motion is translated into the movement of a pointer against a scale. Alternatively, if an electrical output is required, the displacement is measured by some form of displacement transducer, which is commonly a potentiometer or LVDT (linear variable differential transformer), or less often a capacitive sensor. In yet another version, the displacement is measured optically.

The three common shapes of Bourdon tube are shown in Figure 14.9. The maximum possible deflection of the free end of the tube is proportional to the angle subtended by the arc through which the tube is bent. For a C-type tube, the maximum value for this arc is somewhat less than 360°. Where greater measurement sensitivity and resolution are required, spiral and helical tubes are used in which the possible magnitude of the arc subtended is limited only by a practical limit on how many turns it is convenient to have in the helix or spiral. However, this increased measurement performance is only gained at the expense of a substantial increase in manufacturing difficulty and cost compared with C-type tubes, and is also associated with a large decrease in the maximum pressure that can be measured.

C-type tubes are available for measuring pressures up to 6000 bar. A typical C-type tube of 25 mm radius has a maximum displacement travel of 4 mm, giving a moderate level of measurement resolution. Measurement inaccuracy is typically quoted as ±1 percent of full-scale deflection. Similar accuracy is available from helical and spiral types, but whilst the measurement resolution is higher, the maximum pressure measurable is only 700 bar. Special versions of the Bourdon tube are also available for measuring low pressures in the range down to 10 mbar.

Because both the Bourdon tube itself and the secondary transducer

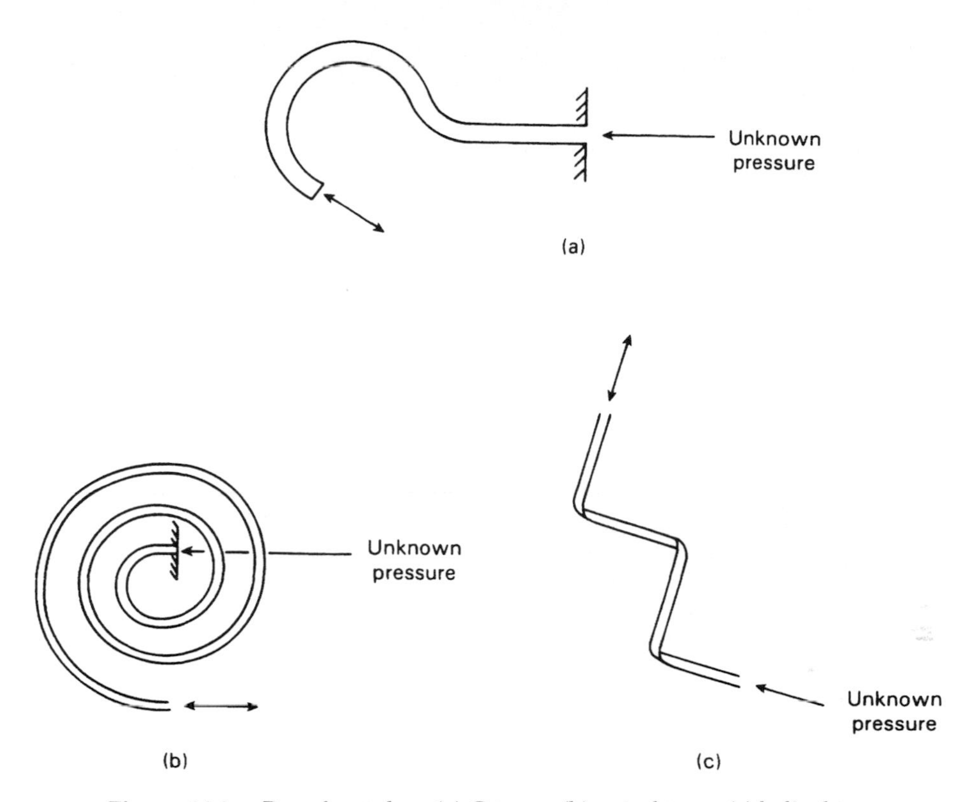

Figure 14.9. Bourdon tubes: (a) C-type ; (b) spiral type; (c) helical type

used to measure the deflection of its end are mechanical in nature, the amount of the change in characteristics with time is dependent upon the operational conditions prevailing. The recalibration frequency necessary must therefore be determined experimentally. Mention should also be made of the practical difficulties of calibrating Bourdon tubes that are to be used for measuring liquid pressures, as discussed more fully later in Section 14.5. In consequence of this, the accuracy limits quoted earlier are only guaranteed when gaseous pressures are being measured. It is possible to achieve these same levels of accuracy when measuring liquid pressures, but only if the gauge can be totally filled with liquid during both calibration and measurement, a condition which is very difficult to fulfil practically.

Resonant Wire Devices

The resonant wire device is a relatively new instrument which has emanated from recent advances in the electronics field. A typical device is shown schematically in Figure 14.10. Wire is stretched across a chamber containing fluid at unknown pressure subjected to a magnetic field. The

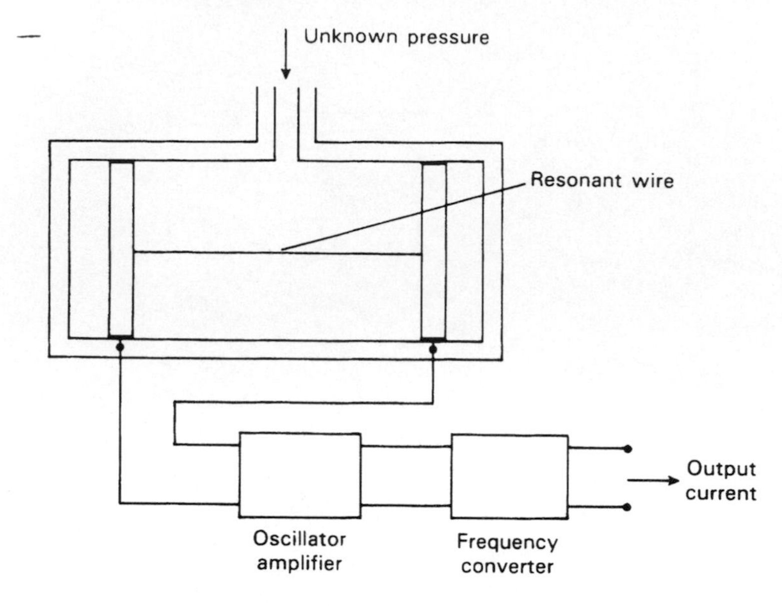

Figure 14.10. Resonant-wire device

wire resonates at its natural frequency according to its tension, which varies with pressure. Thus pressure is calculated by measuring the frequency of vibration of the wire using electronics integrated into the cell. Such devices are highly accurate, with an inaccuracy of ± 0.2 percent of full-scale reading being typical, and they are particularly insensitive to ambient condition changes.

Because of their newness, little practical experience has been gained of the long-term operation of these devices. Comment on the required recalibration frequency must therefore be a little guarded. However, there is little within the instrument which seems likely to suffer characteristics drift and therefore the initial recommendation would be to carry out calibration checks annually.

Thermocouple Gauge

The thermocouple gauge is one of the group of gauges working on the thermal conductivity principle. The Pirani and Thermistor gauges also belong to this group. At low pressure, the kinematic theory of gases predicts a linear relationship between pressure and thermal conductivity. Thus measurement of thermal conductivity gives an indication of pressure.

Figure 14.11 shows a sketch of a thermocouple gauge. Operation of the gauge depends on the thermal conduction of heat between a thin, hot metal strip in the centre and the cold outer surface of a glass tube (which is normally at room temperature). The metal strip is heated by passing a current through it and its temperature is measured by a thermocouple. The

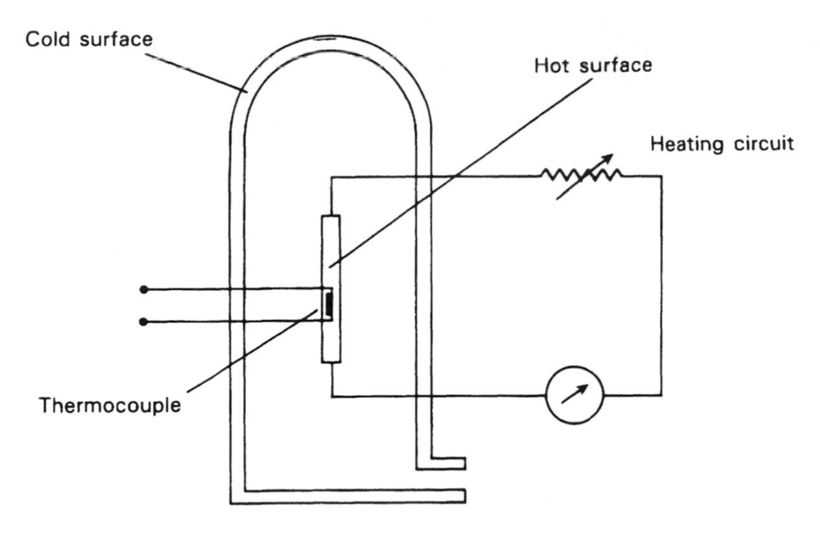

Figure 14.11. Thermocouple gauge

temperature measured depends on the thermal conductivity of the gas in the tube and hence on its pressure. A source of error in this instrument is the fact that heat is also transferred by radiation as well as conduction. This is of a constant magnitude, independent of pressure, and so can be measured and corrected for. However, it is usually more convenient to design for low radiation loss by choosing a heated element with low emissivity. Thermocouple gauges are typically used to measure pressures in the range 10^{-4} mbar up to 1 mbar.

The characteristics of the gauge vary with the nature of the gas whose pressure is being measured. Therefore, it must be calibrated separately for every application.

Pirani Gauge

A typical form of Pirani gauge is shown in Figure 14.12. This is similar to a thermocouple gauge but has a heated element which consists of four coiled tungsten wires connected in parallel. Two identical tubes are normally used, connected in a bridge circuit as shown in Figure 14.13, with one containing the gas at unknown pressure and the other evacuated to a very low pressure. Current is passed through the tungsten element, which attains a certain temperature according to the thermal conductivity of the gas. The resistance of the element changes with temperature and causes an imbalance of the measurement bridge. Thus the Pirani gauge avoids the use of a thermocouple to measure temperature (as in the thermocouple gauge) by effectively using a resistance thermometer as the heated element. Such gauges cover the pressure range 10^{-5} mbar to 1 mbar.

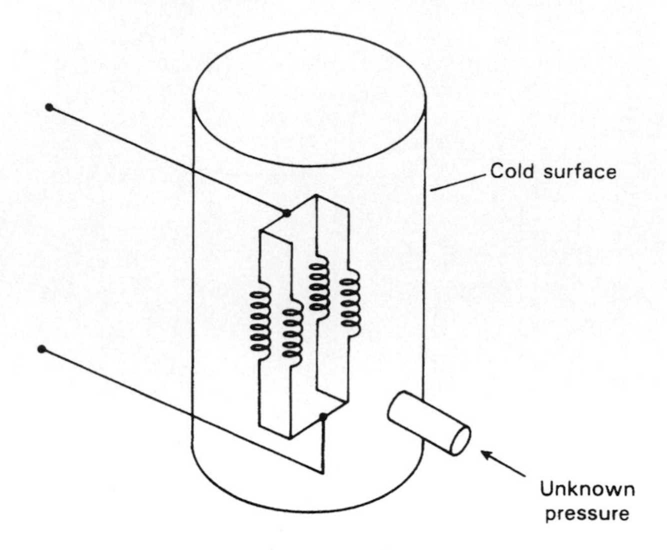

Figure 14.12. Pirani gauge

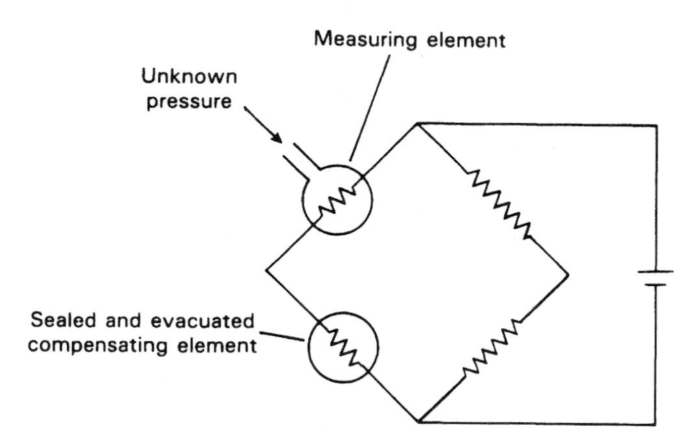

Figure 14.13. Wheatstone bridge circuit to measure output of Pirani gauge

As with the thermocouple gauge, the Pirani gauge must be calibrated separately for each different application.

Thermistor Gauge

Thermistor gauges operate on identical principles to the Pirani gauge but use semiconductor materials for the heated elements instead of metals. The normal pressure range covered is 10^{-4} mbar to 1 mbar. Again, the gauge must be calibrated separately for each different application.

Ionization Gauge

The ionization gauge is a special type of instrument used for measuring very low pressures in the range 10^{-13} to 10^{-3} bar. Gas of unknown pressure is introduced into a glass vessel containing free electrons discharged from a heated filament, as shown in Figure 14.14. Gas pressure is determined by measuring the current flowing in an anode–cathode system within the vessel. This current is proportional to the number of ions per unit volume, which in turn is proportional to the gas pressure. Measurement uncertainty varies between ±1 percent at the top end of the measurement range to ±4 percent at the lower end. No characteristics drift is anticipated in the instrument but calibration checks should be carried out annually nevertheless.

High-Pressure-Measuring Devices

The normal instrument used for measuring high pressures consists of a coil of manganin wire enclosed in a sealed, paraffin-oil filled, flexible bellows, as shown in Figure 14.15 . The unknown pressure is applied to one end of the bellows which transmits the pressure to the coil. The resistance is linearly proportional to pressure and therefore the magnitude of the applied pressure can be determined by measuring the coil resistance. Typical measurement inaccuracy is ±0.5 percent.

Wire made from a gold–chrome alloy is sometimes used in place of manganin wire. This has a lower temperature coefficient but also a lower pressure sensitivity.

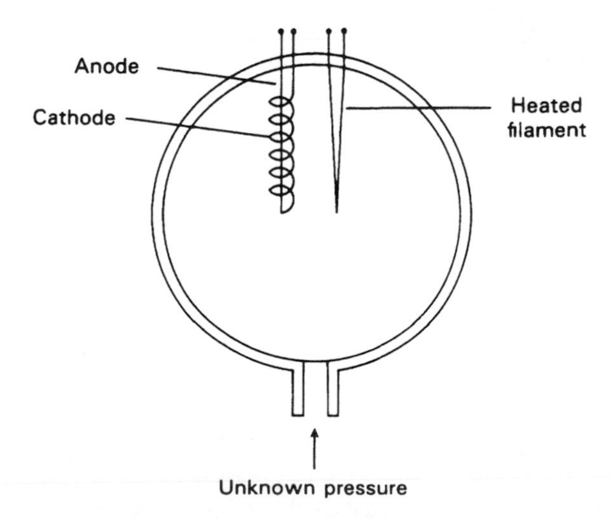

Figure 14.14. Ionization gauge

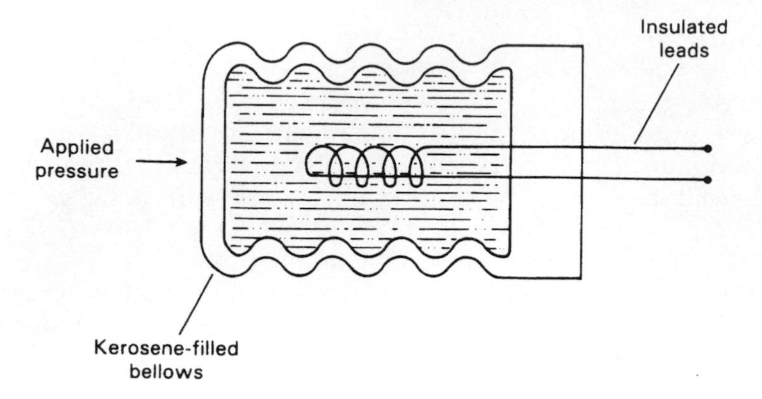

Figure 14.15. High-pressure measurement – wire coil in bellows

Intelligent Pressure Transducers

Adding microprocessor power to pressure transducers brings about substantial improvements in their characteristics. Measurement sensitivity improvement, extended measurement range, compensation for hysteresis and other non-linearities, and correction for ambient temperature and pressure changes are just some of the facilities offered by intelligent pressure transducers. Inaccuracies of only ± 0.1 percent can be achieved with piezoresistive-bridge silicon devices for instance.

Some recent microprocessor-based pressure transducers make use of novel techniques of displacement measurement. For example, both diaphragm and helical Bourdon tube devices are now available which use an optical method of displacement measurement of the form shown in Figure 14.16. In this, the motion is transmitted to a vane which progressively shades one of two monolithic photodiodes that are exposed to infra-red radiation. The second photodiode acts as a reference, enabling the microprocessor to compute a ratio signal which is linearized and is available as either an analogue or digital measurement of pressure. The typical measurement inaccuracy is ± 0.1 percent.

14.2 INTRODUCTION TO CALIBRATION

Pressure is a quantity which is derived from the fundamental quantities of force and area, and is usually measured in terms of the force acting on a known area. In the mid-range of pressures from 0.1 mbar to 20 bar, U-tube manometers, dead-weight gauges and barometers are used to calibrate pressure-measuring instruments. The recently developed vibrating cylinder gauge also provides a very accurate reference standard over part of this range. Above 20 bar, a gold-chrome alloy resistance instrument is normally used. For low pressures in the range of 10^{-1} to 10^{-3} mbar, both the McLeod gauge and various forms of micromanometer are used as a pressure-

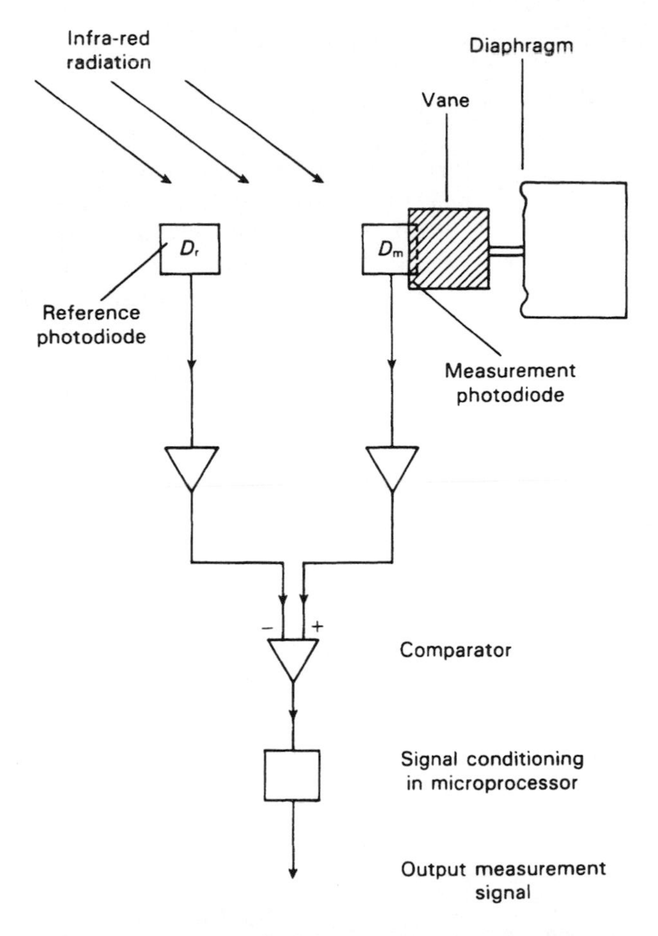

Figure 14.16. One type of intelligent pressure-measuring instrument

measuring standard. At even lower pressures below 10^{-3} mbar, a pressure-dividing technique is used to establish calibration. This involves setting up a series of orifices of accurately known pressure ratio and measuring the upstream pressure with a McLeod gauge or micromanometer.

The limits of accuracy with which pressure can be measured by presently known techniques are as follows:

10^{-7} mbar $\pm$ 4%

10^{-5} mbar $\pm$ 2%

10^{-3} mbar $\pm$ 1%

10^{-1} mbar $\pm$ 0.1%

1 bar $\pm$ 0.001%

10^{4} bar $\pm$ 0.1%

14.3 PRESSURE CALIBRATION INSTRUMENTS

Dead-weight Gauge (Pressure Balance)

The dead-weight gauge or pressure balance, as shown in Figure 14.17, is a null-reading type of measuring instrument in which weights are added to the piston platform until the piston is adjacent to a fixed reference mark, at which time the downward force of the weights on top of the piston is balanced by the pressure exerted by the fluid beneath the piston. The fluid pressure is therefore calculated in terms of the weight added to the platform and the known area of the piston. The device is tedious to use but finds wide application as a reference instrument against which other pressure-measuring instruments are calibrated in the mid-range of pressures.

Special precautions are necessary in the manufacture and use of dead-weight gauges. Friction between the piston and cylinder must be reduced to a very low level, otherwise a significant measurement error will result. Friction reduction is accomplished by designing for a small clearance gap between the piston and cylinder, by machining the cylinder to a slightly greater diameter than the piston. The piston and cylinder are also designed so that they can be turned relative to one another which reduces friction still further. Unfortunately, as a result of the small gap between the piston and cylinder, there is a finite flow of fluid past the seals. This produces a viscous shear force which partly balances the dead-weight on the platform. A theoretical formula exists for calculating the magnitude of this shear force, suggesting that exact correction can be made for it. In practice, however, the piston deforms under pressure and alters the piston/cylinder gap, and so the shear force calculation and correction can only be approximate.

In spite of these difficulties, the instrument gives a typical measurement inaccuracy of only ±0.01 percent. It is normally used for calibrating pressures in the range of 20 mbar up to 20 bar. However, special versions can measure pressures down to 0.1 mbar or up to 7000 bar.

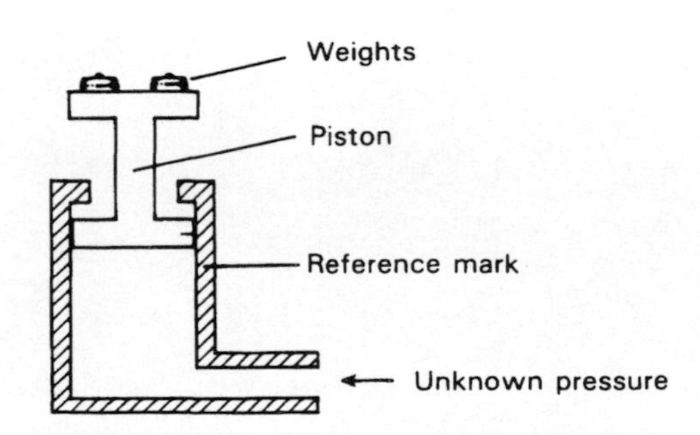

Figure 14.17. Dead-weight gauge

U-tube Manometer

Besides its use for normal process measurements, the U-tube manometer is also used as a reference instrument for calibrating instruments measuring mid-range pressures. Although it is a deflection rather than null type of instrument, it manages to achieve similar degrees of measurement accuracy to the dead-weight gauge, because of the error sources noted in the latter. The major source of error in U-tube manometers arises out of the difficulty in estimating the meniscus level of the liquid column accurately. There is also a tendency for the liquid level to creep up the tube by capillary action, which creates an additional source of error.

U-tubes for measuring high pressures become unwieldy because of the long lengths of liquid column and tube required. Consequently, U-tube manometers are normally used only for calibrating pressures at the lower end of the mid-pressure range.

Barometers

The most commonly used type of barometer for calibration duties is the Fortin barometer. This is a highly accurate instrument which provides measurement inaccuracy levels of between ±0.03 percent of full-scale reading and ±0.001 percent of full-scale reading depending on the measurement range. To achieve such levels of accuracy, the instrument has to be used under very carefully controlled conditions of lighting, temperature and vertical alignment. It must also be manufactured to exacting standards and is therefore very expensive to buy. Corrections have to be made to the output reading according to the ambient temperature, local value of gravity and atmospheric pressure. Because of its expense and the difficulties in using it, the barometer is not normally used for calibration other than as a primary reference standard at the top of the calibration chain.

Vibrating Cylinder Gauge

The vibrating cylinder gauge, shown in Figure 14.18, acts as a reference standard instrument for calibrating pressure measurements up to 3.5 bar. It consists of a cylinder in which vibrations at the resonant frequency are excited by a current-carrying coil. The pressure-dependent oscillation frequency is monitored by a pick-up coil, and this frequency measurement is converted to a voltage signal by a microprocessor and signal conditioning circuitry contained within the package. By evacuating the space on the outer side of the cylinder, the instrument is able to measure the absolute pressure of the fluid inside the cylinder. Measurement errors are less than 0.005 percent over the absolute pressure range up to 3.5 bar.

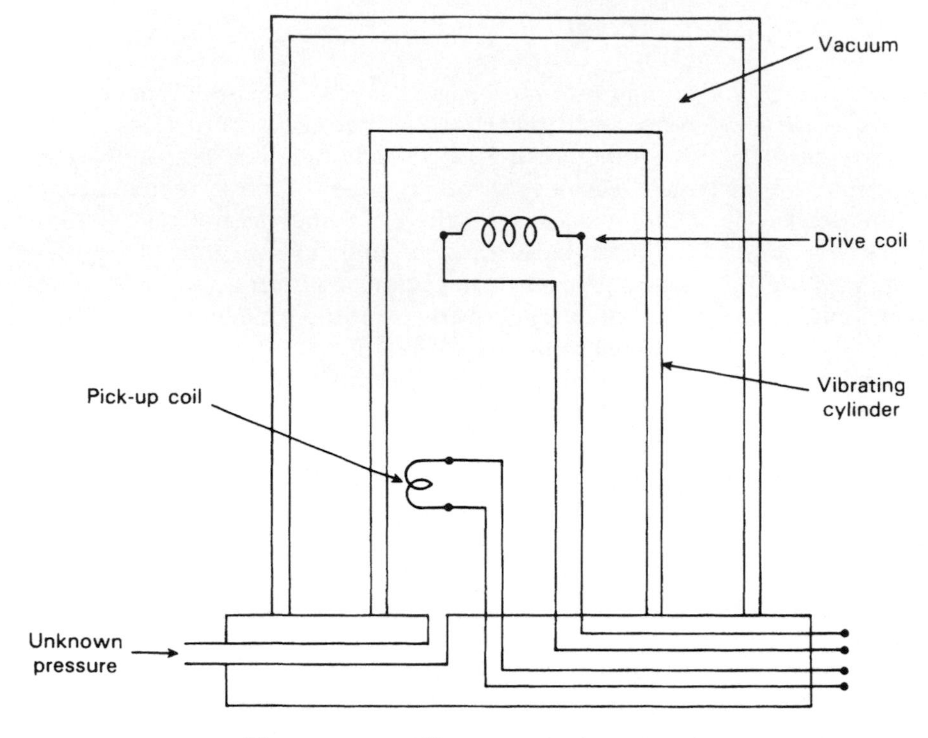

Figure 14.18. Vibrating cylinder gauge

Gold-chrome Alloy Resistance Instruments

For measuring pressures above 7000 bar, an instrument based on measuring the resistance change of a metal coil as the pressure varies is used, and the same type of instrument is also used for calibration purposes. As stated in Section 14.1, such instruments can either use manganin or gold-chrome alloys for the coil. Gold–chrome has a significantly lower temperature coefficient (i.e. its pressure–resistance characteristic is less affected by temperature changes), and is therefore the normal choice for calibration instruments in spite of its higher cost. An inaccuracy of only ± 0.1 percent is achievable in such devices.

McLeod Gauge

The McLeod gauge, shown in Figure 14.19, is used for the calibration of instruments that are designed to measure pressures below 0.1 mbar. Its principle of operation is to compress low resistance fluid to a higher pressure which can be measured by manometer techniques. In essence, the gauge can be visualized as a U-tube manometer containing mercury, sealed at one end, and where the bottom of the U can be blocked at will. To operate

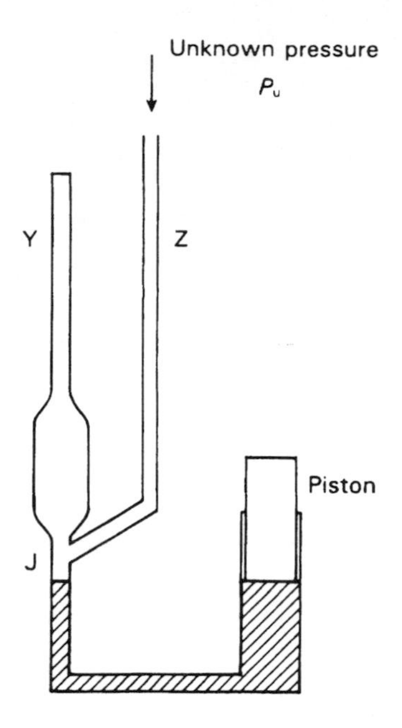

Figure 14.19. McLeod gauge

the gauge, the piston is first withdrawn, causing the level of mercury in the lower part of the gauge to fall below the level of the junction J between the two tubes in the gauge marked Y and Z. Fluid at unknown pressure P_u is then introduced via the tube marked Z, from where it also flows into the tube marked Y, of cross-sectional area A. Next, the piston is pushed in, moving the mercury level up to block the junction J. At the stage where J is just blocked, the fluid in tube Y is at pressure P_u and contained in a known volume V_u. Further movement of the piston compresses the fluid in tube Y and this process continues until the mercury level in tube Z reaches a zero mark. Measurement of the height (h) above the mercury column in tube Y then allows calculation of the unknown pressure according to

$$P_u = \frac{Ah^2 d}{V_u - Ah} \tag{14.1}$$

where d is the mass density of mercury.

The compressed volume V_c is often very much smaller than the original volume, in which case equation (14.1) approximates to:

$$P_u = \frac{Ah^2 dg}{V_u} \text{[for } Ah \ll V_u\text{]} \tag{14.2}$$

Although the lowest measurement uncertainty achievable with McLeod gauges is ±1 percent, this is still better than other gauges can achieve in the low-pressure range. The McLeod gauge can measure low pressures directly down to 10^{-4} mbar. To measure lower pressures, it is necessary to use pressure-dividing techniques as described in Section 14.2.

Ionization Gauge

The ionization gauge can calibrate instruments measuring low pressures across their whole range, and is therefore preferable to the McLeod gauge which can only calibrate instruments at specific points within their measuring range. The ionization gauge has a straight-line relationship between output reading and pressure. Unfortunately, its inherent accuracy is inadequate and it is only usable as a reference instrument if specific points on its output characteristic are calibrated against a McLeod gauge.

Micromanometers

Micromanometers are instruments which work on the manometer principle but are specially designed to minimize capillary effects and meniscus reading errors. The type of micromanometer which is most accurate as a calibration standard down to pressures of 10^{-3} mbar is the centrifugal micromanometer shown schematically in Figure 14.20. In this, a rotating disc serves to amplify a reference pressure, with the speed of rotation being adjusted until the amplified pressure just balances the unknown pressure. This null position is detected by observing when oil droplets sprayed into a glass chamber cease to move. Measurement accuracy is ±1 percent.

Other types of micromanometer also exist [1], which give similar levels of

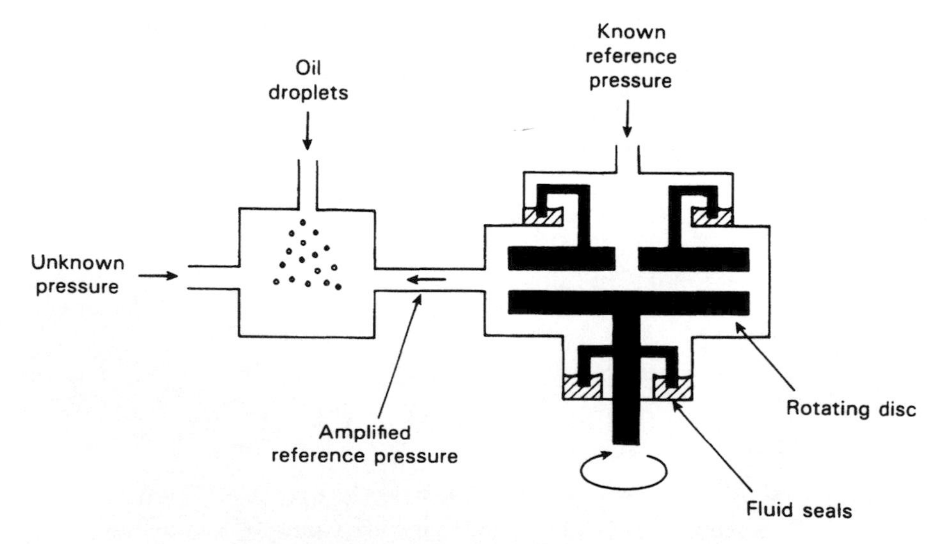

Figure 14.20. Centrifugal micromanometer

accuracy, but only at somewhat higher pressure levels. These can be used as calibration standards at pressures up to 50 mbar.

14.4 CALIBRATION FREQUENCY

Pressure-measuring instruments can be divided into three distinct classes in respect of the required frequency of calibration. The first class includes resonant wire devices, ionization gauges and high-pressure measuring instruments (those working on the principle of resistance change with pressure). These are all very stable and unlikely to suffer from drift in characteristics with time. All forms of manometer are similarly stable, although in the longer term small errors can develop in these through volumetric changes in the glass. Therefore, only annual calibration checks are recommended for this class of instruments, unless of course something happens to the instrument which puts its calibration into question.

The second class includes devices like diaphragms, bellows and Bourdon tubes. All of these contain an elastic element and displacement transducer that measures movement of the element. Both the elastic element and displacement transducer are mechanical in nature, and such instruments can suffer changes in characteristics for a number of reasons. One factor is the characteristics of the operating environment and the degree to which the instrument is exposed to it. Another reason is the amount of mishandling it receives. These parameters are entirely dependent upon the particular application the instrument is used in, and the frequency with which it is used and exposed to the operating environment. A suitable calibration frequency can therefore only be determined on an experimental basis.

The third class of instrument from the calibration-requirements viewpoint is the range of devices working on the thermal conductivity principle, such as the thermocouple gauge, Pirani gauge and thermistor gauge. These instruments have characteristics which vary with the nature of the gas being measured, and must therefore be calibrated each time that they are used.

14.5 CALIBRATION PROCEDURES

Pressure calibration requires the output reading of the instrument being calibrated to be compared with the output reading of a reference standard instrument when the same pressure is applied to both. This necessitates designing a suitable leakproof seal to connect the pressure-measuring chambers of the two instruments.

The calibration of pressure transducers used for process measurements often has to be carried out *in-situ* in order to avoid serious production delays. Such devices are often remote from the nearest calibration laboratory and to transport them there for calibration would take an unacceptably long time. Because of this, portable reference instruments have been developed for calibration at this level in the calibration chain. These use a standard air supply connected to an accurate pressure regulator to provide a range of reference pressures. An inaccuracy of ±0.025 percent is achieved when calibrating mid-range pressures in this manner. Calibration at higher levels in the calibration chain must of course be carried out in a proper calibration laboratory maintained in the correct manner. Irrespective of where calibration is carried out, however, several special precautions are necessary when using the various instruments described in Section 14.3.

U-tube manometers must have their vertical alignment carefully set up before use. Particular care must also be taken to ensure that there are no temperature gradients between the two halves of the tube. Such temperature differences would cause local variations in the specific weight of the manometer fluid, resulting in measurement errors. Correction must also be made for the local value of g (acceleration due to gravity). These comments apply similarly to the use of other types of manometer and micromanometer.

The existence of one potentially major source of error in *Bourdon tube* pressure measurement has not been widely documented and few manufacturers of Bourdon tubes make any attempt to warn users of their products of it. The problem is concerned with the relationship between the fluid being measured and the fluid used for calibration. The pointer of Bourdon tubes is normally set at zero during manufacture, using air as the calibration medium. If however a different fluid, especially a liquid, is subsequently used in a Bourdon tube, the fluid in the tube will cause a non-zero deflection according to its weight compared with air, resulting in a reading error of up to 6 percent of full-scale deflection.

This can be avoided by calibrating the Bourdon tube with the fluid to be measured instead of with air. Alternatively, correction can be made according to the calculated weight of the fluid in the tube. Unfortunately, difficulties arise with both of these solutions if air is trapped in the tube, since this will prevent the tube being filled completely by the fluid. Then, the amount of fluid actually in the tube, and its weight, will be unknown. To avoid this problem, at least one manufacturer now provides a bleed facility in the tube which allows measurement uncertainties of less than 0.1 percent to be achieved.

When using a *McLeod gauge*, care must be taken to ensure that the measured gas does not contain any vapour. This would be condensed during the compression process, causing a measurement error. A further recommendation is the insertion of a liquid–air cold trap between the gauge and the instrument being calibrated to prevent the passage of mercury vapour into the latter.

REFERENCES

1. Benedict, R. P., *Fundamentals of temperature, pressure and flow measurement*, John Wiley & Sons (1984).

CHAPTER 15

Mass, Force and Torque Calibration

15.1 REVIEW OF MASS MEASUREMENT

Mass describes the quantity of matter which a body contains. One way in which it can be measured is to compare the gravitational force on the body with the gravitational force on another body of known mass, using a beam balance, weigh beam, pendulum scale or electromagnetic balance, in a procedure known as 'weighing'. Alternative mass-measurement techniques are to use either a spring balance or a load-cell. Of these, the electronic load-cell has definite advantages and is the preferred instrument nowadays in more and more industrial applications.

Beam Balance (Equal-arm Balance)

In the beam balance, shown in Figure 15.1, standard masses are added to a pan on one side of a pivoted beam until the magnitude of the gravitational force on them balances the magnitude of the gravitational force on the unknown mass acting at the other end of the beam. This equilibrium position is indicated by a pointer which moves against a calibrated scale.

Instruments of this type are capable of measuring a wide span of masses. Those at the top and of the range can typically measure masses up to 1000 g whereas those at the bottom end of the range can measure masses of less than 0.01 g. Measurement resolution can be as good as 1 part in 10^7 of the full-scale reading if the instrument is designed and manufactured very carefully. The lowest measurement inaccuracy figure attainable is ± 0.002 percent.

One serious disadvantage of this type of instrument is its lack of ruggedness. Continuous use and the inevitable shock loading which will occur from time to time both cause damage to the knife-edges, leading to problems in measurement accuracy and resolution, as discussed in Section

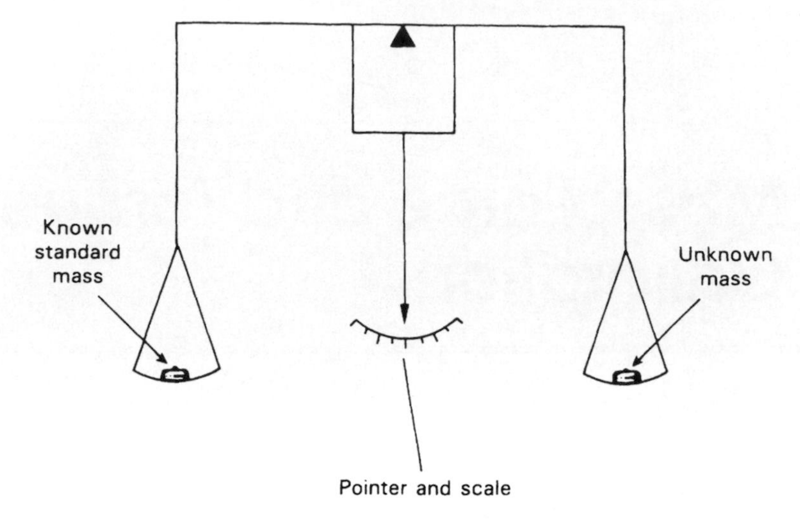

Figure 15.1. Beam balance

15.5. A further problem in industrial use is the relatively long time needed to make each measurement. For these reasons, the beam balance is normally reserved as a calibration standard and is not used in day-to-day production environments.

Weigh Beam

The weigh beam, sketched in two alternative forms in Figure 15.2, operates on similar principles to the beam balance but is much more rugged. In the first form, standard masses are added to balance the unknown mass and fine adjustment is provided by a known mass which is moved along a notched, graduated bar until the pointer is brought to the null, balance point. The alternative form has two or more graduated bars (three bars shown in Figure 15.2). Each bar carries a different standard mass and these are moved to appropriate positions on the notched bar to balance the unknown mass. Versions of these instruments are used to measure masses up to 50 tonne.

Pendulum Scale

The pendulum scale, sketched in Figure 15.3, is another instrument that works on the mass-balance principle. The unknown mass is put on a platform which is attached by steel tapes to a pair of cams. Downward motion of the platform, and hence rotation of the cams, under the influence of the gravitational force on the mass, is opposed by the gravitational force acting on two pendulum-type masses attached to the cams. The amount of rotation of the cams when the equilibrium position is reached is determined

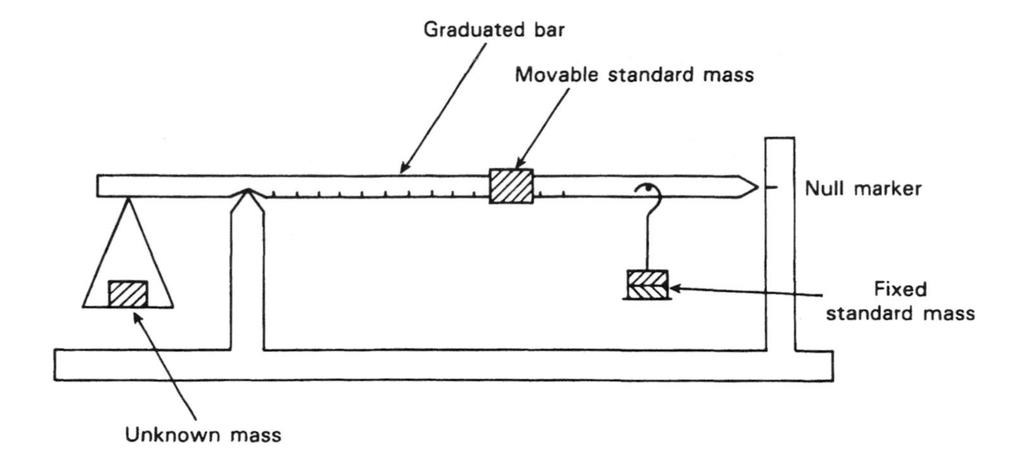

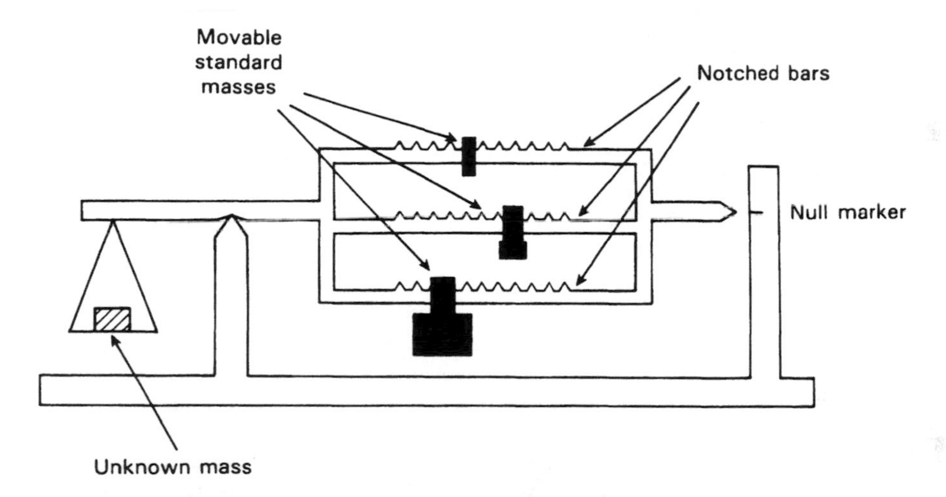

Figure 15.2. Two alternative forms of weigh beam

by the deflection of a pointer against a scale. The shape of the cams is such that this output deflection is linearly proportional to the applied mass.

This instrument is particularly useful in some applications because it is a relatively simple matter to replace the pointer and scale system by a rotational displacement transducer which gives an electrical output. Various versions of the instrument can measure masses in the range between 1 kg and 500 t, with a typical measurement inaccuracy of ±0.1 percent.

One potential source of difficulty with the instrument is oscillation of the weigh platform when the mass is applied. Where necessary, in instruments measuring larger masses, dashpots are incorporated into the cam system to damp out such oscillations. A further possible problem can arise, mainly when measuring large masses, if the mass is not placed centrally on the platform. This can be avoided by designing a second platform to hold the mass which is hung from the first platform by knife-edges. This lessens the criticality of mass placement.

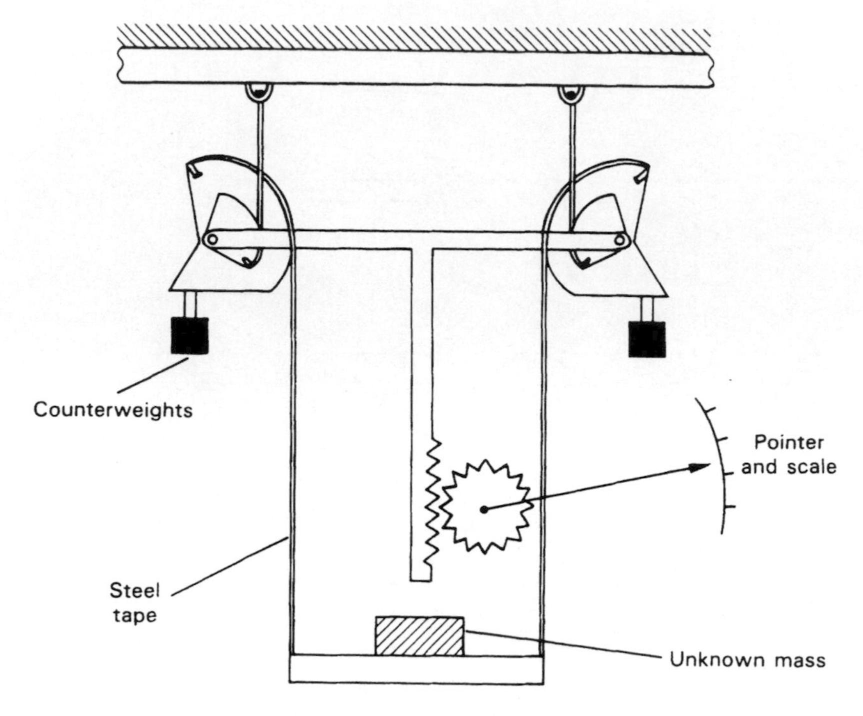

Figure 15.3. Pendulum scale

Electromagnetic Balance

The electromagnetic balance uses the torque developed by a current-carrying coil suspended in a permanent magnetic field to balance the unknown mass against the known gravitational force produced on a standard mass, as shown in Figure 15.4. A light source and detector system is used to determine the null balance point. The voltage output from the light detector is amplified and applied to the coil, thus creating a servosystem in which the deflection of the coil in equilibrium is proportional to the applied force. Its advantages over beam balances, weigh beams and pendulum scales include its smaller size, its insensitivity to environmental changes and the electrical form of its output.

Spring Balance

The spring balance is a well-known instrument for measuring mass. The mass is hung on the end of a spring and the deflection of the spring due to the downwards gravitational force on the mass is measured against a scale. Because the characteristics of the spring are very susceptible to environmental changes, measurement accuracy is usually relatively poor. If,

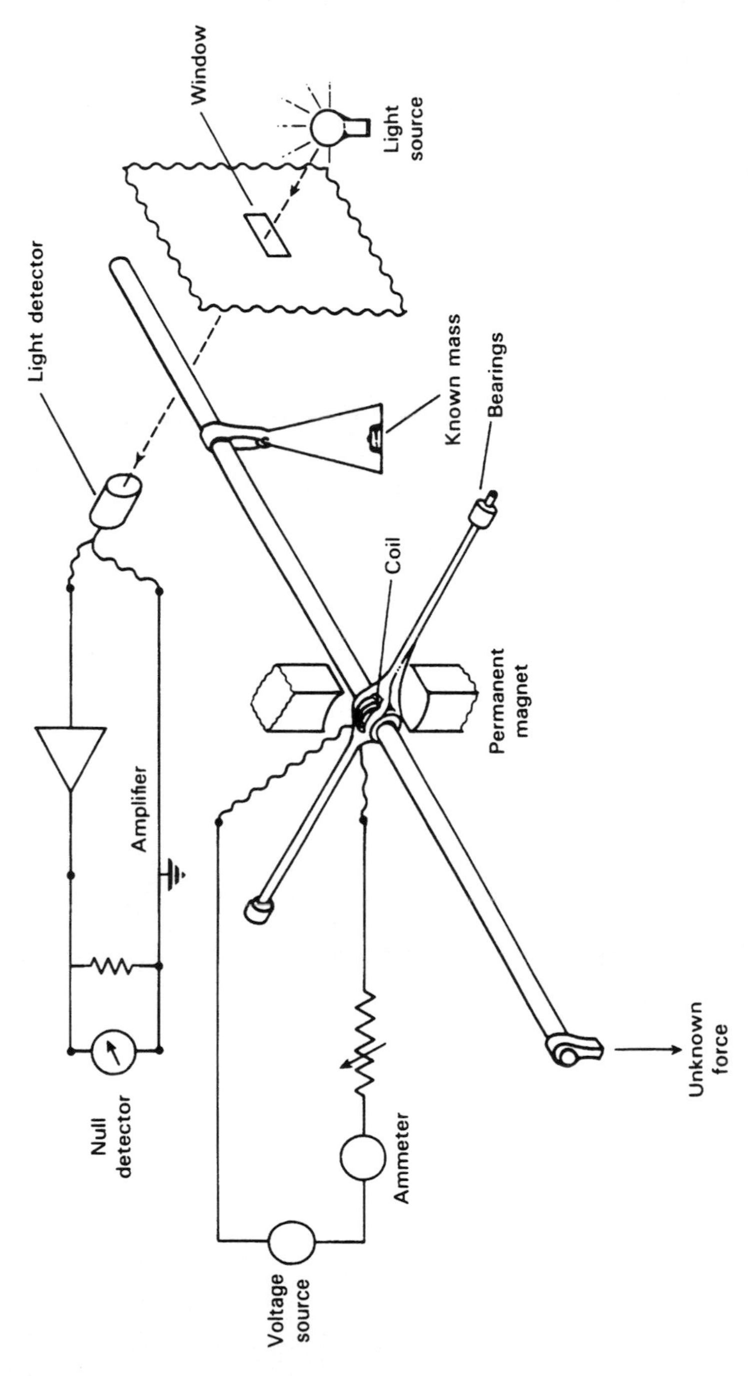

Figure 15.4. Electromagnetic balance

however, compensation is made for the changes in spring characteristics, then a measurement inaccuracy less than ±0.2 percent is achievable. According to the design of the instrument, masses between 0.5 kg and 10 t can be measured.

Electronic Load-cell (Electronic Balance)

Electronic load-cells, also known as electronic balances, have significant advantages over most other forms of mass-measuring instrument and so are the preferred type for most industrial applications. These advantages include relatively low cost, wide measurement range, tolerance of dusty and corrosive environments, remote measurement capability, tolerance of shock loading and ease of installation.

The electronic load cell uses the physical principle that a force applied to an elastic element produces a measurable deflection. The elastic elements used are specially shaped and designed, some examples are shown in

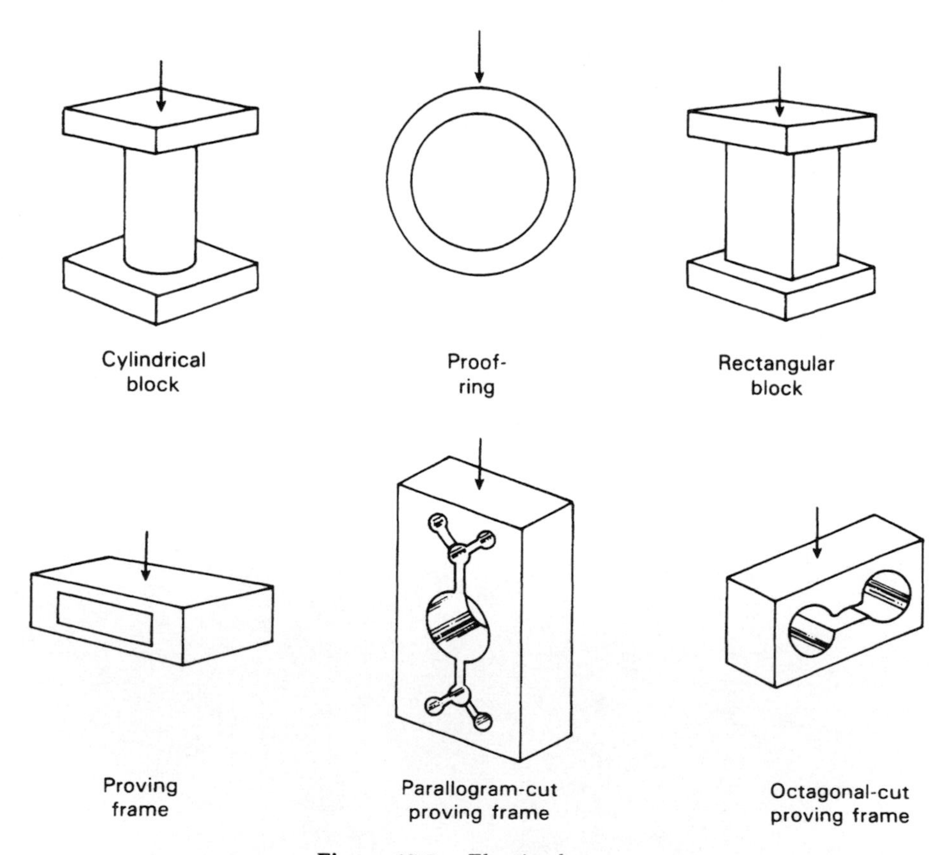

Figure 15.5. Elastic elements

Figure 15.5. The design aims are to obtain a linear output relationship between the applied force and the measured deflection and to make the instrument insensitive to forces which are not applied directly along the sensing axis. Various types of displacement transducer are used to measure the deflection of the elastic elements.

One problem that can affect the performance of this class of instrument is the phenomenon of creep. Creep describes the permanent deformation that an elastic element undergoes after it has been under load for a period of time. This can lead to significant measurement errors in the form of a bias on all readings if the instrument is not recalibrated from time to time. However, careful design and choice of materials can largely eliminate the problem.

The best accuracy in elastic force transducers is obtained by those instruments which use strain gauges to measure displacements of the elastic element, with an inaccuracy figure less than ±0.05 percent of full-scale reading being obtainable. Instruments of this type are used to measure masses over a very wide range between 0 and 3.106 kg. The measurement capability of an individual instrument designed to measure masses at the bottom end of this range would typically be 0.1–5 kg, whereas instruments designed for the top of the range would have a typical measurement span of 10–3000 t.

Elastic force transducers based on differential transformers (LVDT's) to measure defections are used to measure masses up to 25 t. Apart from having a lower maximum measuring capability, they are also inferior to strain-gauge-based instruments in terms of their ±0.2 percent inaccuracy figure. Their major advantage is their longevity and almost total lack of maintenance requirements.

The final type of displacement transducer used in this class of instrument is the piezoelectric device. Such instruments are used to measure masses in the range of 0 to 106 kg. Piezoelectric crystals replace the specially designed elastic member normally used in this class of instrument, allowing the device to be physically small. As discussed previously, such devices can only measure dynamically changing forces because the output reading results from an induced electrical charge whose magnitude leaks away with time. The fact that the elastic element consists of the piezoelectric crystal means that it is very difficult to design such instruments to be insensitive to forces applied at an angle to the sensing axis. Therefore, special precautions have to be taken in applying these devices. Although such instruments are relatively cheap, their lowest inaccuracy is ±1 percent of the full-scale reading, and they also have a high temperature coefficient.

All types of load-cell-based electronic balances normally contain several load-cells, as illustrated in Figure 15.6. Such arrangements are dictated by the need for stability in the mechanical construction. Commonly, either three or four load-cells are used in the balance, with the output mass measurement being formed from the sum of the outputs of each cell. Where appropriate, the upper platform can be replaced by a tank for weighing liquids, powders etc.

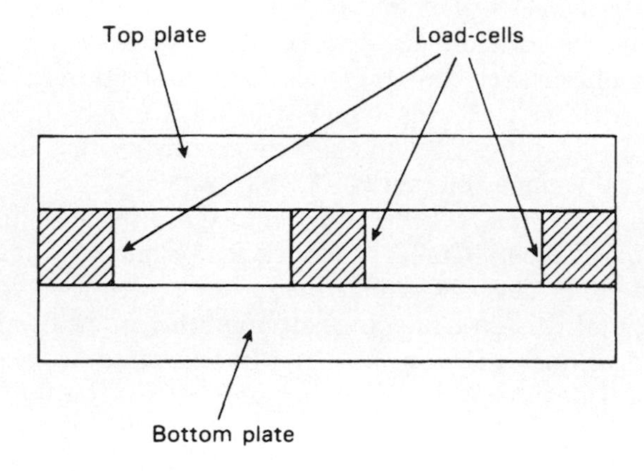

Figure 15.6. Load-cell-based electronic balance

Pneumatic/Hydraulic Load-cells

Alternative forms of load cell also exist which work on either pneumatic or hydraulic principles and convert mass measurement into a pressure measurement problem. A pneumatic load cell is shown schematically in Figure 15.7. Application of a mass to the cell causes deflection of a diaphragm acting as a variable restriction in a nozzle–flapper mechanism. The output pressure measured in the cell is approximately proportional to the magnitude of the gravitational force on the applied mass. The instrument requires a flow of air at its input of around 0.25 m^3/h at a pressure of 4 bar. Standard cells are available to measure a wide range of masses. For measuring small masses, instruments are available with a full-scale reading of 25 kg, whilst at the top of the range, instruments with a full-scale reading of 25 t are obtainable. Inaccuracy is typically ±0.5 percent of full scale in pneumatic load-cells.

The alternative, hydraulic load-cell is shown in Figure 15.8 . In this, the gravitational force due to the unknown mass is applied, via a diaphragm, to oil contained within an enclosed chamber. The corresponding increase in oil

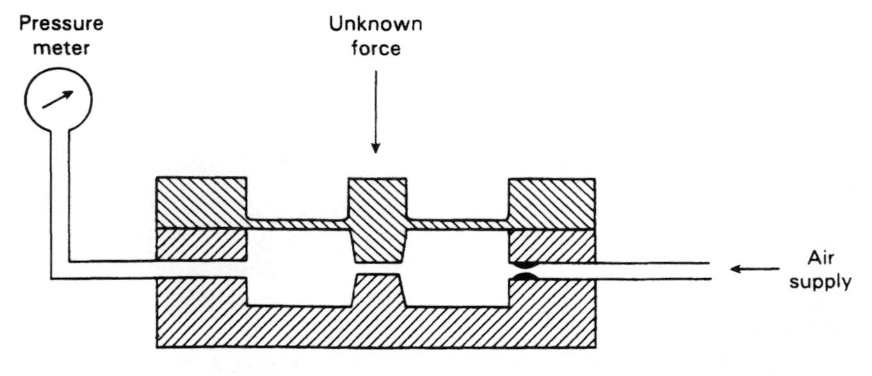

Figure 15.7. Pneumatic load-cell

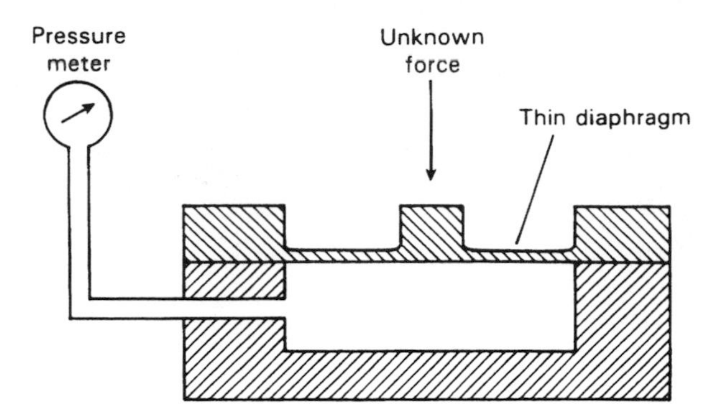

Figure 15.8. Hydraulic load-cell

pressure is measured by a suitable pressure transducer. These instruments are designed for measuring much larger masses than pneumatic cells, with a load capacity of 500 t being common. Special units can be obtained to measure masses as large as 50 000 t. Besides their much greater measuring range, hydraulic load-cells are much more accurate than pneumatic cells, with an inaccuracy figure of ±0.05 percent of full scale being typical, although to obtain this level of accuracy, correction for the local value of g (acceleration due to gravity) is necessary. A measurement resolution of ±0.02 percent is attainable.

Intelligent Load-cells

Intelligent load-cells are formed by adding a microprocessor to a standard cell. This brings no improvement in accuracy because the load-cell is already a very accurate device. What it does produce is an intelligent weighing system which can compute total cost from the measured weight, using stored cost-per-unit-weight information, and provide an output in the form of a digital display. Cost-per-weight figures can be prestored for a large number of substances, making the instrument very flexible in its application.

15.2 REVIEW OF FORCE MEASUREMENT

If a force of magnitude, F, is applied to a body of mass, M, the body will accelerate at a rate, A, according to the equation:

$$F = M \cdot A$$

The standard unit of force is the newton, this being the force which will produce an acceleration of one metre per second squared in the direction of the force when it is applied to a mass of one kilogram. One way of measuring an unknown force is therefore to measure the acceleration when it is applied to a body of known mass.

An alternative technique is to measure the variation in the resonant frequency of a vibrating wire as it is tensioned by an applied force.

Use of Accelerometers

The technique of applying a force to a known mass and measuring the acceleration produced can be carried out using any type of accelerometer. Unfortunately, the method is of very limited practical value because, in most cases, forces are not free entities but are part of a system (from which they cannot be decoupled) in which they are acting on some body which is not free to accelerate. However, the technique can be of use in measuring some transient forces, and also for calibrating the forces produced by thrust motors in space vehicles.

Vibrating Wire Sensor

This instrument, illustrated in Figure 15.9, consists of a wire which is kept vibrating at its resonant frequency by a variable-frequency oscillator. The resonant frequency of a wire under tension is given by

$$f = \frac{0.5}{L}\sqrt{\left[\frac{M}{T}\right]}$$

where M is the mass per unit length of the wire, L is the length of the wire, and T is the tension due to the applied force, F.

Thus, measurement of the output frequency of the oscillator allows the force applied to the wire to be calculated.

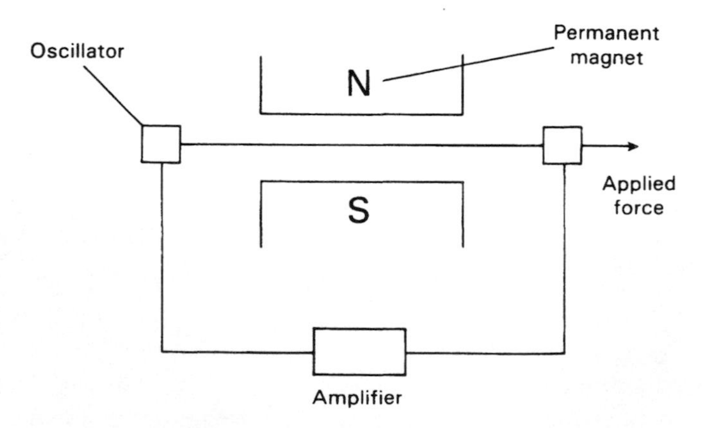

Figure 15.9. Vibrating wire sensor

15.3 REVIEW OF TORQUE MEASUREMENT

Measurement of applied torques is of fundamental importance in all rotating bodies to ensure that the design of the rotating element is adequate to prevent failure under shear stresses. Torque measurement is also a necessary part of measuring the power transmitted by rotating shafts.

Four methods of measuring torque exist:

1. measuring the reaction force in cradled shaft bearings,
2. the 'Prony brake' method,
3. measuring the strain produced in a rotating body owing to an applied torque and
4. optical torque measurement

The first three of the above methods are traditional ones and the fourth is a new technique made possible by recent developments in electronics and fibre-optic technology.

Reaction Forces in Shaft Bearings

Any system involving torque transmission through a shaft contains both a power source and a power absorber where the power is dissipated. The magnitude of the transmitted torque can be measured by cradling either the power source or the power absorber end of the shaft in bearings and then measuring the reaction force F, and the arm length L, as shown in Figure 15.10(a) and (b). The torque is then calculated as the simple product, $F \cdot L$. Pendulum scales are very commonly used for measuring the reaction force. Inherent errors in the method are bearing friction and windage torques.

Prony Brake

The principle of the Prony brake is illustrated in Figure 15.11. It is used to measure the torque in a rotating shaft and consists of a rope wound round the shaft, one end of the rope being attached to a spring balance and the other end carrying a load in the form of a standard mass, m. If the measured force in the spring balance is F_s, then the effective force, F_e, exerted by the rope on the shaft is given by:

$$F_e = mg - F_s$$

If the radius of the shaft is R_s and that of the rope is R_r, then the effective radius, R_e, of the rope and drum with respect to the axis of rotation of the shaft is given by:

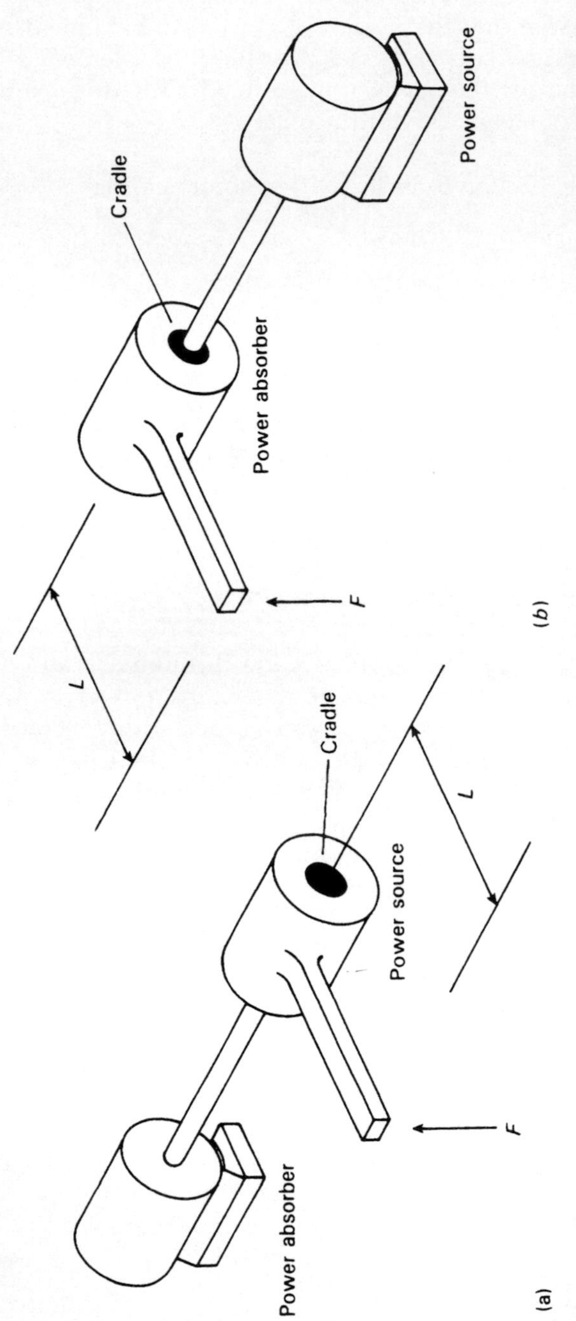

Figure 15.10. Measuring reaction forces in cradled shaft bearings: (a) cradled power source; (b) cradled power absorber

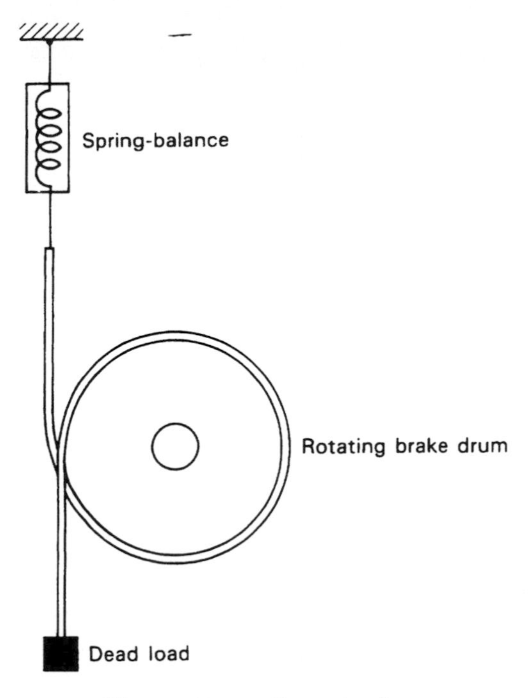

Figure 15.11. Prony brake

$$R_e = R_s + R_r$$

The torque in the shaft, T, can then be calculated as:

$$T = F_e \cdot R_e$$

Whilst this is a well-known method of measuring shaft torque, a lot of heat is generated because of friction between the rope and shaft, and water cooling is usually necessary.

Measurement of Induced Strain

Measuring the strain induced in a shaft due to an applied torque has been the most common method used for torque measurement in recent years. It is a very attractive method because it does not disturb the measured system by introducing friction torques in the same way as the last two methods described. The method involves bonding four strain gauges onto the shaft, connected in a direct-current bridge circuit, as shown in Figure 15.12. The output from the bridge circuit is a function of the strain in the shaft and hence of the torque applied. It is very important that the positioning of the strain gauges on the shaft is precise, and the difficulty in achieving this makes the instrument relatively expensive.

The technique is ideal for measuring the stalled torque in a shaft before

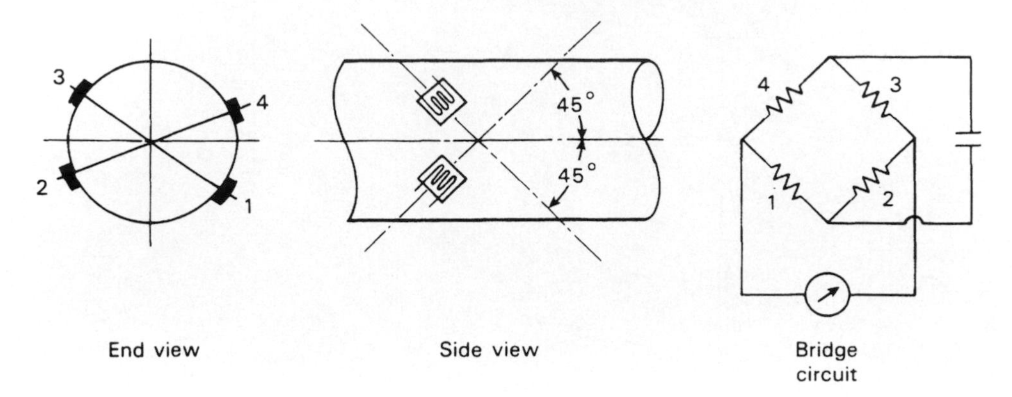

Figure 15.12. Position of torque-measuring strain gauges on shaft

rotation commences. However, a problem is encountered in the case of rotating shafts because a suitable method then has to be found for making the electrical connections to the strain gauges. One solution to this problem found in many commercial instruments is to use a system of slip rings and brushes, although this increases the cost of the instrument still further.

Optical Torque Measurement

Optical techniques for torque measurement have become available recently with the development of laser diodes and fibre-optic light-transmission systems. One such system is shown in Figure 15.13. Two black and white striped wheels are mounted at either end of the rotating shaft and are set in

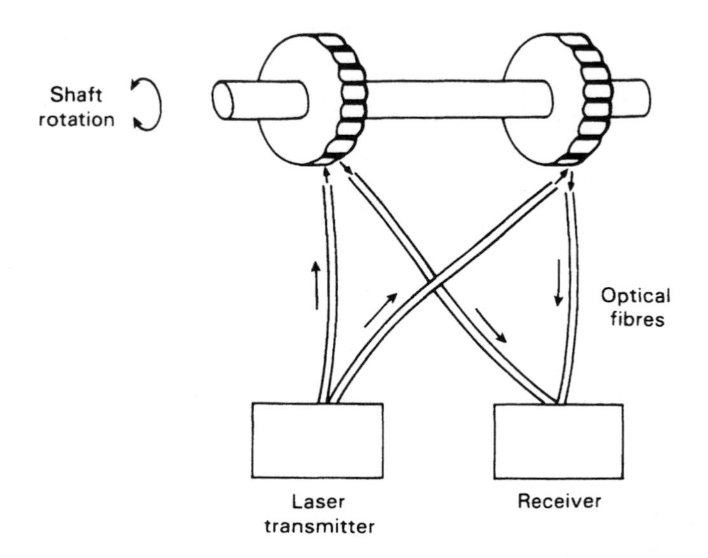

Figure 15.13. Optical torque measurement

alignment when no torque is applied to the shaft. Light from a laser diode light source is directed by a pair of fibre-optic cables onto the wheels. The rotation of the wheels causes pulses of reflected light and these are transmitted back to a receiver by a second pair of fibre-optic cables. Under zero-torque conditions, the two pulse-trains of reflected light are in phase with each other.

If torque is then applied to the shaft, the reflected light is modulated. Measurement by the receiver of the phase difference between the reflected pulse-trains therefore allows the magnitude of torque in the shaft to be calculated. The cost of such instruments is relatively low, and an additional advantage in many applications is their small physical size.

15.4 INTRODUCTION TO CALIBRATION

A complication which arises in the calibration of mass-, force- and torque-measuring instruments is the variability in the value of g (the acceleration due to gravity). Apart from instruments like the beam balance and pendulum scale which directly compare two masses, all other instruments have an output reading which depends on the value of g.

The value of g is given by Helmert's formula:

$$g = 980.6 - 2.6 \cos \Phi - 0.000\,309h$$

where Φ is the latitude and h is the altitude in metres.

It can be seen from this formula that g varies with both latitude and altitude. At the equator ($\cos \Phi = 0°$) $g = 978.0$, whereas at the poles ($\cos \Phi = 90°$) $g = 983.2$. In Britain, a working value of 980.7 is normally used for g, and very little error can normally be expected when using this value. Where necessary, the exact value of g can be established by measuring the period and length of a pendulum.

A further difficulty which arises in calibrating instruments in this category is the upwards force generated by the air medium in which the instruments are tested and used. According to Archimedes' principle, when a body is immersed in a fluid (air in this case), there is an upwards force proportional to the volume of fluid displaced. Even in pure mass-balance instruments, an error is introduced because of this unless both the body of unknown mass and the standard mass have the same density. This error is given by:

$$E = \frac{SG_a}{SG_u} - \frac{SG_a}{SG_m}$$

where SG_a is the specific gravity of air, SG_u is the specific gravity of the substance being measured and SG_m is the specific gravity of the standard mass.

Fortunately, the maximum error due to this (which occurs when weighing low density liquids such as petrol) will not exceed 0.2 percent. Therefore, in most circumstances, the error due to air-buoyancy can be neglected. However, for calibrations at the top of the calibration tree, where the highest levels of accuracy are demanded, either correction must be made for this factor or it must be avoided by carrying out the calibration in vacuum conditions.

15.5 CALIBRATION INSTRUMENTS AND EQUIPMENT

The primary requirement in mass calibration is the maintenance of a set of standard masses against which the mass to be calibrated can be compared [1]. Beam balances, weigh beams, pendulum scales, electromagnetic balances and load-cells are all used in various circumstances to effect this comparison.

Beam Balance

The instrument used most commonly for calibrating masses is the beam balance. This is used for calibrations in the range between 10 mg and 1 kg. The measurement resolution and accuracy of such instruments depends on the quality and sharpness of the knife-edge that the pivot is formed from. For high measurement resolution, friction at the pivot must be as close to zero as possible and hence a very sharp and clean knife-edge pivot is demanded. The two halves of the beam on either side of the pivot are normally of equal length and are measured from the knife-edge. Any bluntness, dirt or corrosion in the pivot can cause these two lengths to become unequal, causing consequent measurement errors. Similar comments apply to the knife-edges on the beam that the two pans are hung from. It is also important that all knife-edges are parallel, otherwise displacement of the point of application of the force over the line of the knife-edge can cause further measurement errors. This last form of error also occurs if the mass is not placed centrally on the pan.

Great care is therefore required in the use of such an instrument but, provided that it is kept in good condition, particularly with regard to keeping the knife-edges sharp and clean, high measurement accuracy is achievable.

Weigh Beam

The weigh beam, when manufactured and maintained at a sufficient standard of accuracy, can be used as a standard for masses up to 50 t.

Pendulum Scale

In the range above 1 kg, up to 500 t, versions of the pendulum scale can be used for mass calibration. Again, such calibration instruments must be kept in good condition, with special attention to the cleanliness and lubrication of moving parts.

Electromagnetic Balance

Various forms of electromagnetic balance exist as alternatives to the last three instruments. Their optical system magnifies motion around the null point, leading to greater accuracy. Consequently, this type of instrument is preferred for higher-level calibrations. The actual degree of accuracy achievable depends on the magnitude of the mass being measured. In the range between 100 g and 10 kg, an inaccuracy of ±0.0001 percent is achievable. Above and below this range, the inaccuracy is worse, increasing to ±0.002 percent measuring 5 t and ±0.03 percent measuring 10 mg.

Proof-ring-based Load-cell

In the range between 150 kg and 2000 t, a proof-ring-based load-cell is used for calibration in which the displacement of the proof-ring is measured by either an LVDT or a micrometer. As the relationship between the applied mass/force and the displacement is not a straight-line one, a force–deflection graph has to be used to interpret the output. Minimum measurement inaccuracy is ±0.1 percent.

Shaft with Known Torque Applied

Torque-measuring instruments are calibrated by applying them to a shaft to which a known torque is applied.

15.6 CALIBRATION FREQUENCY

The primary requirement in mass calibration is a set of standard masses for comparison. Provided that these do not suffer any obvious damage, annual calibration checks are quite adequate.

The required recalibration frequency for instruments used to effect comparison of standard masses is somewhat more difficult to define as it depends on the conditions of use. Therefore, in the case of the beam balance, weigh beam, pendulum scale and electromagnetic balance, practical measurement of the rate of degradation in measurement accuracy is necessary in order to determine a suitable recalibration frequency.

Both elastic-element-based load-cells and (especially) spring balances suffer from elastic deformation of the active element. Again, practical measurement of the rate of accuracy degradation is required but, as a general rule, recalibration of such instruments will have to be carried out much more frequently than for mass-comparison instruments.

REFERENCES

1. NAMAS Document NIS 6: *Traceability – weighing equipment and weight*, NAMAS Executive, National Physical Laboratory, Middlesex, UK (1984).

CHAPTER 16

Dimension Calibration

16.1 REVIEW OF DIMENSION-MEASURING INSTRUMENTS

Dimension measurement includes the measurement of the length, height, depth (of holes and slots) and angles of components. For many such measurements, it is necessary to have a reference flat plane on which components being measured are located. This is provided by a surface plate or table as described below.

The range of instruments available for measuring length includes the steel tape (to 30 m), ultrasonic rule (to 10 m), steel rule (to 1 m), standard callipers (to 600 mm), vernier callipers (to 200 mm) and micrometer (to 50 mm). Gauge blocks and length bars, although primarily intended for calibration duties, are also used for the measurement of larger lengths when very high accuracy is required. Height and depth are measured by the height gauge, depth gauge or dial gauge. Angles are normally measured by either an angle protractor, a bevel protractor or a spirit level. Further information on dimension measurement can be found in references [1, 2].

Surface Planes and Tables

A flat and level reference plane is often an essential component in dimension measurement, especially where the line of measurement and the dimension being measured are not coincident. Such reference planes are available in a range of standard sizes, the smallest having nominal dimensions of 100 mm × 160 mm and the largest 1600 mm × 2500 mm. The smaller sizes exist as a surface plate resting on a supporting table, whereas the larger sizes take the form of free-standing tables, as shown in Figure 16.1. Larger sizes have a projection at the edge to facilitate the clamping of components. They are normally used in conjunction with box cubes and vee blocks (see Figure 16.2) which locate components in a fixed position. The

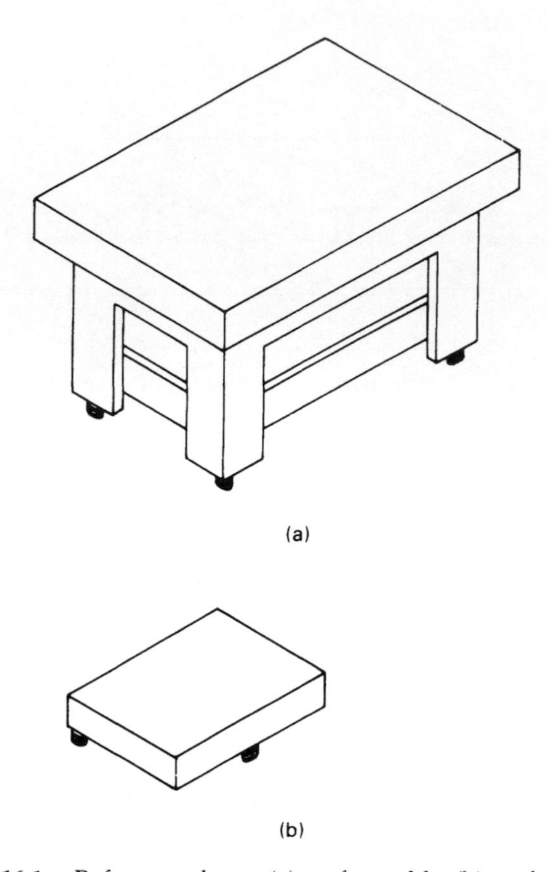

Figure 16.1. Reference planes: (a) surface table; (b) surface plate

plate or table normally has three feet†, with provision for levelling to make the surface exactly horizontal. In modern tables, granite has tended to supersede iron as the preferred material for the plate, although iron plates are still available. Granite is ideal for this purpose as it does not corrode, is dimensionally very stable and does not form burrs when damaged. Iron plates, on the other hand, are prone to rusting and susceptible to damage: this results in burrs on the surface which interfere with measurement procedures.

Both forms of plate are available in four standard grades, from grade 0 to grade 3. Grade 0 is the best and is used for calibration purposes. Grades 1 to 3 are recommended for inspection, marking out and lower-grade marking out respectively. Flatness is defined in terms of the distance between two parallel planes which just contain all points in the surface. Standards of flatness, as defined by BS 817 [3], vary according to the size of the plate or table. For a 2 m × 1 m table, the maximum permitted deviations from flatness for grades 0 to 3 are 8.5 μm, 17 μm, 34 μm, 68 μm respectively.

---

† Plates with dimensions of 1 m × 1.6 m and larger often have more than three feet.

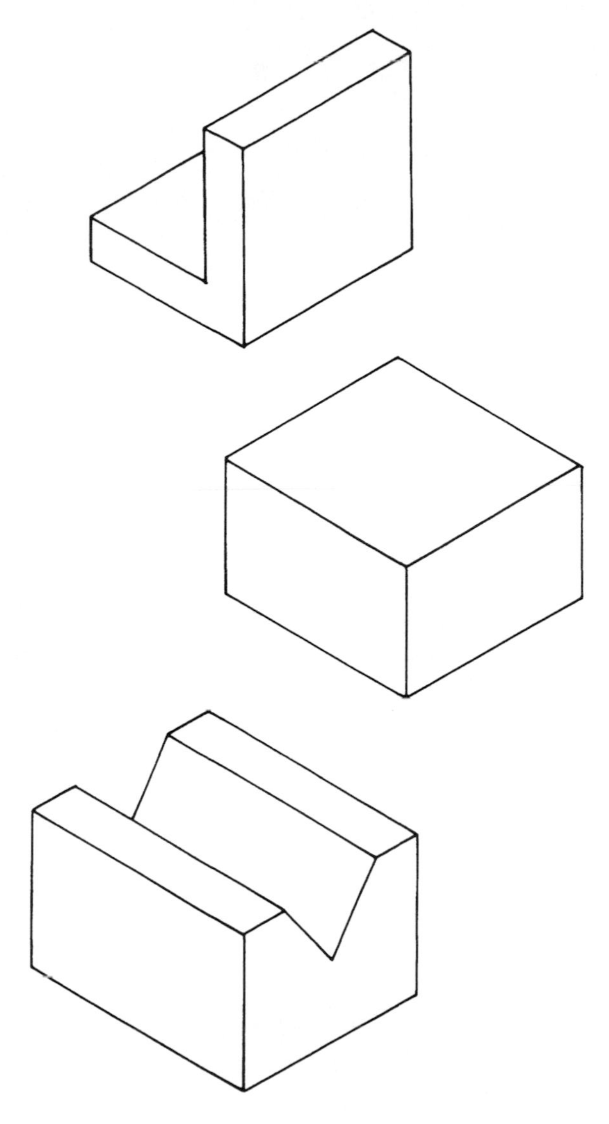

Figure 16.2. Box cubes and vee blocks

Steel Rule

The steel rule is undoubtedly the simplest instrument available for measuring length. Rules are manufactured according to quality standards (see [4]) which define the ruling accuracy, squareness of the end and edge straightness. The best rules have rulings at 0.05 mm intervals and a measurement resolution of 0.02 mm. When used by placing the rule against an object, the measurement accuracy is much dependent upon the skill of the human measurer and, at best, the inaccuracy is likely to be at least ±0.5 percent.

Steel Tape

Retractable steel tapes are another well-known instrument. The end of the tape is usually provided with a flat hook which is loosely fitted so as to allow for automatic compensation of the hook thickness when the rule is used for both internal and external measurements. Again, measurement accuracy is governed by human skill, but, with care, the measurement inaccuracy can be made to be as low as ±0.01 percent. Formal specifications for measuring tapes are defined in references [5, 6].

Ultrasonic Rule

The ultrasonic rule is a recent addition to the measurement scene. It consists of an ultrasonic energy source, an ultrasonic energy detector and battery-powered electronic circuitry housed within a hand-held box, as shown in Figure 16.3. Both source and detector often consist of the same type of piezoelectric crystal excited at a typical frequency of 40 kHz. Energy travels from the source to a target object and is then reflected back into the detector. The time of flight of this energy is measured and this is converted into a distance reading by the enclosed electronics. Maximum measurement inaccuracy of ±1 percent of the full-scale reading is claimed. This is only a modest level of accuracy which is of little use for most engineering measurements. However, it is sufficient for such purposes as the measuring of rooms by estate agents prior to producing sales literature, where the ease and speed of making measurements is of great value.

A fundamental problem in the use of ultrasonic energy of this type is the limited measurement resolution (7 mm) imposed by the 7 mm wavelength of sound at this frequency. Further problems are caused by the variation in the speed of sound with humidity (variations of ±0.5 percent possible) and the temperature-induced variation of 0.2 percent per °C. Therefore, the conditions of use must be carefully controlled if the claimed accuracy figure is to be met.

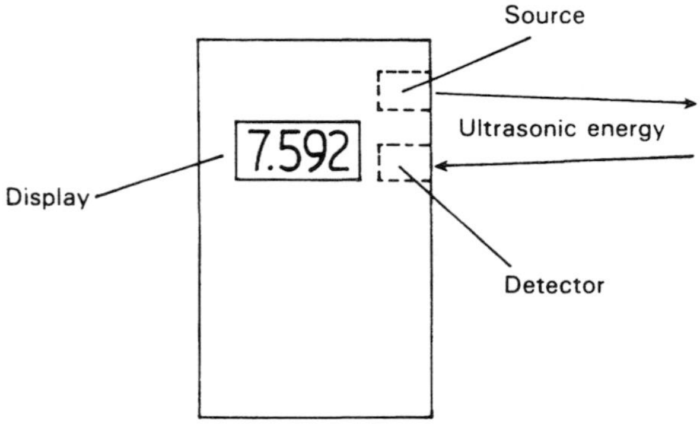

Figure 16.3. Ultrasonic rule

Standard Calliper

Figure 16.4 shows two types of standard calliper. These are used to transfer the measured dimension from the workpiece to a steel rule. This avoids the necessity of aligning the end of the rule exactly with the edge of the workpiece and reduces the measurement inaccuracy by a factor of two. In the basic calliper, careless use can allow the setting of the calliper to be changed during transfer from the workpiece to the rule. Hence, the spring-loaded type, which prevents this happening, is preferable.

Vernier Calliper

Vernier callipers [7, 8], shown in Figure 16.5, are effectively the combination of standard callipers with a steel rule. The main body of the instrument includes the main scale with a fixed anvil at one end. This carries a sliding anvil which is provided with a second, vernier scale. This second scale is shorter than the main scale and is divided into units which are slightly smaller than the main scale units but related to them by a fixed factor. Determination of the point where the two scales coincide enables very

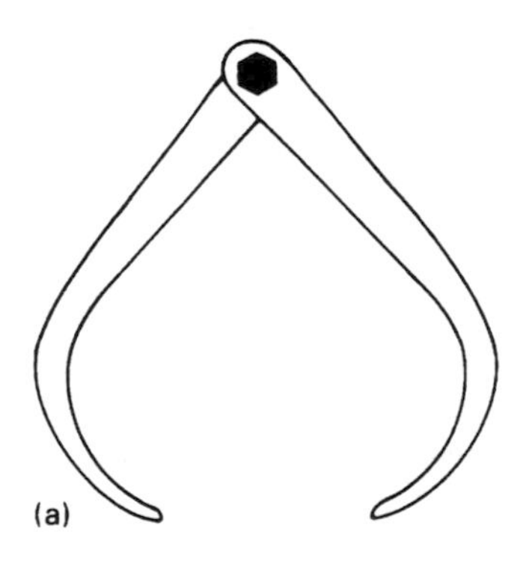

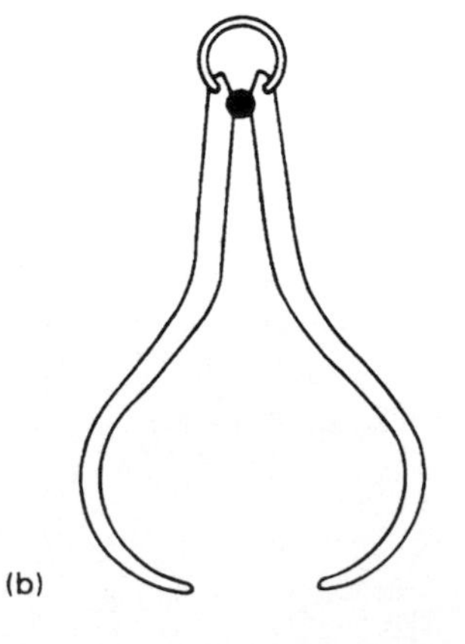

Figure 16.4. Callipers: (a) standard type; (b) spring-loaded type

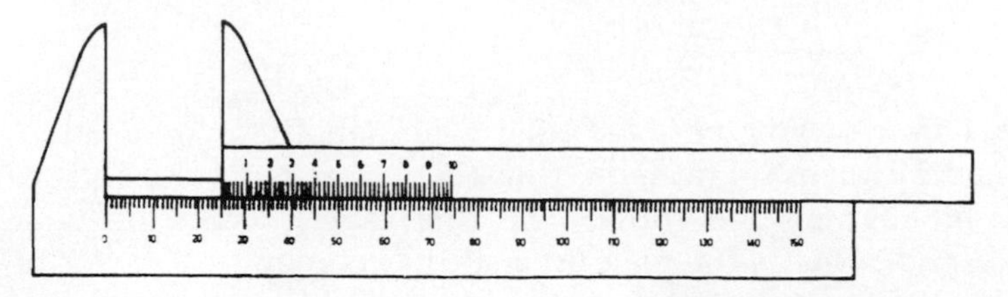

Figure 16.5. Vernier calliper

accurate measurements to be made, with typical inaccuracy levels down to ±0.01 percent.

Figure 16.6 shows details of a typical combination of main and vernier scales. The main scale is ruled in 1 mm units. The vernier scale is 49 mm long and divided into 50 units, thereby making each unit 0.02 mm smaller than the main scale units. Each group of five units on the vernier scale thus differs from the main scale by 0.1 mm and the numbers marked on the scale thus refer to these larger units of 0.1 mm. In the particular position shown in Figure 16.6, the zero on the vernier scale is indicating a measurement between 8 and 9 mm. Both scales coincide at a position of 6.2 (large units). This defines the interval between 8 and 9 mm to be $6.2 \times 0.1 = 0.62$ mm, i.e. the measurement is 8.62 mm.

Intelligent digital callipers are now available which give a measurement resolution of 0.01 mm and an accuracy of ±0.03 mm. These have automatic compensation for wear, and hence calibration checks have to be very infrequent. In some such instruments, the digital display can be directly interfaced to an external computer monitoring system.

Micrometers

In the standard micrometer, shown in Figure 16.7(a), measurement is made between two anvils, one fixed and one which is moved along by the rotation of an accurately machined screw thread. One complete rotation of the screw typically moves the anvil by a distance of 0.5 mm. Such movements of the anvil are measured using scale marked with divisions every 0.5 mm along the barrel of the instrument. A scale marked with 50 divisions is etched

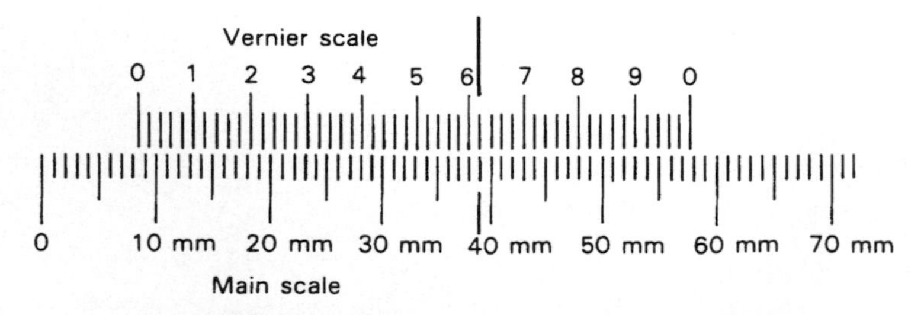

Figure 16.6. Details of vernier calliper scale

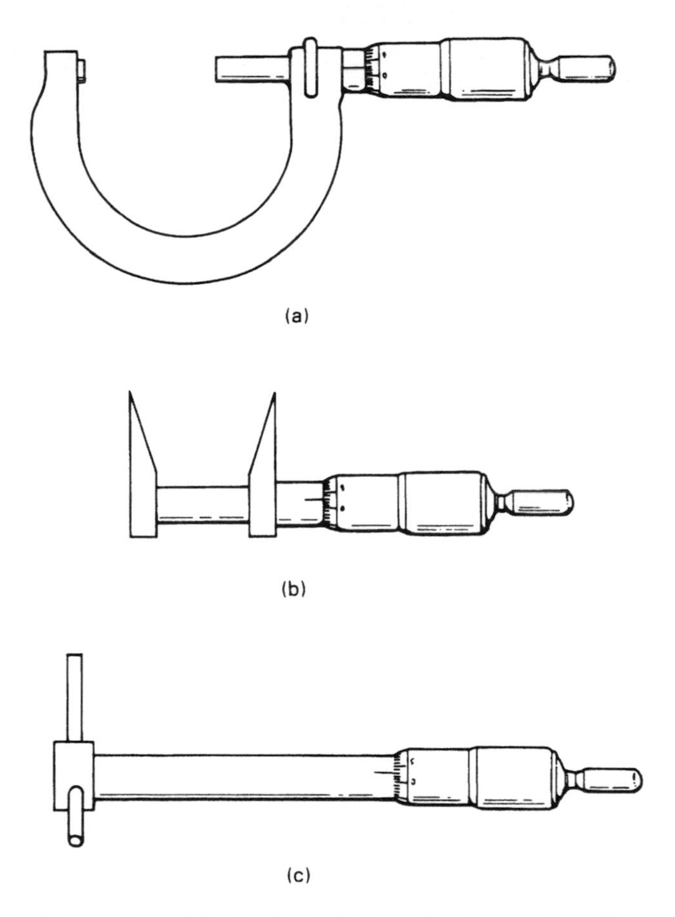

Figure 16.7. Micrometers: (a) standard (external) micrometer; (b) internal micrometer; (c) bore micrometer

around the circumference of the spindle holder: each division therefore corresponds to an axial movement of 0.01 mm. Assuming that the user is able to judge the position of the spindle on this circumferential scale against the datum mark to within one fifth of a division, a measurement resolution of 0.002 mm is possible. Such instruments are manufactured to high standards of accuracy, and typical specifications [9, 10] might be; flatness of anvils, ±0.001 mm; parallelism of anvils, ±0.003 mm; screw traverse, ±0.003 mm; alignment error, 0.05 mm max.

The most common measurement ranges are either 0–25 mm or 25–50 mm, with inaccuracy levels down to ±0.003 percent. However, a whole family of micrometers is available, where each has a measurement span of 25 mm, but with the minimum distance measured varying from 0 mm up to 575 mm. Thus, the last instrument in this family measures the range from 575 to 600 mm. Some manufacturers also provide micrometers with two or more interchangeable anvils, which extends the span measurable with one instrument to between 50 mm and 100 mm according to the number of anvils supplied. Therefore, an instrument with four anvils might for instance measure the range from 300 mm to 400 mm, if appropriate changes are made to the anvils.

An alternative form of micrometer [11] (see Figure 16.7(b)) is able to measure internal dimensions such as hole diameters. In the case of measuring holes, micrometers are inaccurate if there is any ovality in the hole, unless the diameter is measured at several points. An alternative solution to this problem is to use a special type of instrument known as a bore micrometer (Figure 16.7(c)). In this, three probes move out radially from the body of the instrument as the spindle is turned. These probes make contact with the sides of the hole at three equidistant points, thus averaging out any ovality.

Intelligent micrometers in the form of the electronic digital micrometer are now available. These have a self-calibration capability and a digital read-out, with a measurement resolution of 0.001 mm (1 micron).

Gauge Blocks (Slip Gauges) and Length Bars

Gauge blocks, also known as slip gauges, consist of rectangular blocks (see Figure 16.8) of hardened steel which have flat and parallel end faces. These faces are machined to very high standards of accuracy in terms of their surface finish and flatness. They are available in boxed sets containing a range of block sizes, which allows any dimension to be constructed by joining together an appropriate number of blocks.

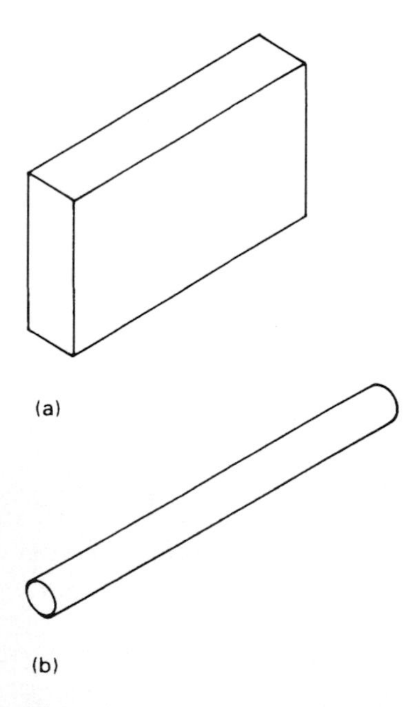

Figure 16.8. (a) Gauge block; (b) length bar

Blocks are joined by *wringing*, a procedure in which the two end faces are rotated slowly against each other. This removes the air film and allows adhesion to develop by intermolecular attraction. Adhesion is so good in fact that, if groups of blocks were not separated within a few hours, the molecular diffusion process would continue to the point where the blocks would be permanently welded together. The typical interblock gap resulting from wringing has been measured as 0.001 μm, which is effectively zero. Thus, any number of blocks can be joined without creating any significant measurement error.

Gauge blocks are available in five standards of accuracy: calibration, 00, 0, 1 and 2. Grades 1 and 2 are used for normal production inspection measurements. The length, end-face-flatness and end-face-parallelism tolerances allowed for gauge blocks are defined by BS 4311† [12].

For a 100 mm long block, the tolerances are

	Length	Flatness	Parallelism
Grade 1	+0.6 μm −0.3 μm	±0.15 μm	±0.25 μm
Grade 2	+1.4 μm −1.0 μm	±0.25 μm	±0.35 μm

The larger positive tolerance for grade 1 and grade 2 blocks is provided to allow for the relatively high level of wear which occurs immediately after a new block is put into use. It should be noted that the length tolerances of blocks wrung together are cumulative. Thus the length tolerance of five grade 2 blocks wrung together is (+7 μm, −5 μm).

A typical set of gauge blocks allows any dimension between 3.0 mm and 200 mm to be built up in steps of 0.0005 mm. By careful choice of the block combination, it is usually possible to construct any dimension using not more than five blocks. Consider the M112/1 set below which consists of the following 112 blocks:

Size or series	Increment	No. of pieces
1.0005	–	1
1.001–1.009	0.001	9
1.01–1.49	0.01	49
0.5–24.5	0.5	49
25–100	25	4
		112

† ISO 3650 [13] defines similar, though not identical, standards.

The procedure for choosing appropriate blocks to make up a particular length is to consider each decimal point in turn, as follows:

1. Choose the thinnest block to satisfy the last decimal place.
2. Choose the next block to satisfy the next-to-last decimal place.
3. Continue in this fashion until the full length required is constructed.

For example to construct a length of 89.4365 mm, the following five blocks would be used:

$$\begin{array}{r} 1.0005 \\ 1.006 \\ 1.43 \\ 11.0 \\ 75.0 \\ \hline 89.4365 \end{array}$$

It is fairly common practice with blocks of grades 0, 1 and 2 to include an extra pair of 2 mm thick blocks in the set, made from wear-resistant tungsten carbide. These are marked with a letter P and are designed to protect the other blocks from wear during use. Where such protector blocks are used, due allowance has to be made for their thickness (4 mm) in calculating the sizes of block needed to make up the required length.

A precaution to be followed when using gauge blocks is to avoid handling them more than is necessary. The length of a bar 100 mm long at 20 °C would increase to 100.02 mm at 37 °C (body temperature). Hence, after wringing bars together, they should be left to stabilize back to the ambient room temperature before use. This wait might need to be several hours if the blocks have been handled to any significant extent.

Where a greater dimension than 200 mm is required, gauge blocks are used in conjunction with length bars. Length bars consist of straight, hardened, high quality steel bars of a uniform 22 mm diameter and in a range of lengths between 100 mm and 1200 mm. They are available in four standards of accuracy, as defined by BS 5317 [14]; reference, calibration, grade 1 and grade 2. Reference and calibration grades have accurately flat end faces, which allows a number of bars to be wrung together to obtain the required standard length. Bars of grades 1 and 2 have threaded ends which allows them to be screwed together. Grade 2 bars are used for general measurement duties, with grade 1 bars being reserved for inspection duties. By combining length bars with gauge blocks, any dimension up to about 2 m can be set up with a resolution of 0.0005 mm.

The length tolerances of a 200 mm length bar are given by:-

$$\text{Grade 1} \quad \begin{matrix} +1.4 \\ -0.6 \end{matrix} \times (0.2 + 0.004L)\ \mu\text{m, i.e.} + 1.4\ \mu\text{m or} - 0.6\ \mu\text{m.}$$

$$\text{Grade 2} \quad {}^{+1.4}_{-0.6} \times (0.4 + 0.006L)\ \mu\text{m, i.e.} + 2.2\ \mu\text{m or} - 1.0\ \mu\text{m}.$$

where L is the length of the bar in mm. As for gauge blocks, the larger positive tolerance for grades 1 and 2 is provided to allow for the high level of wear occurring immediately after new bars are put into use. These length tolerances are only valid if the bar is horizontal and at 20 °C. Horizontal alignment is obtained by mounting the bars at their *Airy* points. The Airy points are at a distance of 0.2117 L from the ends of the bar and are marked by circumferential lines.

Height and Depth Gauges

The height gauge [15], shown in Figure 16.9, effectively consists of a vernier calliper mounted on a flat base. Measurement inaccuracy levels down to ±0.015 percent are possible. The depth gauge, (Figure 16.9), is a further variation on the standard vernier calliper principle, which has the same measurement accuracy capabilities as the height gauge.

In practice, certain difficulties can arise in the use of these instruments where either the base of the instrument is not properly located on the measuring table or where the point of contact between the moving anvil and the workpiece is uncertain. In such cases, a dial gauge, which has a clearly defined point of contact with the measured object, is used in conjunction with the height or depth gauge to avoid these possible sources of error.

These instruments can also be obtained in intelligent versions which give a digital display and have self-calibration capabilities.

Dial Gauge

The dial gauge [16, 17], shown in Figure 16.10, consists of a spring-loaded probe which drives a pointer around a circular scale via rack-and-pinion gearing. Typical measurement resolution is 0.01 mm. When used to measure the height of objects, it is clamped in a retort stand and a measurement taken of the height of the unknown component. Then, it is put in contact with a height gauge (Figure 16.11) which is adjusted until the reading on the dial gauge is the same. At this stage, the height gauge is set to the height of the object. The dial gauge is also used in conjunction with the depth gauge in an identical manner. (Gauge blocks can be used instead of height/depth gauges in such measurement procedures if greater accuracy is required).

Protractors

The simplest form of protractor is the angle type shown in Figure 16.12(a). This consists of two straight edges, one of which is able to rotate with

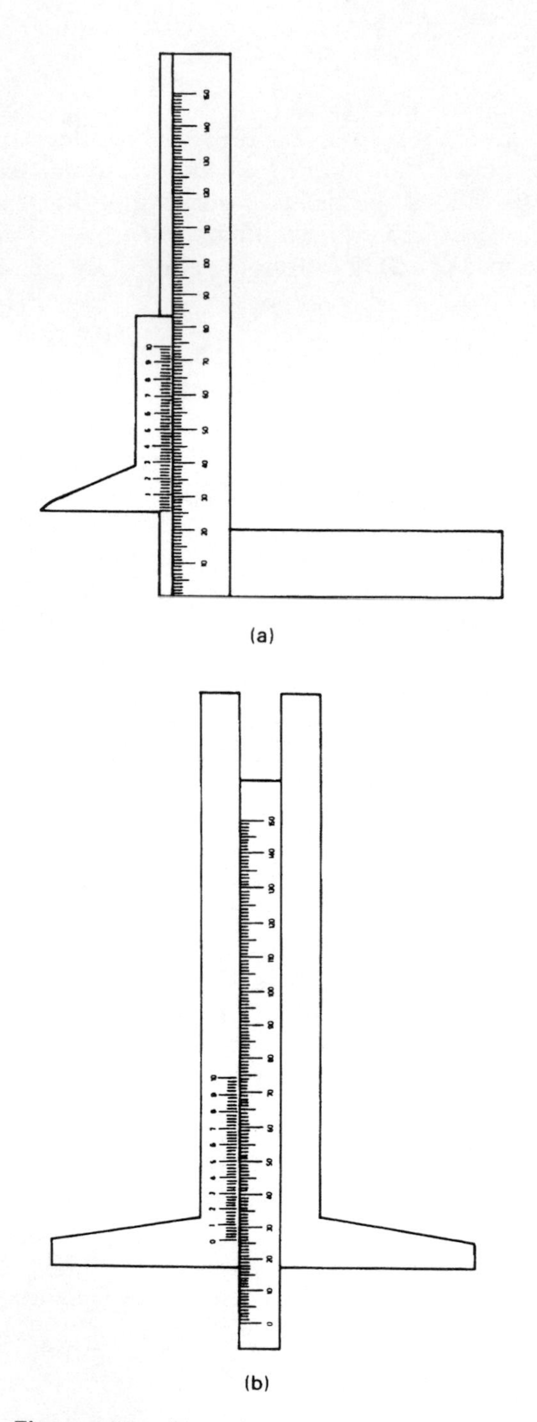

Figure 16.9. (a) Height gauge; (b) depth gauge

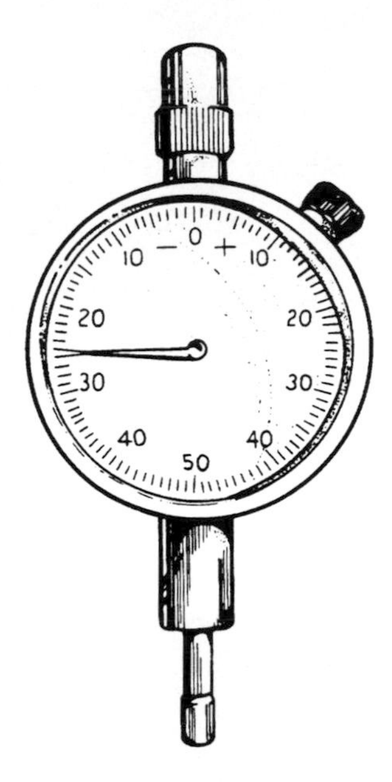

Figure 16.10. Dial gauge

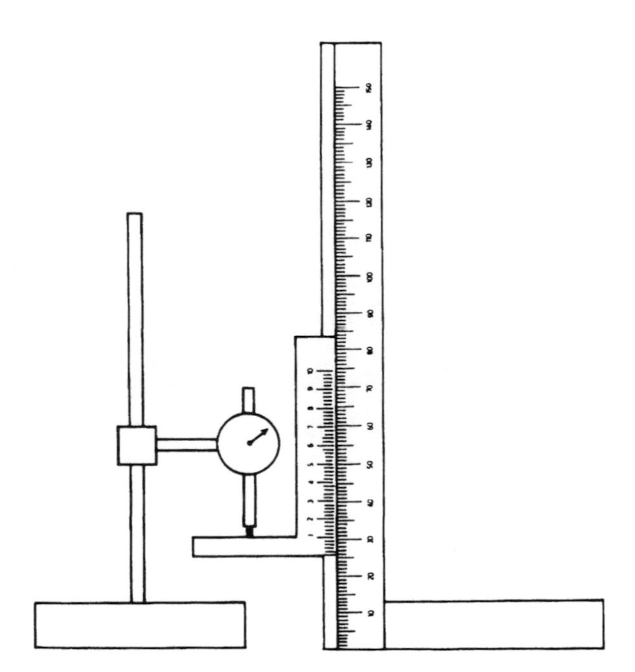

Figure 16.11. Use of dial gauge in conjunction with height gauge

respect to the other. A circular scale attached to one of the straight edges rotates inside a fixed circular housing attached to the other straight edge. The relative angle between the two straight edges in contact with the component being measured is determined by the position of the moving

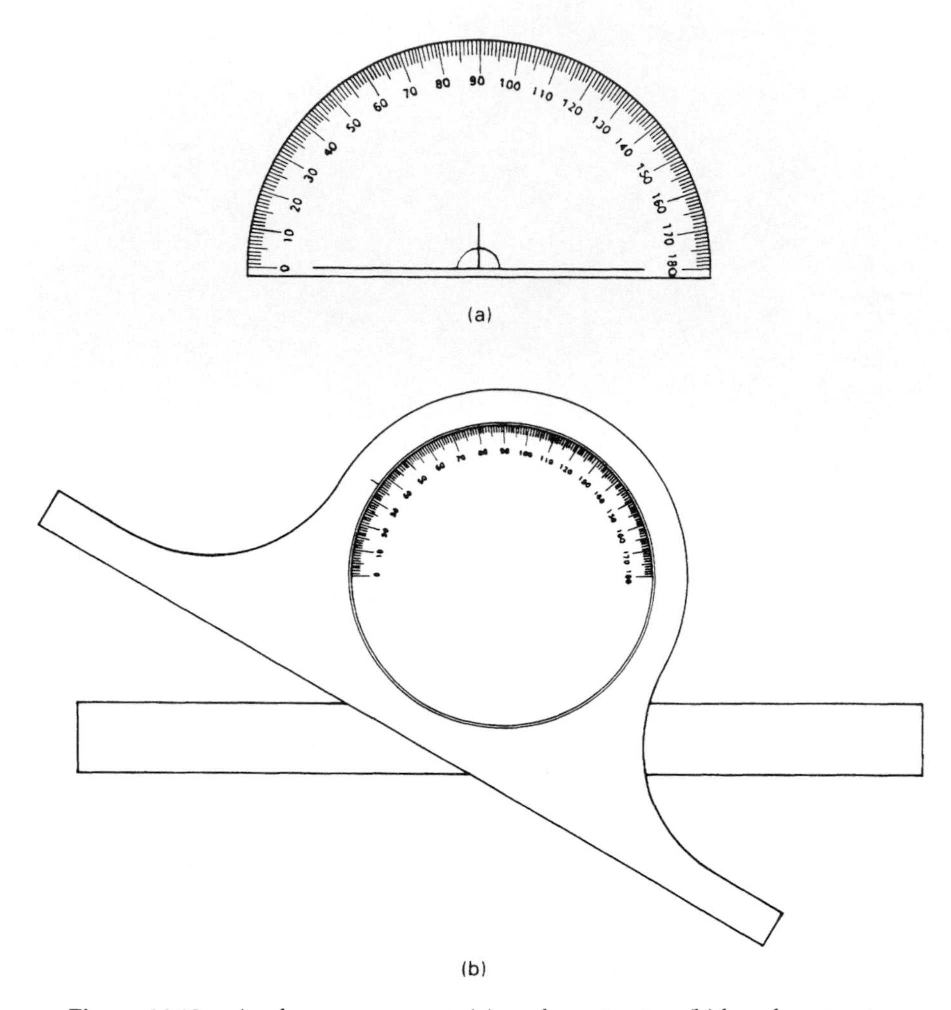

Figure 16.12. Angle measurement: (a) angle protractor; (b) bevel protractor

scale with respect to a reference mark on the fixed housing. With this type of instrument, measurement inaccuracy is at least ±1 percent.

A more accurate instrument is the bevel protractor [18] shown in Figure 16.12(b). This is of the same form as the angle protractor, but it has a vernier scale on the fixed housing. This allows the inaccuracy level to be reduced to ±10 minutes of arc.

Spirit Level

The spirit level shown in Figure 16.13 is an alternative angle-measuring instrument. It consists of a standard spirit level attached to a rotatable circular scale which is mounted inside an accurately machined square frame. When placed on the sloping surfaces of components, rotation of the

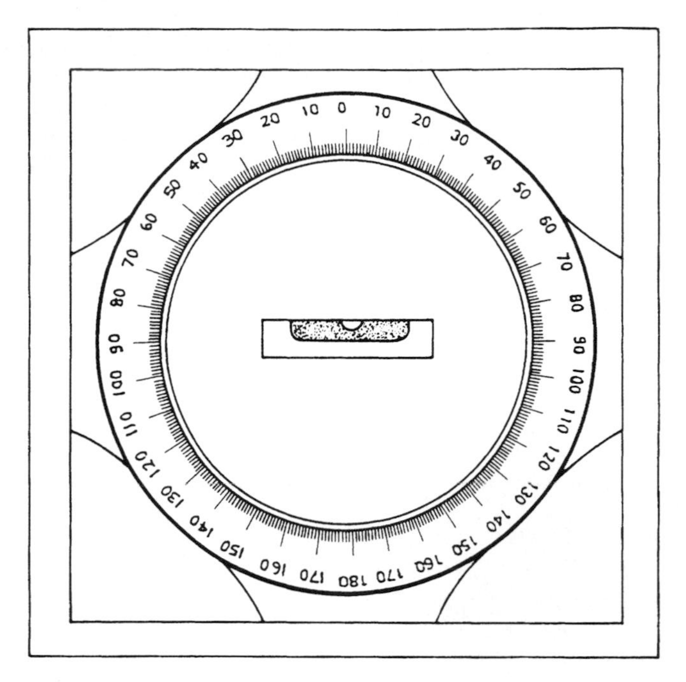

Figure 16.13. Angle-measuring spirit level

scale to centralize the bubble in the spirit level allows the angle of slope to be measured. Again, measuring inaccuracies down to ±10 minutes of arc are possible if a vernier scale is incorporated in the instrument [19].

Electronic Spirit Level

Electronic spirit levels contain a pendulum whose position is sensed electrically. With such instruments, measurement resolutions as good as 0.2 seconds of arc are possible.

16.2 'USER' CALIBRATION

When dimension measurements are being made as part of a quality control system, checks on the way that the human operator is using equipment are just as important as calibration checks on the instruments themselves. Such 'calibration' of the equipment user is very important because there are so many ways in which human-induced measurement errors can be introduced. The scope for human-induced errors is far greater than that which exists when measuring most other physical quantities.

The golden rule in dimension measurement is that the line of measurement and the line of the dimension being measured should be coincident. In the case of steel rules and tapes, the greatest potential source of user-induced error is failure to position the rule squarely across the dimension being measured. Parallax error is also possible if the user does not position the rule and read it from directly above. Callipers and micrometers are less susceptible to these types of error but it still remains sensible to carry out periodic checks on the way in which these instruments are being used, verifying in particular that measurements are being made squarely.

16.3 CALIBRATION EQUIPMENT

The equipment needed for working standard dimension calibration are a set of gauge blocks, a set of end bars, a reference plane (horizontal table) and a sine bar for calibrating angle-measuring instruments. Optically flat glass plates are also needed for some procedures which involve testing the flatness and parallelism of metal faces.

Calibration is performed by comparing measurements made on the reference plane with the unknown instrument and the standard. Reference standards for length consist of either gauge blocks alone or gauge blocks and end bars together for testing larger dimensions. A high-magnification comparator is required to make this comparison. At the very highest levels of calibration of gauge blocks themselves, interferometric methods are used as a primary reference standard. Such calibration services are normally provided by National Standards Organizations (e.g. National Physical Laboratory in the UK) or specialized calibration companies.

Setting gauges also deserve mention in respect of their use for initial setting up of new internal micrometers and subsequent calibration checks of these instruments.

Reference Plane

A reference plane is an essential component in dimension calibration. Surface plates and tables of flatness grade 0 or 1 are used according to the level of calibration. They are constructed either of graphite or iron, as discussed in Section 16.1.

Proper procedures for care and use of reference planes used for calibration duties are defined in BS 817. These require the surface to be wiped clean of dust at frequent intervals, to be used carefully to avoid damage and to be protected by a purpose-designed cover when not in use. Use in conditions of circulating air under constant temperature control, avoiding direct sunlight and draughts, is recommended.

The reference plane itself must be subjected to periodic calibration checks to test its flatness and horizontal alignment. The variation gauge (Figure 16.14) is available for measuring flatness. This has three fixed feet in contact with the flat surface and a floating foot whose position is indicated on a dial gauge. However, flatness measurement is normally entrusted to specialist companies who maintain equipment to carry out such checks, which

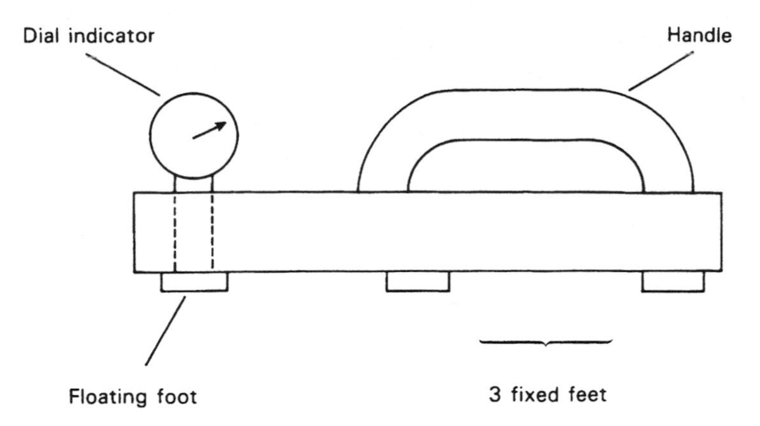

Figure 16.14. Variation gauge

include computational power to simplify the analysis of surface measurements.

Gauge Blocks (Slip Gauges) and Length Bars

Gauge blocks are available in five standards of accuracy; Calibration, 00, 0, 1 and 2. The permitted tolerances on length, flatness and end-face parallelism for the top three grades of bar, as defined by BS 4311 [12], are as follows for a 100 mm block:

	Length	Flatness	Parallelism
Calibration grade	±0.5 μm	±0.05 μm	±0.1 μm
Grade 00	±0.15 μm	±0.05 μm	±0.1 μm
Grade 0	±0.25 μm	±0.1 μm	±0.15 μm

The standard procedure in calibrating gauge blocks themselves is to compare them against gauge blocks of the next highest accuracy standard. Thus grade 2 is calibrated against grade 1, grade 1 against grade 0, etc. In performing this comparison, it is very important to check the flatness and degree of parallelism of the opposing end faces of each block, as these latter two parameters are just as important as length in determining the possible error in total length when a number of blocks are wrung together. Above grade 0, interferometric methods provide better reference standards for length than 'calibration standard' gauge blocks, with inaccuracy levels down to ±0.03 μm being possible. Because of this, 'calibration standard' blocks are only provided for checking the flatness and parallelism of gauge 0 blocks, with length being checked by interferometric methods. However, for the situations where such alternative reference length standards are not available, special grade 00 blocks are available. These have the same standards of flatness and parallelism as 'calibration grade' blocks but they have a tighter length tolerance. The calibration chains for gauge blocks are summarized in Figure 16.15 .

Length bars are available in four standards of accuracy [14]; reference, calibration, grade 1 and grade 2. As for gauge blocks, the standard calibration procedure is to compare them against length bars of the next highest accuracy standard. Thus, grade 2 bars are calibrated against grade 1 bars, grade 1 against 'calibration' grade, etc. Reference and calibration grades are calibrated themselves by interferometric methods, and inaccuracy levels down to ± 0.5 μm for a 1 m bar are achievable. Reference and calibration grades have accurately flat end faces, allowing a number of bars to be wrung together to obtain the required standard length. By combining length bars with gauge blocks, any dimension up to about 2 m can be set up with a resolution of 0.0005 mm.

The length tolerances of a 200 mm length bar are given by

Grade 1 $\quad {}^{+1.4}_{-0.6} \times (0.2 + 0.004L)$ μm, i.e. $+1.4$ μm or -0.6 μm.

Reference grade $\quad (0.05 \pm 0.0015L)$ μm, i.e. ± 0.35 μm.

Calibration grade $\quad (0.1 \pm 0.003L)$ μm, i.e. ± 0.65 μm.

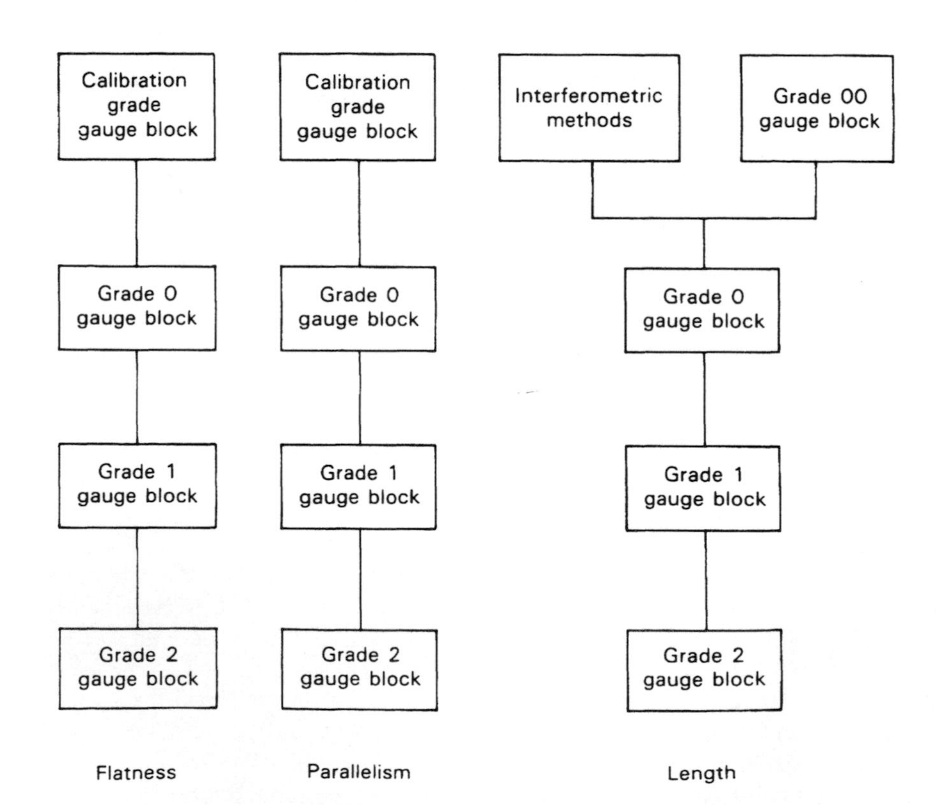

Figure 16.15. Calibration chains for gauge blocks

These length tolerances are only valid if the bar is horizontal and at 20 °C. Horizontal alignment is obtained by mounting the bars at their *Airy* points. The Airy points are at a distance of 0.2117 L from the ends of the bar and are marked by circumferential lines. Reference and calibration grades of bar must only be used in a standards laboratory where the temperature is accurately controlled at 20 °C [20].

Setting Gauges

Setting gauges are used for setting up new internal micrometers and recalibrating those in use. For testing instruments measuring up to 50 mm., the setting gauge consists of a 25 mm diameter steel disc (with ±0.001 mm tolerance on its diameter). For testing larger-range instruments, the setting gauge consists of either a flat-ended or a spherical-ended steel rod, with a length tolerance which varies from ±0.002 mm on a 125 mm long gauge to ±0.006 mm on a 575 mm long gauge.

Sine Bar

The sine bar [21], shown in Figure 16.16, is used in conjunction with gauge blocks for the calibration of angle-measuring equipment. It consists of a rectangular-section piece of steel into which a pair of rollers are located. The distance between the centres of the rollers is known very accurately and is commonly set to be 250 mm or some multiple thereof. By setting the bar up with one roller on a reference table and the other resting on a pile of gauge blocks, any angle can be set up, given by (see figure 16.16)

$$\theta = \sin^{-1}(H/L)$$

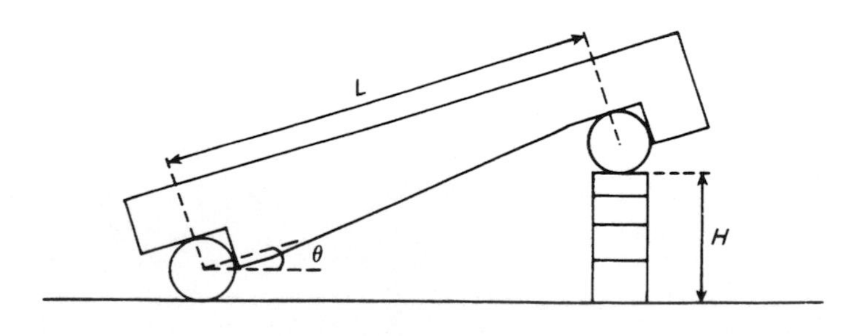

Figure 16.16. Sine bar

16.4 CALIBRATION PROCEDURES

The general principles of calibration are to set up a series of standard lengths using gauge blocks and/or length bars and to compare the output reading of the instrument being calibrated against the correct value when the instrument is set to the standard lengths. This comparison is carried out on a surface plane or table and various ancillary blocks (see Figure 16.2) and clamps are used to hold the set of gauge blocks and length bars in a manner appropriate to the needs of the instrument being calibrated. Various other special comments about the calibration of particular instruments can be made as follows.

New micrometers are set up with a setting gauge and this instrument may often also be suitable for recalibrating workshop micrometers. For calibration of working and reference standard micrometers, gauge blocks are used. When using gauge blocks for calibration, it is important that measurement accuracy is checked at intermediate positions of the thimble as well as for full revolutions. A set of gauge blocks set up to give the following dimensions in turn would satisfy this requirement: 2.5 mm, 5.1 mm, 7.7 mm, 12.9 mm, 15.0 mm, 17.6 mm, 20.2 mm, 22.8 mm, 25.0 mm.

As well as checking length-measurement accuracy, calibration checks on micrometers must include checks on the flatness and parallelism of the anvil faces. Flatness is checked by bringing the anvil into contact with an optically flat glass plate. The degree of non-flatness is indicated by the colour and number of interference bands on the surface. Parallelism of the measuring faces is tested by moving about an optical flat between them and observing the changes in the number of interference bands.

If calibration checks show up errors in length measurement, micrometers are provided with a means of adjustment to take up small amounts of wear in the screw thread. They also usually have some means of rotating the body scale so that the zero mark can be reset.

In the calibration of height gauges, depth gauges and dial gauges, comparison with reference piles of gauge blocks is effected by a special form of dial gauge which has a measurement resolution of 1 μm. In other procedures, suitable combinations of gauge blocks are wrung together and then clamped using special end-fittings. These provide the standard reference lengths used in the calibration of micrometers and callipers.

A reference surface-plane is also required for the purpose of calibrating angle-measuring instruments. The calibration instrument, a sine bar, is used in conjunction with a pile of gauge blocks to set up a series of standard angles. The output reading of the instrument under test can then be compared with the correct value as the instrument is set to each of these angles in turn. Again, various blocks and clamps are often used on the surface plane to assist in this procedure.

16.5 CALIBRATION FREQUENCY

As with so many other types of measuring instrument, experimentally derived knowledge is needed to define suitable calibration frequencies for dimension-measuring equipment. The required frequency depends largely upon their rate and conditions of use. However, whatever recalibration frequency is defined, a vigilant watch must be kept for signs of physical damage to the instrument at intermediate times. Instruments are often abused and mistreated, for instance by using the ends of rulers to open tins and using vernier callipers as a spanner. Whenever such misuse is suspected, the instrument must be withdrawn immediately for calibration checks to be carried out.

REFERENCES

1. Anthony, D. M., *Engineering Metrology*, Pergamon (1986).
2. Hume, K. J., *Engineering Metrology*, McDonald, (1970).
3. BS 817: *Surface plates*, British Standards Institution, London (1988).
4. BS 4372: *Engineers' steel measuring rules*, British Standards Institution, London (1968).
5. ISO 8322: *Building Construction – Measuring Instruments – Procedures for determining accuracy in use. Part 2 – Measuring Tapes*, International Organization for Standards, Geneva (1989).
6. BS 4035: *Linear measuring instruments for use in building and civil engineering constructional works. Steel measuring tapes, steel bands and retractable steel pocket rules*, British Standards Institution, London (1966).
7. BS 887: *Precision vernier callipers*, British Standards Institution, London (1982).
8. ISO 6906: *Vernier callipers reading to 0.02 mm*, International Organization for Standards, Geneva (1984).
9. BS 870: *External micrometers*, British Standards Institution, London (1950).
10. ISO 3611: *Micrometer callipers for external measurement*, International Organization for Standards, Geneva (1978).
11. BS 959: *Internal micrometers (including stick micrometers)*, British Standards Institution, London (1950).
12. BS 4311 (Parts 1 and 2): *Metric gauge blocks*, British Standards Institution, London (1993).
13. ISO 3650: *Gauge blocks*, International Organization for Standards, Geneva (1978).
14. BS 5317: *Metric length bars and their accessories*, British Standards Institution, London (1976).
15. BS 1643: *Vernier height gauges*, British Standards Institution, London (1983).
16. BS 907: *Dial gauges for linear measurements*, British Standards Institution, London (1965).
17. ISO/R 463: *Dial gauges reading in 0.01 mm, 0.001 inch and 0.0001 inch*, International Organization for Standards, Geneva (1965).
18. BS 1685: *Bevel protractors (mechanical and optical)*, British Standards Institution, London (1951).
19. BS 958: *Spirit levels for use in precision engineering*, British Standards Institution, London (1968).

20. NAMAS Document B 7002: *Length bars – calibration procedures and uncertainties*, NAMAS Executive, National Physical Laboratory, Middlesex, UK (1982).
21. BS 3064: *Metric sine bars and sine tables (excluding compound tables)*, British Standards Institution, London (1978).

CHAPTER 17

Volume Flow Rate Calibration

Volume flow rate measurement is extremely important in all the process industries. It is used for quantifying the flow of all materials that are in a gaseous, liquid or semi-liquid slurry form and carried in pipes.

17.1 REVIEW OF INSTRUMENTS MEASURING VOLUME FLOW RATE

A wide range of instruments have been developed for measuring volume flow rate and they can be divided into the following classes:

1. Differential pressure meters
2. Variable-area meters
3. Positive displacement meters
4. Turbine flowmeters
5. Electromagnetic flowmeters
6. Vortex shedding flowmeters
7. Gate-type meters
8. Ultrasonic flowmeters
9. Cross-correlation flowmeters
10. Laser–Doppler flowmeters

The number of relevant factors to be considered when specifying a flowmeter for a particular application is very large. These include the temperature and pressure of the fluid, its density, viscosity, chemical properties and abrasiveness, whether it contains particles, whether it is a liquid or gas, etc.

The required performance factors of accuracy, measurement range, acceptable pressure drop, output signal characteristics, reliability and service life must also be assessed.

Differential Pressure Meters

Differential pressure meters involve the insertion of some device into a fluid-carrying pipe which causes an obstruction and creates a pressure difference on either side of the device. The orifice plate is very much the most common of such devices and accounts for 95 percent of all differential pressure instruments. Other devices in this category include the Venturi tube, the flow nozzle, the Dall flow tube and the Pitot tube. When such a restriction is placed in a pipe, the velocity of the fluid through the restriction increases and the pressure decreases. The volume flow rate is then proportional to the square root of the pressure difference across the obstruction. The manner in which this pressure difference is measured is important. Measuring the two pressures with different instruments and calculating the difference between the two measurements is not satisfactory because of the large measurement error which can arise when the pressure difference is small. The normal procedure is therefore to use a diaphragm-based differential pressure transducer.

All applications of this method of flow measurement assume that flow conditions upstream of the obstruction device are in steady state, and a certain minimum length of straight run of pipe ahead of the flow measurement point is specified to ensure this. The minimum lengths required for various pipe diameters are specified in British Standards tables (and also in alternative but equivalent national standards used in other countries), but a useful rule of thumb widely used in the process industries is to specify a length of 10 times the pipe diameter. If physical restrictions make this impossible to achieve, special flow smoothing vanes can be inserted immediately ahead of the measurement point.

Flow-restriction-type instruments are popular because they have no moving parts and are therefore robust, reliable and easy to maintain. One disadvantage of this method is that the obstruction causes a permanent loss of pressure in the flowing fluid. The magnitude and hence importance of this loss depends on the type of obstruction element used, but where the pressure loss is large, it is sometimes necessary to recover the lost pressure by an auxiliary pump further down the flow line. This class of device is not normally suitable for measuring the flow of slurries as the tappings into the pipe to measure the differential pressure are prone to blockage, although the Venturi tube can be used to measure the flow of dilute slurries.

Figure 17.1 illustrates approximately the way in which the flow pattern is interrupted when an orifice plate is inserted into a pipe. The other obstruction devices also have a similar effect to this. Of particular interest is the fact that the minimum cross-sectional area of flow occurs not within the obstruction but at a point downstream of it. Knowledge of the pattern of pressure

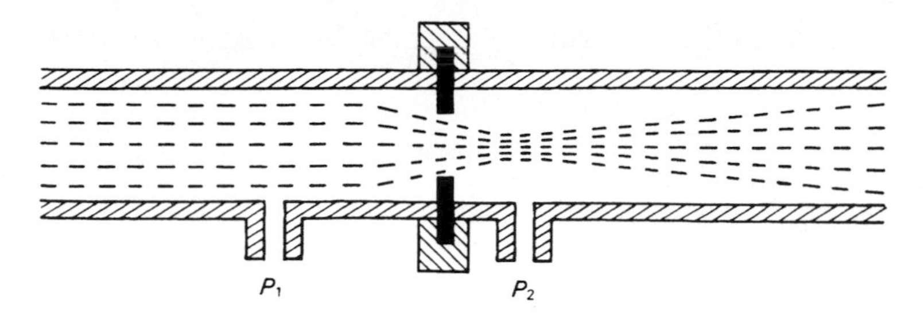

Figure 17.1. Profile of flow across orifice plate

variation along the pipe, as shown in Figure 17.2, is also of importance in using this technique of flow measurement. This shows that the point of minimum pressure coincides with the point of minimum cross-section flow, a little way downstream of the obstruction. Figure 17.2 also shows that there is a small rise in pressure immediately before the obstruction. It is therefore important not only to position the instrument measuring P_2 exactly at the point of minimum pressure, but also to measure the pressure P_1 at a point upstream of the point where the pressure starts to rise before the obstruction.

In the absence of any heat transfer mechanisms, and assuming frictionless flow of an incompressible fluid through the pipe, the theoretical volume flow rate of the fluid, Q, is given by [1]

$$Q = \frac{A_2}{\sqrt{[1 - (A_2/A_1)^2]}} \cdot \sqrt{[2(P_1 - P_2)/d]} \qquad (17.1)$$

where A_1 and P_1 are the cross-sectional area and pressure of the fluid flow before the obstruction, A_2 and P_2 are the cross-sectional area and pressure

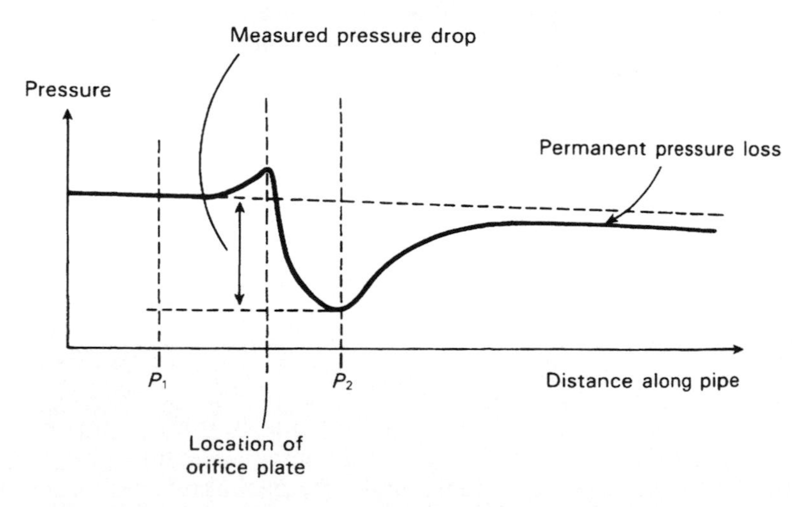

Figure 17.2. Pattern of pressure variation along pipe obstructed by an orifice plate

of the fluid flow at the narrowest point of the flow beyond the obstruction, and d is the fluid density.

Equation (17.1) is never applicable in practice for several reasons. Firstly, frictionless flow is never achieved. However, in the case of turbulent flow through smooth pipes, friction is low and it can be adequately accounted for by a variable called the *Reynold's number*, which is a measurable function of the flow velocity and the viscous friction. The other reasons for the non-applicability of equation (17.1) are that the initial cross-sectional area of the fluid flow is less than the diameter of the pipe carrying it, and that the minimum cross-sectional area of the fluid is less than the diameter of the obstruction. Therefore, neither A_1 nor A_2 can be measured. These problems are taken account of by modifying equation (17.1) to the following:

$$Q = \frac{C_D \cdot A_2'}{\sqrt{[1 - (A_2'/A_1')^2]}} \cdot \sqrt{[2(P_1 - P_2)/d]} \tag{17.2}$$

where A_1' and A_2' are the pipe diameters before and at the obstruction and C_D is a constant, known as the discharge coefficient, which accounts for the Reynold's number and the difference between the pipe and flow diameters.

Before equation (17.2) can be evaluated, the discharge coefficient must be calculated. As this varies between each measurement situation, it would appear at first sight that the discharge coefficient must be determined by practical experimentation in each case. However, provided that certain conditions are met, standard tables can be used to obtain the value of the discharge coefficient appropriate to the pipe diameter and fluid involved.

It is particularly important in applications of flow restriction methods to choose an instrument whose range is appropriate to the magnitudes of flow rate being measured. This requirement arises because of the square-root type of relationship between the pressure difference and the flow rate, which means that as the pressure difference decreases, the error in flow-rate measurement can become very large. In consequence, restriction-type flowmeters are only suitable for measuring flow rates between of 30 percent and 100 percent of the instrument range.

Orifice Plate

The orifice plate is a metal disc with a hole in it, as shown in Figure 17.3, inserted into the pipe carrying a flowing fluid. This hole is normally concentric with the disc. Over 50 percent of the instruments used in industry for measuring volume flow rate are of this type. The use of the orifice plate is so widespread because of its simplicity, cheapness and availability in a wide range of sizes. However, the best accuracy obtainable with this type of obstruction device is only ±2 percent and the permanent pressure loss caused in the flow is very high, being between 50 percent and 90 percent of the pressure difference $(P_1 - P_2)$ in magnitude. Other problems with the orifice plate are a gradual change in the discharge coefficient

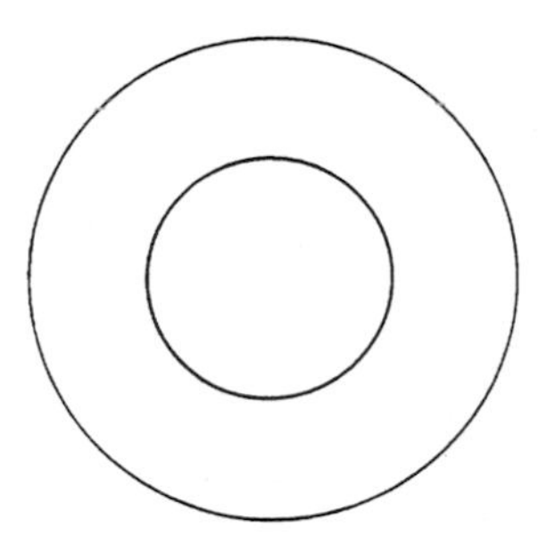

Figure 17.3. Orifice plate

over a period of time as the sharp edges of the hole wear away, and a tendency for any particles in the flowing fluid to stick behind the hole and gradually build up and reduce its diameter. The latter problem can be minimized by using an orifice plate with an eccentric hole. If this hole is close to the bottom of the pipe, solids in the flowing fluid tend to be swept through, and build-up of particles behind the plate is minimal.

A very similar problem arises if there are any bubbles of vapour or gas in the flowing fluid when liquid flow is involved. These also tend to build up behind an orifice plate and distort the pattern of flow. This difficulty can be avoided by mounting the orifice plate in a vertical run of pipe.

Flow Nozzle

The form of a flow nozzle is shown in Figure 17.4. This is not prone to solid particles or bubbles of gas in a flowing fluid sticking in the flow restriction, and so in this respect it is superior to the orifice plate. Its useful working life is also greater because it does not get worn away in the same way as an orifice plate. These factors contribute to giving the instrument a greater measurement accuracy and to a need for calibration at less frequent intervals. However, as the engineering effort involved in fabricating a flow nozzle is greater than that required to make an orifice plate, the instrument is somewhat more expensive. In terms of the permanent pressure loss imposed on the measured system, the flow nozzle is very similar to the orifice plate. A typical application of the flow nozzle is in measurement of steam flow.

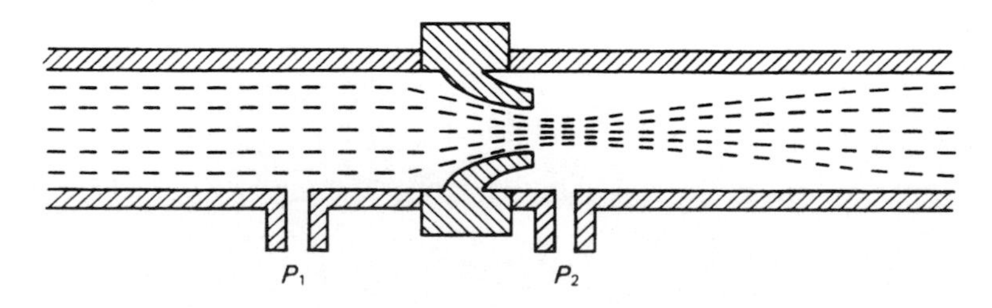

Figure 17.4. Flow nozzle

Venturi

The venturi is a precision-engineered tube of a special shape, as shown in Figure 17.5. It is a very expensive instrument but offers very good accuracy and imposes a permanent pressure loss on the measured system of only 10–15 percent of the pressure difference ($P_1 - P_2$) across it. The smooth internal shape of this type of restriction means that it is unaffected by solid particles or gaseous bubbles in the flowing fluid, and in fact can even cope with dilute slurries. It has almost no maintenance requirements and its working life is very long.

Dall Flow Tube

The Dall flow tube, shown in Figure 17.6, consists of two conical reducers inserted into the fluid-carrying pipe. It has a very similar internal shape to the venturi, except that it lacks a throat. This construction is much easier to manufacture than a venturi (which requires complex machining), and this gives the Dall flow tube an advantage in cost over the venturi, although the measurement accuracy obtained is not quite as good. Another advantage of the Dall flow tube is its shorter length, which makes the engineering task of inserting it into the flow line easier. The Dall tube has one further operational advantage, in that the permanent pressure loss imposed on the system is only about 5 percent of the measured pressure difference

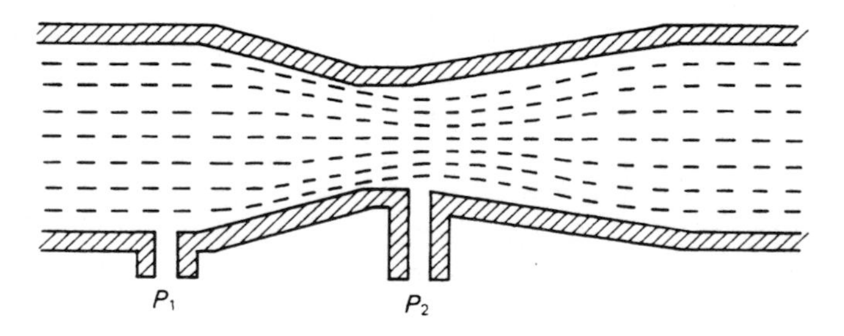

Figure 17.5. Venturi

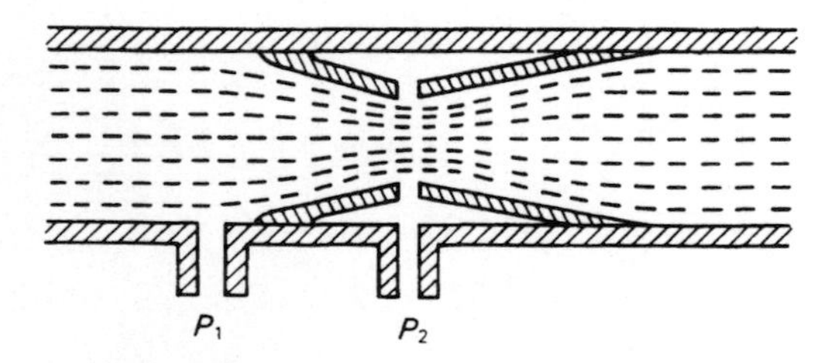

Figure 17.6. Dall flow tube

$(P_1 - P_2)$, and thus is only half that due to a venturi. In other respects, the two instruments are very similar, with their low maintenance requirement and long life.

Pitot Tube

The Pitot-static tube is mainly used for making temporary measurements of flow, although it is also used in some instances for permanent flow monitoring. The instrument depends on the principle that a tube placed with its open end in a stream of fluid, as shown in Figure 17.7, will bring to rest that part of the fluid which impinges on it, and the loss of kinetic energy will be converted to a measurable increase in pressure inside the tube.

The flow velocity can be calculated from the formula:

$$v = C \cdot \sqrt{(2g[P_1 - P_2])}$$

The constant C, known as the Pitot tube coefficient, is a factor which corrects for the fact that not all fluid incident on the end of the tube will be brought to rest: a proportion will slip around it according to the design of the tube.

Having calculated v, the volume flow rate can then be calculated by multiplying v by the cross-sectional area of the flow pipe, A. Inferring the volume flow rate from measurement of the flow velocity at one point in the fluid obviously requires the flow profile to be very uniform. If this condition is not met, multiple Pitot tubes can be used to measure velocities across the whole pipe cross-section.

Pitot tubes have the advantage that they cause negligible pressure loss in the flow. They are also cheap, and the installation procedure consists of the very simple process of pushing them down a small hole drilled in the flow-carrying pipe.

Their main failing is that the lowest measurement uncertainty achievable is usually about ±5 percent, and sensitive pressure-measuring devices are needed to achieve even this limited level of accuracy, as the pressure difference created is very small. More recently, measurement capabilities with an uncertainty down to ±1 percent have been claimed for specially designed Pitot tubes [2].

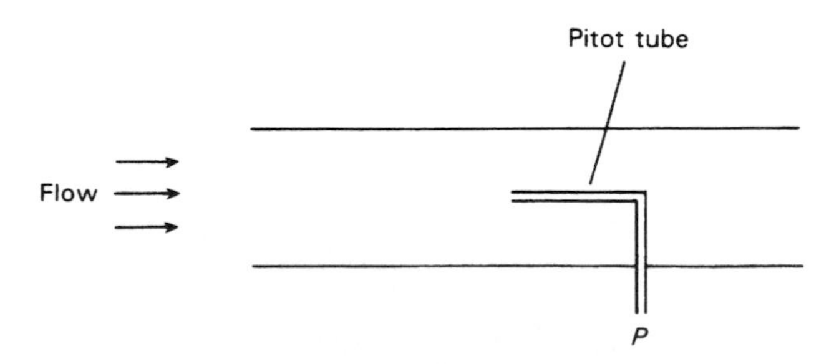

Figure 17.7. Pitot tube

Variable Area Flowmeters

In this class of flowmeter, the differential pressure across a variable aperture is used to adjust the area of the aperture. The aperture area is then a measure of the flow rate. This type of instrument only gives a visual indication of flow rate, and so is of no use in automatic control schemes. However, it is reliable, cheap and used extensively throughout industry, accounting for 20 percent of all flowmeters sold.

In its simplest form, shown in Figure 17.8, the instrument consists of a tapered glass tube containing a float which takes up a stable position where its submerged weight is balanced by the upthrust due to the differential pressure across it. The position of the float is a measure of the effective annular area of the flow passage and hence of the flow rate. The accuracy of the cheapest instruments is only ±3 percent, but more expensive versions offer measurement accuracies as high as ±0.2 percent. The normal measurement range is between 10 percent and 100 percent of the full-scale reading for any particular instrument.

Positive-Displacement Flowmeters

All positive-displacement meters operate by using mechanical divisions to displace discrete volumes of fluid successfully. Although this principle of operation is common, many different mechanical arrangements exist for putting the principle into practice. All versions of positive-displacement meter are low-friction, low-maintenance and long-life devices, although they do impose a small permanent pressure loss on the flowing fluid. Low friction is especially important when measuring gas flows, and meters with special mechanical arrangements to satisfy this requirement have been developed.

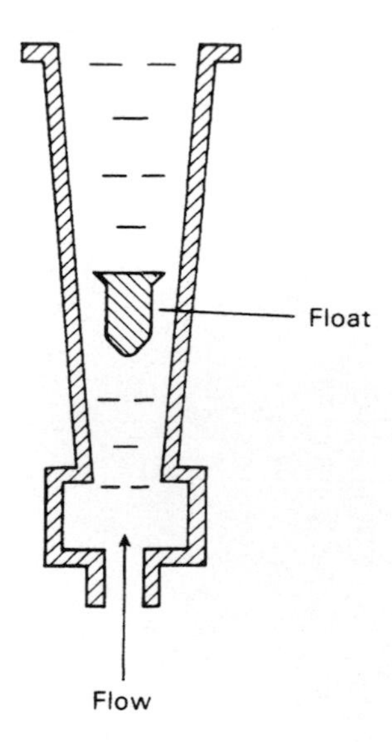

Figure 17.8. Variable-area flowmeter

The rotary piston meter is the most common type of positive-displacement meter, and this is illustrated in Figure 17.9. This uses a cylindrical piston which is displaced around a cylindrical chamber by the flowing fluid. Rotation of the piston drives an output shaft. This can either be used with a pointer-and-scale system to give a visual flow reading or it can be converted into an electrical output signal.

Positive-displacement flowmeters account for nearly 10 percent of the total number of flowmeters used in industry. Such devices are used in large numbers for metering domestic gas and water consumption and they are also capable of measuring the flow of high-viscosity fluids. The cheapest instruments have an accuracy of about ± 1.5 percent, but the accuracy in more expensive ones can be as good as ± 0.2 percent. These higher quality instruments are used extensively within the oil industry, as such an application can justify their high cost.

Turbine Meters (Inferential Meters)

A turbine or inferential flowmeter consists of a multi-bladed wheel mounted in a pipe along an axis parallel to the direction of fluid flow in the pipe, as shown in Figure 17.10. The flow of fluid past the wheel causes it to rotate at a rate which is proportional to the volume flow rate of the fluid. This rate of rotation is measured by constructing the flowmeter so that it behaves as a variable reluctance tachogenerator. This is achieved by fabricating the turbine blades from a ferromagnetic material and placing a permanent magnet and coil inside the meter housing. A voltage pulse is induced in the coil as each blade on the turbine wheel moves past it, and if these pulses are measured by a pulse counter, the pulse frequency and hence the flow rate can be deduced. Provided that the turbine wheel is mounted on low-friction bearings, measurement uncertainty can be as low as ± 0.1 percent. However, turbine flowmeters are less rugged and reliable

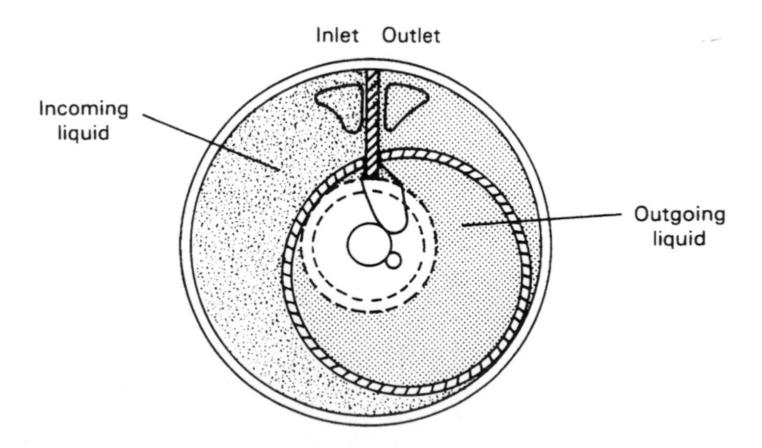

Figure 17.9. Rotaty piston form of positive displacement flowmeter

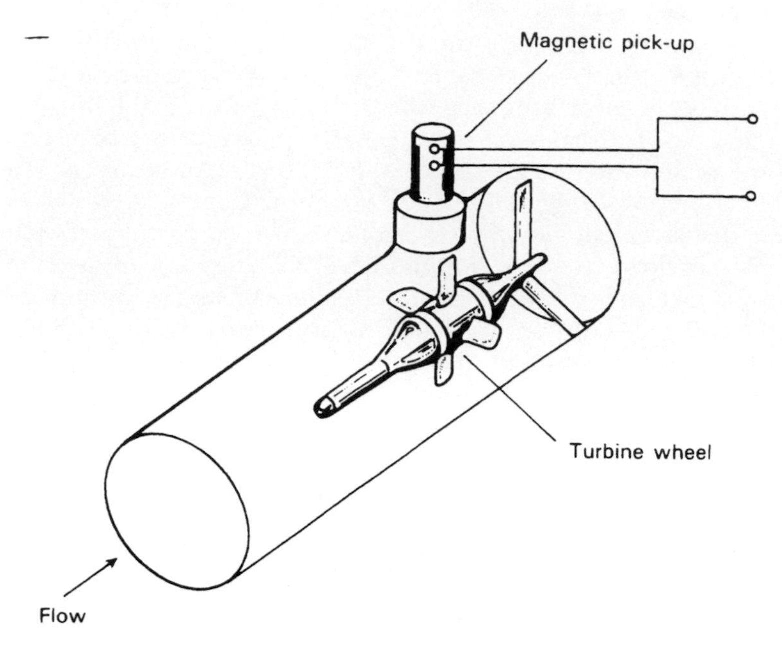

Figure 17.10. Turbine flowmeter

than flow-restriction-type instruments, and are badly affected by any particulate matter in the flowing fluid. Bearing wear is a particular problem and they also impose a permanent pressure loss on the measured system.

Turbine meters are particularly prone to large errors when there is any significant second phase in the fluid measured. For instance, using a turbine meter calibrated on pure liquid to measure a liquid containing 5 percent air produces a 50 percent measurement error. As an important application of the turbine meter is in the petrochemical industries, where gas–oil mixtures are common, special procedures are being developed to avoid such large measurement errors. The most promising approach is to homogenize the two gas–oil phases prior to flow measurement [3].

Turbine meters have a similar cost and market share to positive-displacement meters, and compete for many applications, particularly in the oil industry. Turbine meters are smaller and lighter than positive-displacement meters and are preferred for low-viscosity, high-flow measurements. Positive-displacement meters are superior, however, in conditions of high viscosity and low flow rate.

Electromagnetic Flowmeters

Electromagnetic flowmeters are limited to measuring the volume flow rate of electrically conductive fluids and account for about 5 percent of the flowmeters used in industry. A reasonable measurement accuracy is given, with uncertainty levels typically around ±1.0 percent, but such instruments are expensive to buy. A further significant expense is the need for careful calibration of each instrument individually during manufacture, as there is considerable variation in the properties of the magnetic materials used. Running costs in terms of electricity consumption are also generally high, although versions with a much lower power consumption (and ±0.5 percent inaccuracy) are now becoming available [4].

The instrument, shown in Figure 17.11, consists of a stainless-steel cylindrical tube, fitted with an insulating liner, which carries the measured fluid. Typical lining materials used are Neoprene, polytetrafluoroethylene (PTFE) and polyurethane. A magnetic field is created in the tube by placing mains-energized field coils either side of it, and the voltage induced in the fluid is measured by two electrodes inserted into opposite sides of the tube. The ends of these electrodes are usually flush with the inner surface of the cylinder. The electrodes are constructed from a material which is unaffected by most types of flowing fluid, such as stainless steel, platinum–iridium alloys, hastelloy, titanium or tantalum. In the case of the rarer metals in this list, the electrodes account for a significant part of the total instrument cost.

By Faraday's law of electromagnetic induction, the voltage, E, induced across a length, L, of flowing fluid moving at velocity v in a magnetic field of flux density B, is given by:

$$E = B \cdot L \cdot v \tag{17.3}$$

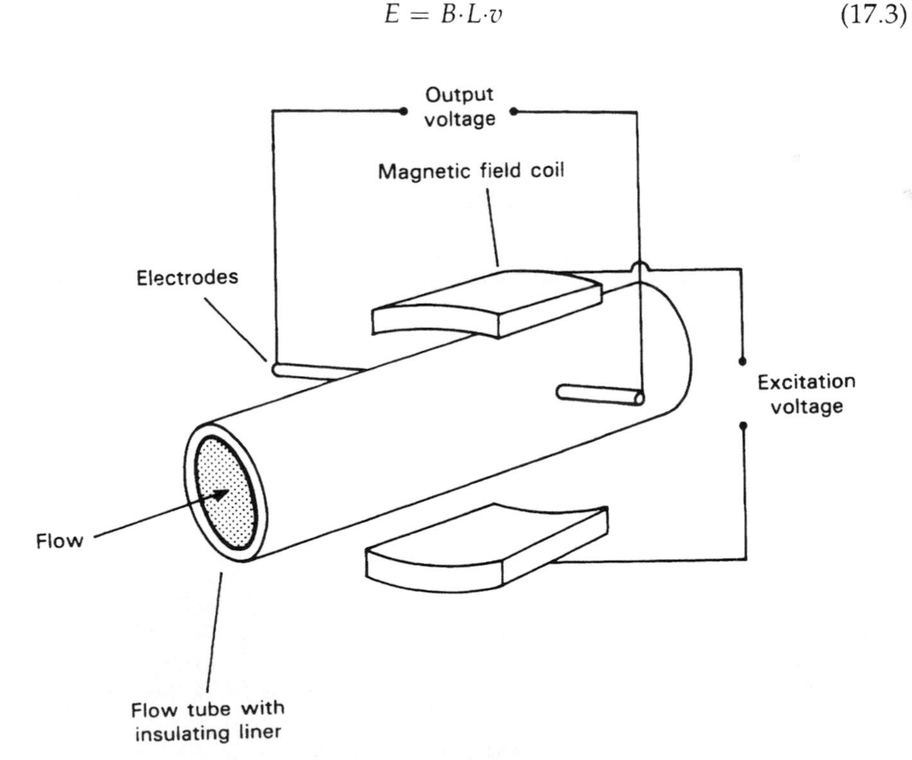

Figure 17.11. Electromagnetic flowmeter

L is the distance between the electrodes, which is the diameter of the tube, and B is a known constant. Hence, measurement of the voltage E induced across the electrodes allows the flow velocity v to be calculated from equation (17.3). Having thus calculated v, it is a simple matter to multiply v by the cross-sectional area of the tube to obtain a value for the volume flow rate. The typical voltage signal measured across the electrodes is 1 mV when the fluid flow rate is 1 m/s.

The internal diameter of a magnetic flowmeter is normally the same as that of the rest of the flow-carrying pipework in the system. Therefore, there is no obstruction to the fluid flow and consequently no pressure loss associated with measurement. Like other forms of flowmeter, the magnetic type requires a minimum length of straight pipework immediately prior to the point of flow measurement in order to guarantee the accuracy of measurement, although a length equal to five pipe diameters is usually sufficient.

Whilst the flowing fluid must be electrically conductive, the method is of use in many applications and is particularly useful for measuring the flow of slurries in which the liquid phase is electrically conductive. It is also suitable for measuring the flow of many types of corrosive fluid. At the present time, magnetic flowmeters account for about 15 percent of the new flowmeters sold and this total is slowly growing. One operational problem is that the insulating lining is subject to damage when abrasive fluids are being handled, and this can give the instrument a limited life.

Current new developments in electromagnetic flowmeters are producing physically smaller instruments and employing better coil designs which reduce electricity consumption. Also, whereas conventional electromagnetic flowmeters require a minimum fluid conductivity of 10 μmho/cm^3, some new versions are now becoming available which can cope with fluid conductivities as low as 1 μmho/cm^3.

Vortex-shedding Flowmeters

Vortex-shedding flowmeters only account for about 1 percent of flowmeters sold at present, but this percentage is likely to grow in the future as the characteristics of these instruments become more generally known. The operating principle of the instrument is based on the natural phenomenon of vortex shedding, created by placing an unstreamlined obstacle (known as a bluff body) in a fluid-carrying pipe, as indicated in Figure 17.12. When fluid flows past the obstacle, boundary layers of viscous, slow-moving fluid are formed along the outer surface. Because the obstacle is not streamlined, the flow cannot follow the contours of the body on the downstream side, and the separate layers become detached and roll into eddies or vortices in the low-pressure region behind the obstacle. The shedding frequency of these alternately shed vortices is proportional to the fluid velocity past the body. Various thermal, magnetic, ultrasonic and capacitive vortex detection techniques are employed in different instruments.

Such instruments have no moving parts, operate over a wide flow range,

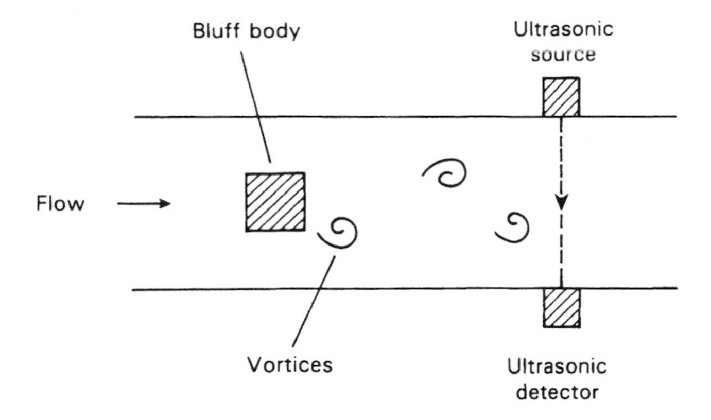

Figure 17.12. Vortex-shedding flowmeter

have a low power consumption and require little maintenance. They can measure both liquid and gas flows and a common inaccuracy figure quoted is ± 1 percent of full-scale reading, though this can be seriously downgraded in the presence of flow disturbances upstream of the measurement point, and a straight run of pipe before the measurement point of fifty pipe diameters is recommended. Another problem with the instrument is its susceptibility to pipe vibrations, although new designs are becoming available which have a better immunity to such vibrations.

Gate-type Meters

The gate meter was the earliest device in this class. It consists of a spring-loaded, hinged flap mounted at right angles to the direction of fluid flow in the fluid-carrying pipe. The flap is connected to a pointer outside the pipe. The fluid flow deflects the flap and pointer and the flow rate is indicated by a graduated scale behind the pointer. The major difficulty with such devices is in preventing leaks at the hinge point.

A variation on this principle is the air-vane meter which measures deflection of the flap by a potentiometer inside the pipe. This is commonly used to measure air flow within automotive fuel-injection systems.

Another device in this class is the target meter. This consists of a circular-disc-shaped flap in the pipe. Fluid flow rate is inferred from the force exerted on the disc, measured by strain gauges bonded to it. This meter is very useful for measuring the flow of dilute slurries but it does not find wide application elsewhere as it has a relatively high cost.

Ultrasonic Flowmeters

The ultrasonic technique of volume flow rate measurement is, like the magnetic flowmeter, a non-invasive method. It is not restricted to conductive fluids, however, and is particularly useful for measuring the flow of corrosive fluids and slurries. A further advantage over magnetic flowmeters is that the instrument is one which clamps on externally to existing pipework rather than being inserted as an integral part of the flow line, as in the case of the magnetic flowmeter. As the procedure of breaking into a pipeline to insert a flowmeter can be as expensive as the cost of the flowmeter itself, the ultrasonic flowmeter has significant cost advantages. Its clamp-on mode of operation has safety advantages in avoiding the possibility of personnel installing flowmeters coming into contact with hazardous fluids such as poisonous, radioactive, flammable or explosive ones. Also, any contamination of the fluid being measured (e.g. food substances and drugs) is avoided. The introduction of this type of flowmeter is a comparatively recent one and its present market share is very small. In view of its distinct advantages, however, its industrial use is likely to increase over the next few years.

Two different types of ultrasonic flowmeter exist which employ distinct technologies, based on Doppler shift and on transit time. In the past, this has not always been readily understood, and has resulted in ultrasonic technology being rejected entirely when one of these two forms has been found to be unsatisfactory in a particular application. This is unfortunate, because the two technologies have distinct characteristics and areas of application, and many situations exist where one form is very suitable and the other is not. To reject both, having only tried out one, is therefore a serious mistake. Rough guidelines about which type to choose could be stated as follows: for clean liquids use transit-time types and for gritty, aerated liquids use Doppler-shift types.

Particular care has to be taken to ensure a stable flow profile in ultrasonic flowmeter applications. It is usual to increase the normal specification of the minimum length of straight pipe-run prior to the point of measurement, expressed as a number of pipe diameters, from a figure of 10 up to 20 or in some cases even 50 diameters. Analysis of the reasons for poor performance in many instances of ultrasonic flowmeter application has shown failure to meet this stable flow-profile requirement to be a significant factor.

Doppler Shift Ultrasonic Flowmeter

The principle of operation of the Doppler-shift flowmeter is shown in Figure 17.13. A fundamental requirement of these instruments is the presence of scattering elements within the flowing fluid which deflect the ultrasonic energy output from the transmitter so that it enters the receiver. These can be provided by either solid particles, gas bubbles or eddies in the flowing fluid. The scattering elements cause a frequency shift between the transmitted and reflected ultrasonic energy, and measurement of this shift enables the fluid velocity to be inferred.

The instrument consists essentially of an ultrasonic transmitter–receiver

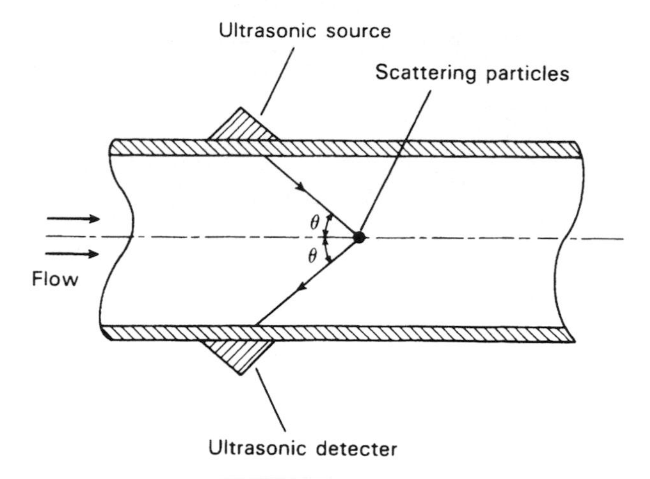

Figure 17.13. Doppler-shift ultrasonic flowmeter

pair clamped onto the outside wall of a fluid-carrying vessel. Ultrasonic energy consists of a train of short bursts of sinusoidal waveforms at a frequency between 0.5 MHz and 20 MHz. This frequency range is described as ultrasonic because it is outside the range of human hearing. The flow velocity, v, is given by

$$v = \frac{c(f_t - f_r)}{2 \cdot f_t \cos(\theta)} \tag{17.4}$$

where f_t and f_r are the frequencies of the transmitted and received ultrasonic waves respectively, c is the velocity of sound in the fluid being measured, and θ is the angle that the incident and reflected energy waves make with the axis of flow in the pipe. Volume flow rate is then readily calculated by multiplying the measured flow velocity by the cross-sectional area of the fluid-carrying pipe.

The electronics involved in Doppler-shift flowmeters is relatively simple and therefore cheap. Ultrasonic transmitters and receivers are also relatively inexpensive, being based on piezoelectric oscillator technology. As all of its components are cheap, the Doppler-shift flowmeter itself is inexpensive. The measurement accuracy obtained depends on many factors such as the flow profile, the constancy of pipe-wall thickness, the number, size and spatial distribution of scatterers, and the accuracy with which the speed of sound in the fluid is known. Consequently, accurate measurement can only be achieved by the tedious procedure of carefully calibrating the instrument for each particular flow measurement application. Otherwise measurement errors can approach ± 10 percent of the reading, and for this reason Doppler-shift flowmeters are often used merely as flow indicators, rather than for accurate quantification of the volume flow rate.

Versions are now available which avoid the problem of variable pipe thickness by being fitted inside the flow pipe, flush with its inner surface. Accuracy of ±0.5 percent is claimed for such devices. Other recent developments are the use of multiple-path ultrasonic flowmeters which use an array of ultrasonic elements to obtain an average velocity measurement. This substantially reduces the error due to non-uniform flow profiles. There is a substantial cost penalty involved in this, however.

Transit-time Ultrasonic Flowmeter

The transit-time ultrasonic flowmeter is an instrument designed for measuring the volume flow rate in clean liquids or gases. It consists of a pair of ultrasonic transducers mounted along an axis aligned at an angle θ with respect to the fluid-flow axis, as shown in Figure 17.14. Each transducer consists of a transmitter–receiver pair, with the transmitter emitting ultrasonic energy which travels across to the receiver on the opposite side of the pipe. These ultrasonic elements are normally piezoelectric oscillators of the same type as used in Doppler-shift flowmeters. Fluid flowing in the pipe causes a time difference between the transit times of the beams travelling upstream and downstream, and measurement of this difference allows the flow velocity to be calculated. The typical magnitude of this time difference is 100 ns in a total transit time of 100 μs, and high precision electronics are therefore needed to measure it. There are three distinct ways of measuring the time shift. These are direct measurement, conversion to a phase change and conversion to a frequency change. The third of these options is particularly attractive as it obviates the need to measure the speed of sound in the measured fluids which the first two methods require. A scheme applying this third option is shown in Figure 17.15. This also multiplexes the transmitting and receiving functions, so that only one ultrasonic element is needed in each transducer.

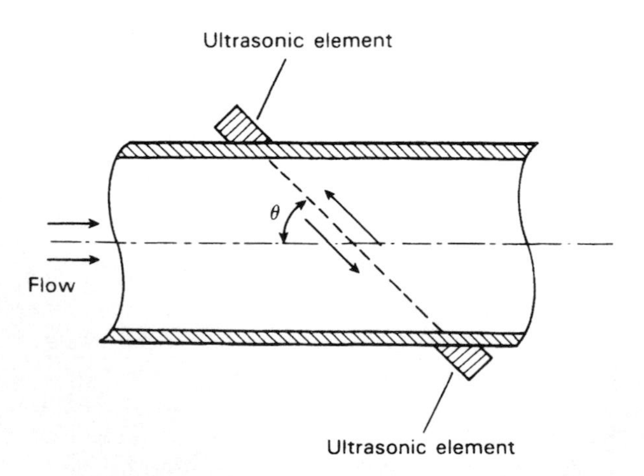

Figure 17.14. Transit-time ultrasonic flowmeter

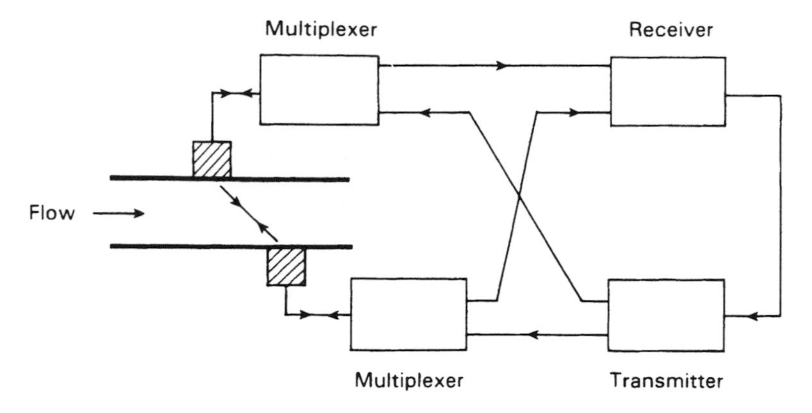

Figure 17.15. Transit-time measurement system

The forward and backward transit times across the pipe, T_f and T_b, are given by

$$T_f = \frac{L}{c + v \cdot \cos(\theta)}; \quad T_b = \frac{L}{c - v \cdot \cos(\theta)}$$

where c is the velocity of sound in the fluid, v is the flow velocity, L is the distance between the ultrasonic transmitter and receiver, and θ is the angle of the ultrasonic beam with respect to the fluid flow axis.

The time difference (TD) is given by:

$$\mathrm{TD} = T_b - T_f = \frac{2 \cdot v \cdot L \cdot \cos(\theta)}{c_2 - v^2 \cdot \cos^2(\theta)}$$

This requires knowledge of c before it can be solved. However, a solution can be found much more simply if the receipt of a pulse is used to trigger the transmission of the next ultrasonic energy pulse. Then, the frequencies of the forward and backward pulse trains are given by

$$F_f = \frac{1}{T_f} = \frac{c - v \cdot \cos(\theta)}{L}; \quad F_b = \frac{1}{T_b} = \frac{c + v \cdot \cos(\theta)}{L}$$

If the two frequency signals are now multiplied together, the resulting beat frequency (BF) is given by

$$BF = F_b - F_f = \frac{2 \cdot v \cdot \cos(\theta)}{L}$$

c has now been eliminated and v can be calculated from a measurement of BF as

$$v = \frac{L \cdot BF}{2 \cdot \cos(\theta)}$$

Transit-time flowmeters are of more general use than Doppler-shift flowmeters, particularly where the pipe diameter involved is large and hence the transit time is consequently sufficiently large to be measured with reasonable accuracy. It is possible then to achieve an accuracy figure of ± 0.5 percent. The instrument costs more than a Doppler-shift flowmeter, however, because of the greater complexity of the electronics needed to make accurate transit-time measurements.

Cross-correlation Flowmeters

Cross-correlation flowmeters are a type of flowmeter which has not yet achieved widespread practical use in industry. Much development work is still going on, and they therefore mainly exist only as prototypes in research laboratories. However, they are included here because their use is likely to become much more widespread in the future. Cost is likely to be similar to that of the larger versions of electromagnetic flowmeter.

Such instruments require some detectable random variable to be present in the flowing fluid. This can take forms such as velocity turbulence and temperature fluctuations. When such a stream of variables is detected by a sensor, the output signal generated consists of noise with a wide frequency spectrum.

Cross-correlation flowmeters use two such sensors placed a known distance apart in the fluid-carrying pipe and cross-correlation techniques are applied to the two output signals from these sensors. This procedure compares one signal with progressively time-shifted versions of the other signal until the best match is obtained between the two wave-forms. If the distance between the sensors is divided by this time shift, a measurement of the flow velocity is obtained. A digital processor is an essential requirement in calculating the cross-correlation function, and therefore the instrument must be properly described as an intelligent one. Measurement uncertainty is typically ± 2 percent.

In practice, the existence of random disturbances in the flow is unreliable, and their detection is difficult. To answer this problem, ultrasonic cross-correlation flowmeters are under development. These use ultrasonic transducers to inject disturbances into the flow and also to detect the disturbances further downstream.

A particular application which is likely to develop for cross-correlation flowmeters is in measuring multiphase flow where the fluid is a mixture of two or more gas, liquid and solid components. Successful measurement of solid–liquid flows has been demonstrated and trials with gas–liquid flows are encouraging [5].

Further information about cross-correlation flowmeters can be found in reference [6].

Laser-Doppler Flowmeter

This instrument gives direct measurements of flow velocity for liquids containing suspended particles flowing in a transparent pipe. Light from a laser is focused by an optical system to a point in the flow. The movement of particles causes a Doppler shift of the scattered light and produces a signal in a photodetector which is related to the fluid velocity. A very wide range of flow velocities between 10 μm/s and 800 m/s can be measured by this technique.

Sufficient particles for satisfactory operation are normally present naturally in most liquid and gaseous fluids, and the introduction of artificial particles is rarely needed. The technique is advantageous in measuring flow velocity directly rather than inferring it from a pressure difference. It also causes no interruption in the flow and, as the instrument can be made very small, it can measure velocity in confined areas. One limitation is that it measures local flow velocity in the vicinity of the focal point of the light beam, which can lead to large errors in the estimation of mean volume flow rate if the flow profile is not uniform. However, this limitation is often used constructively in applications of the instrument where the flow profile across the cross-section of a pipe is determined by measuring the velocity at a succession of points.

Intelligent Flowmeters

All the usual benefits associated with intelligent instruments are potentially applicable to many types of flowmeter. One general benefit in flowmeters is extension of the measurement range for a particular instrument whilst maintaining a good standard of accuracy. However, the availability of intelligent flowmeters in the market place is currently very limited.

Intelligent differential-pressure-measuring instruments can be used to good effect in conjunction with obstruction-type flow transducers. One immediate benefit of this in the case of the commonest flow-restriction device, the orifice plate, is to extend the lowest flow measurable with acceptable accuracy down to 20 percent of the maximum flow value.

In positive-displacement meters, intelligence allows compensation for thermal expansion of meter components and temperature-induced viscosity changes. Correction for variations in flow pressure is also provided for.

Intelligent electromagnetic flowmeters are also available, and these have a self-diagnosis and self-adjustment capability. The rangeability is 30:1 and quoted accuracy is ±0.5 percent. It is also normal to include a non-volatile memory to protect constants used for correcting for environmental inputs, etc., against power supply failures.

Intelligent turbine meters are able to detect their own bearing wear and also report deviations from initial calibration due to blade damage, etc. Some versions also have self-adjustment capability.

The trend is now moving towards total flow computers which can process inputs from almost any type of transducer. Such devices allow user input of parameters like specific gravity, fluid density, viscosity, pipe diameters, thermal expansion coefficients, discharge coefficients, etc. Auxiliary inputs from temperature transducers are also catered for. After processing the raw flow transducer output using this additional data, flow computers are able to produce measurements of flow to a very high degree of accuracy.

17.2 INTRODUCTION TO CALIBRATION

Calibration of volume flow-measuring instruments is a relatively expensive procedure even when only a moderate level of accuracy is demanded. Where high accuracy is required, the cost of calibration can be very great indeed. Therefore, in quality control systems that involve flow measurements, it is particularly important to establish exactly what accuracy level is needed so that the calibration system instituted does not cost more than necessary. In some cases, such as handling valuable fluids or where there are legal requirements as in petrol pumps, high accuracy levels (e.g. error $\leqslant 0.1$ percent) are justified. In other situations, however, such as in measuring additives to the main stream in a process plant, only low levels of accuracy are needed (e.g. error ≈ 5 percent).

It is the normal practice to calibrate process flow-measuring instruments on-site as far as possible. This ensures that calibration is performed under the actual flow conditions, which are difficult or impossible to reproduce exactly in a laboratory. This is necessary because the accuracy of flow measurement is greatly affected by the flow conditions and characteristics of the flowing fluid. On account of this, it is also standard practice to repeat flow calibration checks until the same reading is obtained in two consecutive tests. However, Berman [7] has suggested that even these precautions are inadequate and that statistical procedures are needed.

The equipment and procedures used for calibration depend on whether gaseous or liquid flows are being measured. Therefore, separate sections are devoted to each of these cases. It must also be stressed that all calibration procedures refer only to flows of single-phase fluids (i.e. liquids or gases). Where a second or third phase is present (i.e. there is a mixture of gas and/or liquid and/or solid), suitably accurate calibration techniques have yet to be established [3].

17.3 CALIBRATION EQUIPMENT AND PROCEDURES FOR LIQUID FLOW

Calibrated Tank

Probably the simplest piece of equipment available for calibrating instruments measuring liquid flow rates is the calibrated tank. This consists of a cylindrical vessel, as shown in Figure 17.16, with conical ends which

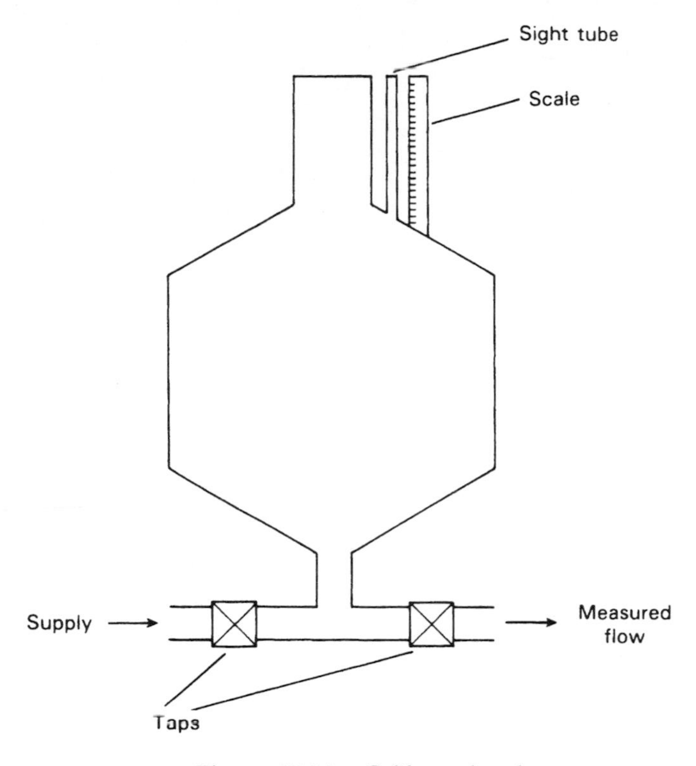

Figure 17.16. Calibrated tank

facilitate draining and cleaning of the tank. A *sight tube* with a graduated scale is placed alongside the final, upper, cylindrical part of the tank and this allows the volume of liquid in the tank to be measured accurately. Flow-rate calibration is performed by measuring the time taken, starting from an empty tank, for a given volume of liquid to flow into the vessel.

Because the calibration procedure starts and ends in zero-flow conditions, it is not suitable for calibrating instruments that are affected by flow acceleration and deceleration characteristics. This therefore excludes instruments like differential-pressure meters (orifice plate, flow nozzle, venturi, Dall flow tube, Pitot tube), turbine flowmeters and vortex-shedding flowmeters. The technique is further limited to the calibration of low-viscosity liquid flows, although lining the tank with an epoxy coating can allow the system to cope with somewhat higher viscosities. The limiting factor in this case is the drainage characteristics of the tank, which must be such that the residue liquid left after draining has an insufficient volume to affect the accuracy of the next calibration.

Pipe Prover

The commonest form of pipe prover is the bidirectional type, shown in Figure 17.17, which consists of a U-shaped tube of metal of accurately

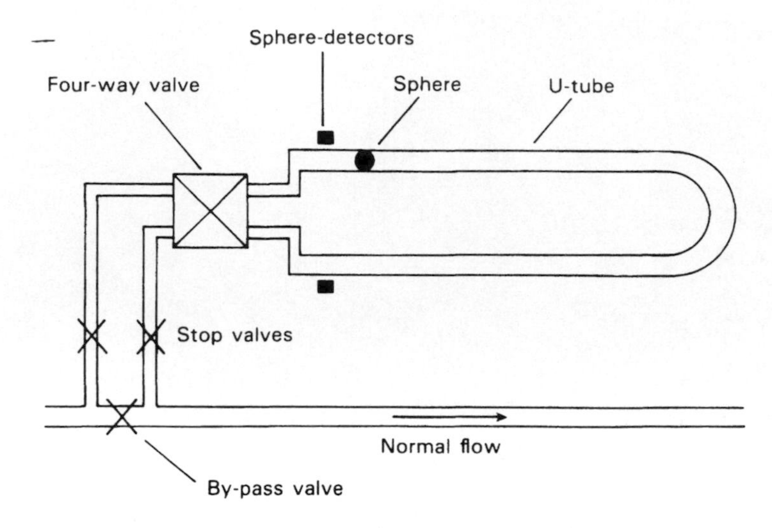

Figure 17.17. Bi-directional pipe prover

known cross-section. The purpose of the U-bend is to give a long flow path within a compact spatial volume. Alternative versions with more than one U-bend also exist to cater for situations where an even longer flow path is required. Inside the tube is a hollow, inflatable sphere which is filled with water until its diameter is about 2 percent larger than that of the tube. As such, the sphere forms a seal with the sides of the tube and acts as a piston. The prover is connected into the existing fluid-carrying pipe network via tappings either side of a bypass valve. A four-way valve at the start of the U-tube allows fluid to be directed in either direction around it. Calibration is performed by diverting flow into the prover and measuring the time taken for the sphere to travel between two detectors in the tube. The detectors are normally of an electromechanical, plunger type.

Unidirectional versions of the above also exists in which fluid only flows in one direction around the tube. A special handling valve has to be provided to return the sphere to the starting point after each calibration, but the absence of a four-way flow control valve makes such devices significantly cheaper than bidirectional types.

Pipe provers are particularly suited to the calibration of pressure-measuring instruments that have a pulse type of output, such as turbine meters. In such cases, the detector switches in the tube can be made to gate the instrument's output pulse counter. This enables not only the basic instrument to be calibrated, but also the ancillary electronics within it at the same time.

The inaccuracy level of such provers can be as low as ± 0.1 percent. This level of accuracy is maintained for high fluid-viscosity levels and also at very high flow rates. Even higher accuracy is provided by an alternative form of prover which consists of a long, straight metal tube containing a metal piston. However, such devices are more expensive than the other types discussed above and their large space requirements also often cause great difficulties.

Compact Prover

The compact prover has an identical operating principle to that of the other pipe provers described above but occupies a much smaller spatial volume. It is therefore used extensively in situations where there is insufficient room to use a larger prover. Many different designs of compact prover exist, operating both in the unidirectional and bidirectional modes, and one such design is shown in Figure 17.18. A common feature of compact provers is an accurately machined cylinder containing a metal piston which is driven between two reference marks by the flowing fluid. The instants at which the reference marks are passed are detected by switches, of optical form in the case of the version shown in Figure 17.18. Provision has to be made within these instruments for returning the piston to the starting point after each calibration and a hydraulic system is commonly used for this. Again, measuring the piston traverse time is made easier if the switches can be made to gate a pulse-train, and therefore compact provers are also most suited to instruments having a pulse-type output, such as turbine meters. Measurement uncertainty levels down to ±0.1 percent are possible.

The main technical difficulty in compact provers is measuring the traverse time, which can be as small as one second. The pulse count from a turbine meter in this time would typically be only about one hundred, making the possible measurement error 1 percent. To overcome this problem, electronic pulse-interpolation techniques have been developed which can count fractions of pulses.

Positive Displacement Meter

High quality versions of the positive-displacement flowmeter can be used as a reference standard in flowmeter calibration. The general principles of

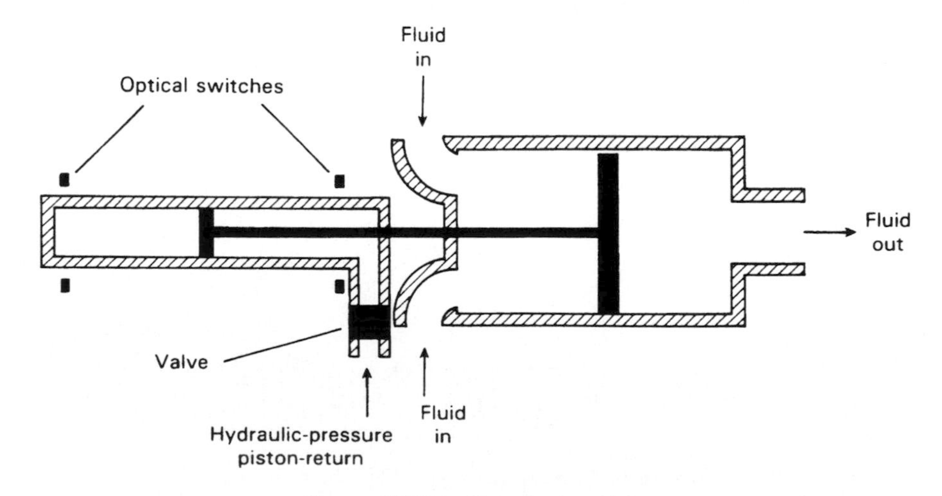

Figure 17.18. Compact prover

these were explained in Section 17.1. Such devices can give measurement inaccuracy levels down to ±0.2 percent.

Gravimetric Method

A variation on the principle of measuring the volume of liquid flowing in a given time is to weigh the quantity of fluid flowing in a given time. Apart from its applicability to a wider range of instruments, this technique is not limited to low-viscosity fluids as any residual fluid in the tank before calibration will be detected by the load-cells and therefore compensated for. In the simplest implementation of this system, fluid is allowed to flow for a measured length of time into a tank resting on load-cells. As before, the stop–start mode of fluid flow makes this method unsuitable for calibrating differential-pressure, turbine and vortex-shedding flowmeters. It is also unsuitable for measuring high flow rates because of the difficulty in bringing the fluid to rest. These restrictions can be overcome by directing the flowing fluid into the tank via diverter valves. In this alternative, it is important that the timing system be carefully synchronized with the operation of the diverter valves.

All versions of gravimetric calibration equipment are less robust than volumetric types and so on-site use is not recommended.

Orifice Plate

A flow line equipped with a certified orifice plate is sometimes used as a reference standard in flow calibration, especially for high flow rates through large-bore pipes. Whilst measurement uncertainty is of the order of ±1 percent at best, this is adequate for calibrating many flow-measuring instruments.

Turbine Meter

Turbine meters are also used as a reference standard for testing flowmeters. Their main application, as for orifice plates, is in calibrating high flow rates through large-bore pipes. Measurement uncertainty down to ±0.2 percent is attainable.

17.4 CALIBRATION EQUIPMENT AND PROCEDURES FOR GASEOUS FLOW

Calibration of gaseous flows poses considerable difficulties compared with calibrating liquid flows. These problems include the lower density of gases,

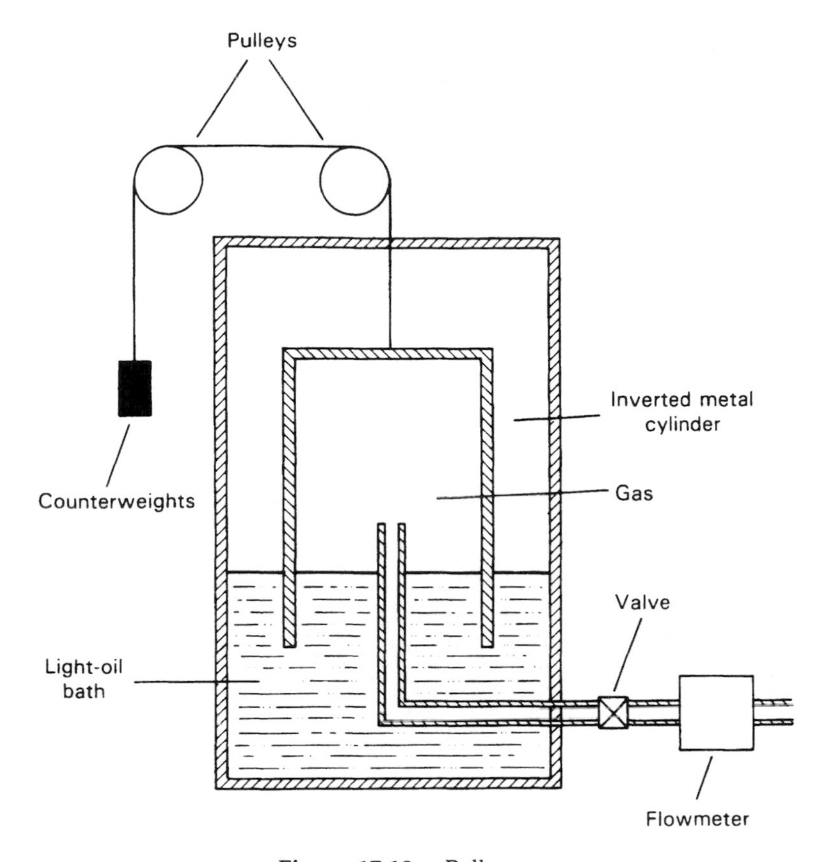

Figure 17.19. Bell prover

their compressibility and the difficulty in establishing a suitable liquid–air interface as utilized in many liquid-flow measurement systems.

In consequence, the main methods of calibrating gaseous flows, as described below, are small in number. Certain other specialized techniques, including the gravimetric method and the pressure–volume–temperature method, are also available, as described elsewhere [8]. These provide primary reference standards for gaseous flow calibration with measurement uncertainty down to ±0.3 percent. However, the expense of the equipment involved is such that it is usually only available in National Standards Laboratories.

Bell Prover

The bell prover consists of a hollow, inverted, metal cylinder suspended over a bath containing light oil, as shown in Figure 17.19. The air volume in the cylinder above the oil is connected, via a tube and a valve, to the flowmeter being calibrated. An air flow through the meter is created by allowing the cylinder to fall downwards into the bath, thus displacing the air contained within it. The flow rate, which is measured by timing the rate of fall of the cylinder, can be adjusted by changing the value of counterweights attached via a low-friction pulley system to the cylinder. This is essentially laboratory-only equipment and therefore on-site calibration is not possible.

Positive-displacement Meter

As for liquid-flow calibration, positive displacement flowmeters can be used for the calibration of gaseous flows with inaccuracy levels down to ±0.2 percent.

Compact Prover

Compact provers of the type used for calibrating liquid flows are unsuitable for application to gaseous flows. However, special designs of compact prover are being developed for gaseous flows [9] and hence such devices may find application in gaseous flow calibration in the future.

17.5 REFERENCE STANDARDS

Traceability of flow-rate calibration to fundamental standards is provided for by reference to primary standards of the separate quantities that flow rate is calculated from. Mass measurements are calibrated by comparison with a copy of the international standard kilogram (see Chapter 15) and time is calibrated by reference to a caesium resonator standard. Volume measurements are calibrated against standard reference volumes which are themselves calibrated gravimetrically using a mass-measurement system traceable to the standard kilogram.

REFERENCES

1. Bentley, J. P., *Principles of measurement systems*, Longman (1975).
2. Britton, C. and Mesnard, D., *A performance summary of averaging pitot-type primaries Measurement and Control*, 15, 341–350 (1982).
3. King, N. W., 'Multi-phase flow measurement at NEL', *Measurement and Control*, 21, 237–239 (1988).
4. Blickley, G. J., 'Magmeters require less power', *Control Engineering*, Aug 65 (1987).
5. Kinghorn, F. C., 'Challenging areas in flow measurement', *Measurement and Control*, 21 229–235 (1988).
6. Medlock, R. S., 'Cross-correlation flow measurement', *Measurement and Control*, 18(8) 293–298 (1985).
7. Berman, M. and Prowse, D. B., 'The statistical properties of the number of

observations used for proving meters', *Measurement and Control*, 00(0) 158–160 (1986).

8. Pursley, W. C., 'The calibration of flowmeters' *Measurement and Control*, 19(5) 37–45 (1986).
9. Bellinga, H. *et al.*, 'Using a piston prover as a primary standard', *Proc. of Flomenko 85, Int. conf. on Flow Measurement*, Melbourne, Australia (1985).

CHAPTER 18

Calibration of Miscellaneous Parameters

18.1 VIBRATION AND SHOCK

Vibration

One common parameter in the quality evaluation of a product is its ability to withstand vibrations up to a certain magnitude, and therefore a test rig has to be set up in which vibrations are excited and measured. Vibration measurement is also necessary to monitor the operation of products that can excite vibrations, such as machines, to check that vibration levels do not exceed limits defined in the quality documentation. Consequently, the calibration of vibration-measuring equipment is an essential part of many quality systems.

Vibration normally consists of linear harmonic motion in which the motion parameters are related by the following formulae:

$$x = X_p \sin(\omega t)$$
$$v = \frac{dx}{dt} = V_p \cos(\omega t) \tag{18.1}$$
$$a = \frac{dv}{dt} = A_p \sin(\omega t + \pi)$$

where X_p, V_p and A_p are the peak values of the displacement, velocity and acceleration, x, v and a are instantaneous values of these quantities and (ωt) is the angular frequency.

Thus, the intensity of vibration can be measured in terms of either displacement, velocity or acceleration. It is apparent from the above equations that displacements are large at low frequencies and that therefore either displacement or velocity transducers are theoretically best for measuring vibration at such frequencies. Accelerometers would only appear to be preferable at high frequencies. However, there are considerable practical difficulties in mounting and calibrating displacement and velocity transducers and they are rarely used. Consequently, vibration is usually measured by accelerometers at all frequencies. The most common type of transducer used is the piezo-accelerometer, which has typical inaccuracy levels of ±2 percent.

When measuring vibration, consideration must be given to the fact that attaching an accelerometer to the vibrating body will significantly affect the vibration characteristics if the body has a low mass. The effect of such 'loading' of the measured system can be quantified by the following equation:

$$a_1 = a_b \left[\frac{m_b}{m_b + m_a} \right]$$

where a_1 is the acceleration of the body with accelerometer attached, a_b is the acceleration of the body without the accelerometer, m_a is the mass of the accelerometer and m_b is the mass of the body.

Such considerations emphasize the advantage of piezo-accelerometers, as these have a lower mass than other forms of accelerometer and so contribute least to this system-loading effect.

As well as an accelerometer, a vibration measurement system requires other elements, as shown in Figure 18.1, to translate the accelerometer output into a recorded signal. The three other necessary elements are a signal conditioning element, a signal analyser and a signal recorder. The signal conditioning element amplifies the relatively weak output signal from the accelerometer and also transforms the high output impedance of the accelerometer to a lower impedance value. The signal analyser then converts the signal into the form required for output. The output parameter may be either displacement, velocity or acceleration and this may be expressed as either the peak value, r.m.s. value or average absolute value. The final element of the measurement system is the signal recorder.

All elements of the measurement system, and especially the signal recorder, must be chosen very carefully to avoid distortion of the vibration wave form. The bandwidth should be such that it is at least a factor of 10 better than the bandwidth of the vibration frequency components at both

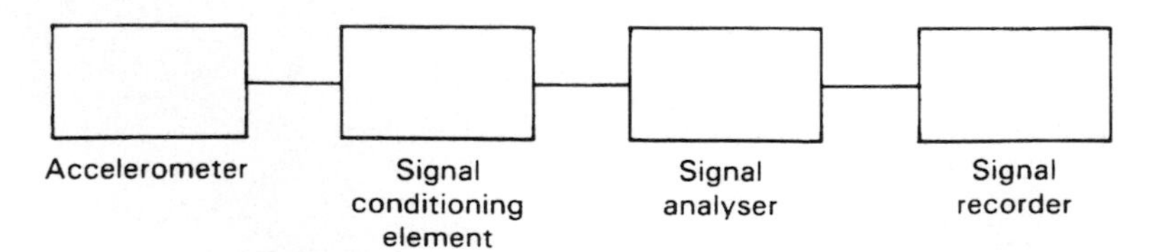

Figure 18.1. Vibration measurement system

ends. Thus its lowest frequency limit should be less than or equal to 0.1 times the fundamental frequency of vibration and its upper frequency limit should be greater than or equal to 10 times the highest significant vibration frequency component.

All elements in the measurement system require calibration, but the element that requires most frequent attention is the vibration transducer, which is normally a piezo-accelerometer. At the lowest level, calibration of the vibration transducer is performed by comparison with a standard accelerometer. In this procedure, both transducers are mounted on a piece of equipment which is able to excite vibrations at a range of frequencies and amplitudes. The standard instrument is most commonly a piezoelectric accelerometer, as this has the best stability and frequency range. If the vibration transducer itself measures acceleration, as is commonly the case, the outputs of the two transducers can be compared directly. Otherwise, the outputs have to be converted to the same forms using equations (18.1) above. Calibration at yearly intervals is usually recommended.

Reference standard accelerometers are similarly used in comparison procedures at all intermediate levels of calibration. Only at the very highest levels of calibration are absolute methods used. The most common absolute methods of calibration involve interferometric techniques which give measurement uncertainty levels of around ±0.5 percent. Such facilities are only usually found in National Standards Laboratories.

Shock

Shock describes a type of motion where a moving body is brought suddenly to rest, often because of a collision. This is very common in industrial situations and usually involves a body being dropped and hitting the floor. The ability of a product to withstand shocks of a certain magnitude is often quality-related and therefore instruments that measure shock, and their calibration, are very important.

Shocks characteristically involve large-magnitude decelerations (e.g. 500 g) which last for a very short time (e.g. 5 ms). An instrument having a very high frequency response is required for shock measurement, and for this reason, piezo-crystal-based accelerometers are commonly used. Again, other elements for analysing and recording the signal are required as shown in Figure 18.1 and described in the last section. A storage oscilloscope is a suitable instrument for recording the output signal, as this allows the time duration as well as the acceleration levels in the shock to be measured. Alternatively, if a permanent record is required, the screen of a standard oscilloscope can be photographed. A further option is to record the output on magnetic tape, which facilitates computerized signal analysis.

Shock calibration requires calibration of all the elements used in the measurement system. Calibration of the piezo-accelerometer is carried out using a shock machine of the form shown in Figure 18.2. The accelerometer is mounted, together with a standard accelerometer, on the anvil of the

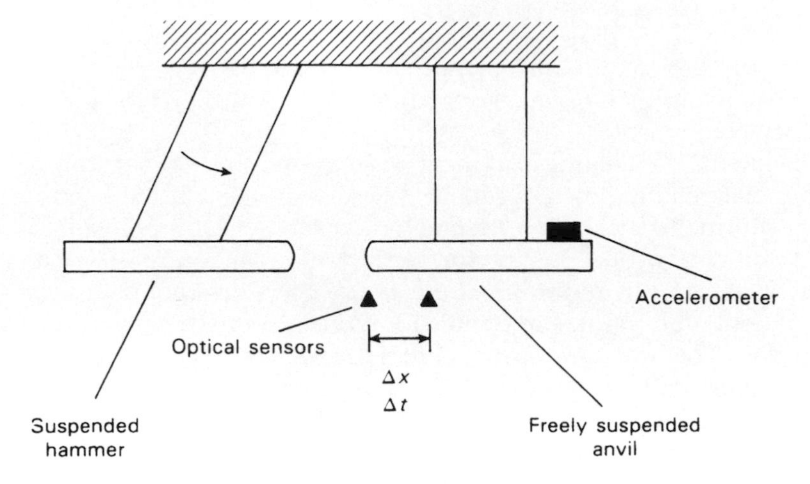

Figure 18.2. Shock machine

machine, and records of the shock signals from the two transducers are recorded on a transient recorder and compared. If optical sensors are used to measure the time Δt taken for the anvil to move a fixed distance Δx, an absolute measurement of acceleration can also be obtained according to the following equation:

$$A = \int_0^{\delta t} a(t)\,\mathrm{d}t = \Delta V = \frac{\Delta x}{\Delta t}$$

where $a(t)$ is the instantaneous acceleration at time t and ΔV is the change in velocity of the anvil as it moves through Δx.

Traceability to fundamental standards is then provided by the procedures described in the last section on vibration measurement.

18.2 VISCOSITY

Viscosity measurement is important in many process industries. In the food industry, the viscosity of raw materials such as dough, batter and ice cream has a direct effect on the quality of the product. Similarly, in other industries such as the ceramics, the quality of raw materials affects the final product quality. Viscosity control is also very important in assembly operations that involve the application of mastics and glue flowing through tubes. Clearly, successful assembly requires such materials to flow through tubes at the correct rate and therefore it is essential that their viscosity is correct.

Viscosity describes the way in which a fluid flows when it is subject to an applied force. Consider an elemental cubic volume of fluid and a shear force F applied to one of its faces of area A. If this face moves a distance L and at a velocity V relative to the opposite face of the cube under the action of F, the shear stress (α) and shear rate (β) are given by

$$\alpha = \frac{F}{A}; \quad \beta = \frac{V}{L}$$

The coefficient of viscosity (μ) is the ratio of shear stress to shear rate, i.e.

$$\mu = \frac{\alpha}{\beta}$$

μ is often described simply as the 'viscosity'. A further term, kinematic viscosity, is also sometimes used, given by:

$$\sigma = \frac{\mu}{\delta}$$

where σ is the kinematic viscosity and δ is the fluid density. To avoid confusion, μ is often known as the dynamic viscosity, to distinguish it from σ. μ is measured in units of poise or $N\,s/m^2$ and σ is measured in units of stokes or m^2/s.

Viscosity was originally defined by Newton, who assumed that it was constant with respect to shear rate. However, it has since been shown that the viscosity of many fluids varies significantly at high shear rates and the viscosity of some varies even at low shear rates. The worst non-Newtonian characteristics tend to occur with emulsions, pastes and slurries. For non-Newtonian fluids, a subdivision into further classes can also be made according to the manner in which the viscosity varies with shear rate, as shown in Figure 18.3.

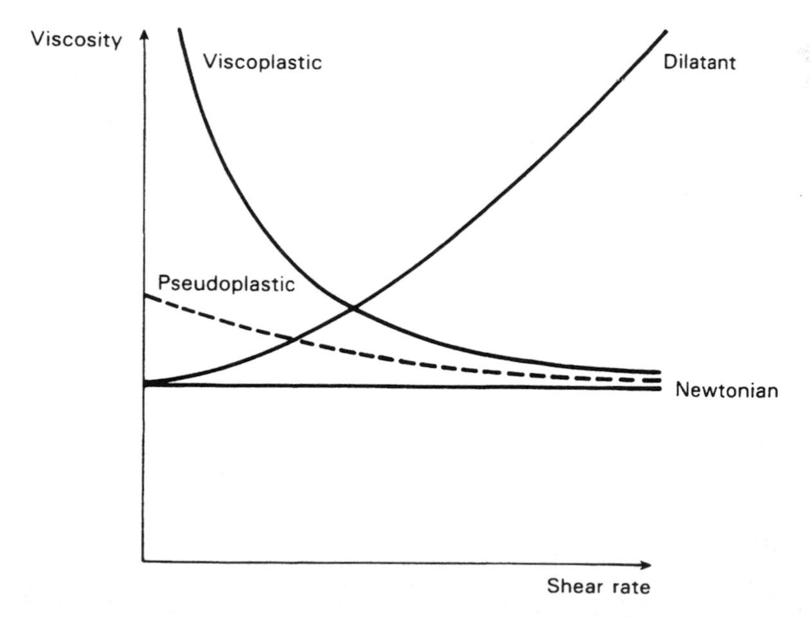

Figure 18.3. Different viscosity/shear-rate relationships

18.2.1 Viscosity Measurement

The relationship between the input variables and output measurement for instruments which measure viscosity normally assumes that the measured fluid has Newtonian characteristics. For non-Newtonian fluids, a correction must be made for shear-rate variations [1]. If such a correction is not made, the measurement obtained is known as the apparent viscosity and this can differ from the true viscosity by a large factor. The true viscosity is often called the absolute viscosity to avoid ambiguity. Viscosity also varies with fluid temperature and density.

Instruments for measuring viscosity work on one of three physical principles:

1. Rate of flow of the liquid through a tube
2. Rate of fall of a body through the liquid
3. Viscous friction force exerted on a rotating body

1. Capillary and Tube Viscometers

These are the most accurate type of viscometer, with typical measurement inaccuracy levels down to ±0.3 percent. Liquid is allowed to flow, under gravity from a reservoir, through a tube of known cross-section. In different instruments, the tube can vary from capillary-sized to a large diameter. The pressure difference across the ends of the tube and the time for a given quantity of liquid to flow are measured, and then the liquid viscosity for Newtonian fluids can be calculated as (in units of poise):

$$\mu = \frac{1.25\pi R^4 PT}{LV} \qquad (18.2)$$

where R is the radius (m) of the tube, L is its length (m), P is the pressure difference (N/m^2) across the ends and V is the volume of liquid flowing in time T (m^3/s). For non-Newtonian fluids, corrections must be made for shear-rate variations [1].

For any given viscometer, R, L and V are constant and equation (18.2) can be written as

$$\mu = KPT \qquad (18.3)$$

where K is known as the viscometer constant.

2. Falling Body Viscometer

The falling-body viscometer is particularly recommended for the measurement of high-viscosity fluids and can give measurement uncertainty levels down to ±1 percent. It involves measuring the time taken for a spherical body to fall a given distance through the liquid. The viscosity for Newtonian fluids is then given by Stoke's formula as (in units of poise)

$$\mu = \frac{R^2 g(\delta_s - \delta_1)}{450V}$$

where R is the radius (m) of the sphere, g is the acceleration due to gravity (m/s^2), δ_s and δ_1 are the specific gravities (g/m^3) of the sphere and liquid respectively and V is the velocity (m/s) of the sphere. For non-Newtonian fluids, correction for the variation in shear rate is very difficult.

3. Rotational Viscometers

Rotational viscometers are relatively easy to use but their measurement inaccuracy is at least ±10 percent. All types have some form of element rotating inside the liquid at a constant rate. One common version has two coaxial cylinders with the fluid to be measured contained between them. One cylinder is driven at a constant angular velocity by a motor and the other is suspended by torsion wire. After the driven cylinder starts from rest, the suspended cylinder rotates until an equilibrium position is reached where the force due to the torsion wire is just balanced by the viscous force transmitted through the liquid. The viscosity (in poise) for Newtonian fluids is then given by:

$$\mu = 2.5G \frac{\left[\frac{1}{R_1^2} - \frac{1}{R_2^2}\right]}{\pi h \Omega}$$

where G is the couple (N·m) formed by the force exerted by the torsion wire and its deflection, R_1 and R_2 are the radii (m) of the inner and outer cylinders, h is the length of the cylinder (m) and Ω is the angular velocity (rad/s) of the rotating cylinder. Again, corrections have to be made for non-Newtonian fluids.

18.2.2 Viscosity Calibration

Viscosity standards are only defined for Newtonian fluids. No such standards exist for non-Newtonian fluids.

Viscosity calibration for Newtonian liquids is carried out using a set of glass capillary tubes with accurately known viscometer constants. These provide reference standards over a range of viscosities. The set is calibrated by a 'step-up' procedure from a calibration on distilled water, using a series of standard liquids. The steps in this procedure are as follows:

A 'master' viscometer is calibrated first using distilled water as the flowing fluid. The 'master' is then used to measure the viscosity of the first standard viscosity liquid. This liquid is then used to calibrate the first standard viscometer. After this, the viscosity of the second standard liquid is measured by the first standard viscometer and then the second standard viscometer is calibrated using the second standard liquid. This procedure is repeated to calibrate the other standard viscometers in the set using the other standard liquids. A more thorough explanation of this procedure and a discussion of the types of standard liquid used can be found in [2].

It is common practice to maintain two sets of such standard viscometers, one as a working set for routine calibrations and one as a reference set for calibration checks on the working set. Higher-level calibration is provided by comparison with a set of standard glass-capillary viscometers maintained by National Standards Organizations. Such organizations also commonly supply a range of standard viscosity liquids as an alternative calibration source. Where standard liquids are kept for calibration, they must be checked periodically to ensure their continuing stability as they age.

As there are no absolute physical quantities that the calibration of viscosity values can be checked against, the various National Standards Organizations carry out cross-checks between each others laboratories. These checks take two forms. One is to transport a reference set of viscometers between laboratories and to compare the viscometer constants measured, with appropriate correction for differences in the value of g (acceleration due to gravity). The other form of check is to compare measurements made on standard liquids at a specified temperature. Precise guidelines for these calibration procedures are given in [3].

In the case of non-Newtonian fluids, as has already been stated, there are no formal viscosity standards. Therefore, the only course of action available for checking viscometers intended to measure non-Newtonian fluids is to calibrate them against standard viscosity Newtonian fluids.

18.3 MOISTURE

There are many quality-related industrial requirements for the measurement of the moisture content of both solids, liquids and gases. The physical properties and storage stability of most solid materials is affected by their water content. There is also a statutory requirement to limit the moisture content in the case of many materials sold by weight. In consequence, the requirement for moisture measurement pervades a large number of industries involved in the manufacture of foodstuffs, pharmaceuticals, cement, plastics, textiles and paper.

Measurement of the water content in liquids is commonly needed for fiscal purposes, but is also often necessary to satisfy statutory requirements. The petrochemical industry has wide-ranging needs for moisture measurement in oil etc. The food industry also needs to measure the water content of products such as beer and milk.

In the case of moisture in gases, the most common measurement is the amount of moisture in air. This is usually known as the humidity level. Humidity measurement and control is an essential requirement in many buildings, greenhouses and vehicles.

As there are several ways in which humidity can be defined, three separate terms have evolved so that ambiguity can be avoided.

- *Absolute humidity* is the mass of water in a unit volume of moist air.
- *Specific humidity* is the mass of water in a unit mass of moist air.
- *Relative humidity* is the ratio of the actual water-vapour pressure in air to the saturation vapour pressure, usually expressed as a percentage.

18.3.1 Industrial Moisture Measurement Techniques

Industrial methods for measuring moisture are based on the variation of some physical property of the material with moisture content. Many different properties can be used and therefore the range of available techniques, as listed below, is large.

Electrical Methods

Measuring the amount of absorption of microwave energy beamed through the material is the most common technique for measuring moisture content and is described in detail in [4, 5]. Microwaves at wavelengths between 1 mm and 1 m are absorbed to a much greater extent by water than most other materials. Wavelengths of 30 mm or 100 mm are commonly used because 'off-the-shelf' equipment to produce these is readily available from instrument suppliers. The technique is suitable for moisture measurement in solids, liquids and gases at moisture-content levels up to 45 percent and measurement uncertainties down to ±0.3 percent are possible.

The *capacitance moisture meter* uses the principle that the dielectric constant of materials varies according to their water content. Capacitance measurement is therefore related to moisture content. The instrument is useful for measuring moisture-content levels up to 30 percent in both solids and liquids, and measurement uncertainty down to ±0.3 percent has been claimed for the technique [6]. Drawbacks of the technique include (a) limited measurement resolution owing to the difficulty in measuring small changes in a relatively large standing capacitance value and (b) difficulties when the sample has a high electrical conductivity. An alternative capacitance charge transfer technique has been reported [7] which overcomes these problems by measuring the charge-carrying capacity of the material.

In this technique, wet and dry samples of the material are charged to a fixed voltage and then simultaneously discharged into charge-measuring circuits.

The *electrical conductivity* of most materials varies with moisture content and this therefore provides another means of measurement. It is cheap and can measure moisture levels up to 25 percent. However, the presence of other conductive substances in the material such as salts or acids affects the measurement.

A further technique is to measure the frequency change in a quartz crystal which occurs as it takes in moisture.

Neutron Moderation

Neutron moderation measures moisture content using a radioactive source and a neutron counter. Fast neutrons emitted from the source are slowed down by hydrogen nuclei in the water, forming a cloud whose density is related to the moisture content. Measurements take a long time because the output density reading may take up to a few minutes to reach steady state, according to the nature of the materials involved. Also, the method cannot be used with any materials that contain hydrogen molecules, such as oils and fats, as these slow down neutrons as well. Specific humidities up to 15 percent (± 1 percent error) can be measured.

Low Resolution Nuclear Magnetic Resonance (NMR)

Low-resolution nuclear magnetic resonance involves subjecting the sample to both a unidirectional and an alternating radio-frequency (r.f.) magnetic field. The amplitude of the uni-directional field is varied cyclically, which causes resonance once per cycle in the coil producing the r.f. field. Under resonance conditions, protons are released from the hydrogen content of the water in the sample. These protons cause a measurable moderation of the amplitude of the r.f. oscillator waveform which is related to the moisture content of the sample. A fuller description of the technique can be found in [8].

Materials with their own natural hydrogen content cannot normally be measured. However, pulsed NMR techniques have been developed which overcome this problem by taking advantage of the different relaxation times of hydrogen nuclei in water and oil. In such pulsed techniques, the dependence on the relaxation time limits the maximum fluid flow rate for which moisture can be measured.

Optical Methods

The *refractometer* is a well-established instrument which is used for measuring the water content of liquids. It measures the refractive index of the liquid, which changes according to the moisture content.

Moisture-related *energy absorption* of near-infra-red light can be used for measuring the moisture content of solids, liquids and gases. At a wavelength of 1.94 μm, energy absorption due to moisture is high, whereas at 1.7 μm, absorption due to moisture is zero. Therefore, measuring absorption at both 1.94 μm and 1.7 μm allows absorption due to components in the material other than water to be compensated for, and the resulting measurement is directly related to energy content. The latest instruments use multiple-frequency infra-red energy and have an even greater capability for eliminating the effect of components in the material, other than water, that absorb energy. Such multi-frequency instruments also cope much better with variations in particle size in the measured material.

In alternative versions of this technique, energy is either transmitted through the material or reflected from its surface. In either case, materials that are either very dark or highly reflective give poor results. The technique is particularly attractive, where applicable, because it is a non-contact method which can be used to monitor moisture content continuously at moisture levels up to 50 percent, with inaccuracy as low as ±0.1 percent in the measured moisture level. A deeper treatment can be found in [9].

Ultrasonic Methods

The presence of water changes the speed of propagation of ultrasonic waves through liquids. The moisture content of liquids can therefore be determined by measuring the transmission speed of ultrasound. This has the inherent advantage of being a non-invasive technique but temperature compensation is essential because the velocity of ultrasound is particularly affected by temperature changes. This method is best suited to measurement of high moisture levels in liquids which are not aerated or of high viscosity. Typical measurement uncertainty is ±1 percent but measurement resolution is very high, with changes in moisture level as small as 0.05 percent being detectable. Further details can be found in [10].

Mechanical Properties

Density changes in many liquids and slurries can be measured and related to moisture content, with good measurement resolution up to 0.2 percent moisture. Moisture content can also be estimated by measuring the moisture-level-dependent viscosity of liquids, pastes and slurries.

18.3.2 Laboratory Techniques for Moisture Measurement

Laboratory techniques for measuring moisture content generally take much longer to obtain a measurement than the industrial techniques described above. However, the measurement accuracy obtained is usually much better.

Water Separation

Various laboratory techniques are available which enable the moisture content of liquids to be measured accurately by separating the water from a sample of the host liquid. Separation is effected by either titration (Karl Fischer technique), distillation (Dean and Stark technique) or a centrifuge. Any of these methods can measure water content in a liquid with measurement uncertainty levels down to ±0.03 percent.

Gravimetric Methods

Moisture content in solids can be measured accurately by weighing the moist sample, drying it and then weighing it again. Great care must be taken in applying this procedure, as many samples rapidly take up moisture again if they are removed from the drier and exposed to the atmosphere before being weighed. Normal procedure is to put the sample in an open container, dry it in an oven and then screw an airtight top onto the container before it is removed from the oven.

Phase-change Methods

The boiling and freezing point of materials is altered by the presence of moisture, and therefore the moisture level can be determined by measuring the phase-change temperature. This technique is used for measuring the moisture content in many food products and in some oil and alcohol products.

Equilibrium Relative Humidity Measurement

This technique involves placing a humidity sensor in close proximity to the sample in an airtight container. The water-vapour pressure close to the sample is related to the moisture content of the sample. The moisture level can therefore be determined from the humidity measurement.

18.3.3 Humidity Measurement

The three major instruments used for measuring humidity in industry are the *electrical hygrometer*, the *psychrometer* and the *dew-point meter*. The dew point meter, described in Section 18.3.4, is the most accurate of these and is used as a calibration standard. A more detailed description of these and other types of hygrometer can be found in [11].

The Electrical Hygrometer

The electrical hygrometer measures the change in capacitance or conductivity of a hygroscopic material as its moisture level changes. Conductivity types use two noble-metal electrodes either side of an insulator coated in a hygroscopic salt such as calcium chloride. Capacitance types have two plates either side of a hygroscopic dielectric such as aluminium oxide.

These instruments are suitable for measuring moisture levels between 15 percent and 95 percent, with a typical measurement uncertainty of ±3 percent. Atmospheric contaminants and operation in saturation conditions both cause characteristics drift, and therefore the recalibration frequency has to be determined according to the conditions of use.

The Psychrometer (Wet and Dry Bulb Hygrometer)

The psychrometer, also known as the *wet and dry bulb hygrometer*, has two temperature sensors, one exposed to the atmosphere and one enclosed in a wet wick. Air is blown across the sensors which causes evaporation and a reduction in temperature in the wet sensor. The temperature difference between the sensors is related to the humidity level. The lowest measurement uncertainty attainable is ±4 percent.

18.3.4 Calibration

Microwave techniques and instruments such as the electrical hygrometer and dew-point meter are accurate enough for working standard calibrations. Otherwise, such calibrations are performed by gravimetric or water-separation laboratory techniques (see Section 18.3.2). For calibration to secondary and primary reference standards, test samples must be used in which the water content is accurately known.

Electrical Hygrometer

The electrical hygrometer, as described in the Section 18.3.3, can be used for first-level calibrations of instruments in regular use providing that it is not subjected to near-saturation conditions and is stored carefully away from possible contaminants.

Microwave Techniques

Measuring the absorption of microwaves, as described in Section 18.3.1, is also accurate enough to be used as a first-level calibration technique for instruments in everyday use, as described in [4, 5]

Dew-point Meter

The elements of the dew-point meter, also known as the dew-point hygrometer, are shown in Figure 18.4. The sample is introduced into a vessel with an electrically cooled mirror surface. The mirror surface is cooled until a light source–light detector system detects the formation of dew on the mirror, and the condensation temperature is measured by a sensor bonded to the mirror surface. The dew point is the temperature at which the sample becomes saturated with water. Therefore, this temperature is related to the moisture level in the sample. A microscope is also provided in the instrument so that the thickness and nature of the condensate can be observed. The instrument is described in greater detail in [12].

Even small levels of contaminants on the mirror surface can cause large changes in the dew-point and therefore the instrument must be kept very clean. When necessary, the mirror should be cleaned with deionized or distilled water applied with a lint-free swab. Any contamination can be detected by a skilled operator, as this makes the condensate look 'blotchy' when viewed through the microscope. The microscope also shows up other potential problems such as large ice crystals in the condensate which cause temperature gradients between the condensate and the temperature sensor.

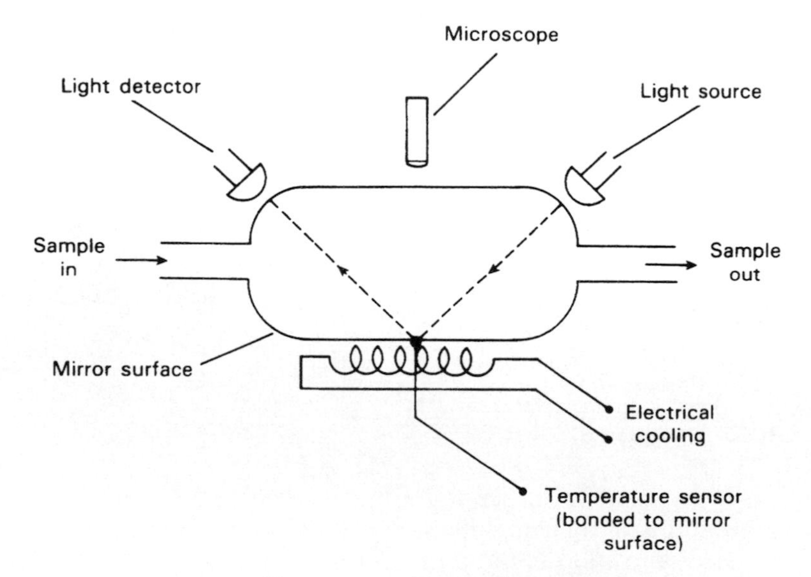

Figure 18.4. Dew-point meter

At suitable intervals, the instrument must be calibrated itself against a test gas containing a known amount of water.

Test Samples

The traceability of moisture measurements to reference standards is achieved through the provision by National Standards Organizations of test samples which have an accurately known water content. The instrument to be calibrated is applied to these samples to see whether its output reading agrees with the known water content.

Test samples created in a company's own calibration laboratory can be used as working standards which are checked against primary reference standards at suitable intervals. This involves making up mixtures of a gas or liquid with water in known proportions. In the case of solids, it is often difficult to make up a test sample with an even distribution of water through it and gravimetric techniques of calibration (see Section 18.3.2) have to be used instead.

18.4 VOLUME

Volume measurement is required in its own right as well as being required as a necessary component in some techniques for the measurement and calibration of other quantities such as volume flow rate and viscosity.

18.4.1 Volume Measurement

The volume of vessels of a regular shape, where the cross-section is circular or oblong in shape, can be readily calculated from the dimensions of the vessel, using instruments as described in Chapter 16. Otherwise, for vessels of irregular shape, either gravimetric techniques or a set of calibrated volumetric measures are required.

In the gravimetric technique, the dry vessel is weighed and then is completely filled with water and weighed again. The alternative technique involves transferring the liquid from the vessel into an appropriate number of volumetric measures taken from a standard-capacity, calibrated set. Each vessel in the set has a mark which shows the volume of liquid contained when the vessel is filled up to the mark. Special care is needed to ensure that the meniscus of the water is in the correct position with respect to the reference mark on the vessel when it is deemed to be full. Normal practice is to set the water level so that the reference mark forms a smooth tangent with the convex side of the meniscus. This is made easier to achieve if the meniscus is viewed against a white background and the vessel is shaded from stray illumination.

18.4.2 Volume Calibration

Either of the following two techniques may be used for calibration purposes, although the gravimetric technique gives the better accuracy [13].

Gravimetric Technique

When the gravimetric technique, as described above, is used for volumetric calibration, certain special procedures should be followed. Firstly, the water used to fill the vessel should be either distilled or deionized. (Where very large vessels are being calibrated and it is impractical to use other than tap water, the difference in water density compared with pure water gives a typical error of 0.02 percent). Secondly, the calibration should be carried out under specified temperature and pressure conditions, and sufficient time should be allowed for the vessel and water to reach a close thermal equilibrium. This latter condition is most easily achieved by maintaining a store of distilled or deionized water in the calibration laboratory, which ensures that the water is automatically at the correct temperature. Lastly, correction may have to be made for changes in air buoyancy, which become significant if calibration is carried out at altitudes greater than 150 m.

Calibrated Measures

The measurement uncertainty using calibrated volumetric measures depends on the number of measures used for any particular measurement. The total error is a multiple of the individual error of each measure, typical values of which are shown in Table 18.1.

As mentioned in the last section, special care must be taken to ensure that the meniscus level is set correctly with respect to the reference mark when each vessel is filled. Calibration traceability is provided by calibrating all volumetric measures gravimetrically at suitable intervals of time.

Table 18.1.

Capacity	Volumetric uncertainty
1 ml	±4%
10 ml	±0.8%
100 ml	±0.2%
1 l	±0.1%
10 l	±0.05%
100 l	±0.02%
1000 l	±0.02%

REFERENCES

1. *The quality Manual*, United Trade Press, 62–106 (1975).
2. Cheng, D. C.-H., 'Viscosity: calibration and standards', *Measurement and Control*, 14, 73–78 (1981).
3. ISO 5725: *Accuracy of measurement test methods and results*, International Organization for Standards, Geneva (1986) (also published by British Standards Institution as BS.ISO 5725).
4. Anderson, J. G., 'Paper moisture measurement using microwaves', *Measurement and Control*, 14 82–84 (1981).
5. Thompson, F., 'Moisture measurement using microwaves', *Measurement and Control*, 14 210–215 (1981).
6. Slight, H. A., 'Further thoughts on moisture measurement', *Measurement and Control*, 14 85–86 (1986).
7. Gimson, C., 'Using the capacitance charge transfer principle for water content measurement', *Measurement and Control*, 14 79–81 (1981).
8. Young, L., 'Moisture measurement using low resolution nuclear magnetic resonance', *Measurement and Control*, 14 54–55 (1981).
9. Benson, I. B., 'Industrial applications of near infrared reflectance for the measurement of moisture', *Measurement and Control*, 22 45–49 (1989).
10. Wiltshire, M. P., 'Ultrasonic moisture measurement', *Measurement and Control*, 22 51–53 (1989).
11. *The quality Manual*, United Trade Press, 180–209 (1975).
12. Pragnell, R. F., modern condensation dewpoint hygrometer', *Measurement and Control*, 22, 74-77 (1989).
13. NAMAS Document NIS 5: *Traceability – volumetric glassware*, NAMAS Executive, National Physical Laboratory, Middlesex, UK (1984).

APPENDIX 1

Tables of Fundamental and Derived SI Units

(a) Fundamental Units

Quantity	Standard Unit	Symbol
length	metre	m
mass	kilogram	kg
time	second	s
electric current	ampere	A
temperature	kelvin	K
luminous intensity	candela	cd
matter	mole	mol

(b) Supplementary Fundamental Units

Quantity	Standard Unit	Symbol
plane angle	radian	rad
solid angle	steradian	sr

(c) Derived Units

Quantity	Standard Unit	Symbol	Derivation formula
area	square metre	m^2	
volume	cubic metre	m^3	
velocity	metre per second	m/s	
acceleration	metre per second squared	m/s^2	
angular velocity	radian per second	rad/s	
angular acceleration	radian per second squared	rad/s^2	
density	kilogram per cubic metre	kg/m^3	
specific volume	cubic metre per kilogram	m^3/kg	
mass flow rate	kilogram per second	kg/s	
volume flow rate	cubic metre per second	m^3/s	
force	newton	N	$kg{\cdot}m/s^2$
pressure	pascal	Pa	N/m^2
torque	newton metre	N·m	
momentum	kilogram metre per second	kg·m/s	
moment of inertia	kilogram metre squared	$kg{\cdot}m^2$	
kinematic viscosity	square metre per second	m^2/s	
dynamic viscosity	newton second per sq metre	$N{\cdot}s/m^2$	
work, energy, heat	joule	J	N·m
specific energy	joule per cubic metre	J/m^3	
power	watt	W	J/s
thermal conductivity	watt per metre Kelvin	W/m·K	
electric charge	coulomb	C	A·s
voltage, e.m.f., pot diff	volt	V	W/A
electric field strength	volt per metre	V/m	
electric resistance	ohm	Ω	V/A
electric capacitance	farad	F	A·s/V
electric inductance	henry	H	V·s/A
electric conductance	siemen	S	A/V
resistivity	ohm metre	Ω·m	
permittivity	farad per metre	F/m	
permeability	henry per metre	H/m	
current density	ampere per square metre	A/m^2	
magnetic flux	weber	Wb	V·s
magnetic flux density	tesla	T	Wb/m^2
magnetic field strength	ampere per metre	A/m	
frequency	hertz	Hz	s^{-1}
luminous flux	lumen	lm	cd·sr
luminance	candela per square metre	cd/m^2	
illumination	lux	lx	lm/m^2
molar volume	cubic metre per mole	m^3/mol	
molarity	mole per kilogram	mol/kg	
molar energy	joule per mole	J/mol	

APPENDIX 2

Conversion Tables Between Imperial, Metric and SI Units

Length

SI Units: mm, m, km — Imperial units: in, ft, mile

	mm	m	km	in	ft	mile
mm	1	10^{-3}	10^{-6}	0.039 370	3.281×10^{-3}	–
m	10^3	1	10^{-3}	39.370	3.2808	6.214×10^{-4}
km	10^6	10^3	1	–	3280.8	0.621 37
in	25.4	0.0254	–	1	0.083 333	–
ft	304.8	0.3048	3.048×10^{-4}	12	1	1.894×10^{-4}
mile	–	1609.3	1.6093	–	5280	1

Area

SI Units: mm^2, m^2, km^2 — Imperial Units: in^2, ft^2, yd^2

	mm^2	m^2	km^2	in^2	ft^2	yd^2
mm^2	1	10^{-6}	–	1.550×10^{-3}	1.076×10^{-5}	1.196×10^{-6}
m^2	106	1	10^{-6}	1550	10.764	1.196
km^2	–	10^6	1	–	–	1.196×10^6
in^2	645.16	6.452×10^{-4}	–	1	6.944×10^{-3}	7.716×10^{-4}
ft^2	92 903	0.092 90	–	144	1	0.1111
yd^2	836 127	0.836 13	8.361×10^{-7}	1296	9	1

Other units: 1 acre = 4840 yd^2
1 hectare = 10 000 m^2 (= 2.471 acres)
1 sq mile = 2.590 km^2

Mass

SI units: g, kg, tonne Imperial units: lb, cwt, ton

	g	kg	tonne	lb	cwt	UK ton
g	1	10^{-3}	10^{-6}	2.205×10^{-3}	1.968×10^{-5}	–
kg	10^3	1	10^{-3}	2.2046	0.019 684	9.842×10^{-4}
tonne	10^6	10^3	1	2204.6	19.684	0.984 21
lb	453.59	0.453 59	4.536×10^{-4}	1	8.929×10^{-3}	4.464×10^{-4}
cwt	50 802	50.802	0.050 802	112	1	0.05
UK ton	–	1016.05	1.016 05	2240	20	1

Density

SI unit: kg/m^3 Metric unit: g/cm^3
Imperial units: lb/ft^3, lb/in^3

	kg/m^3	g/cm^3	lb/ft^3	lb/in^3
kg/m^3	1	10^{-3}	0.062 428	3.605×10^{-5}
g/cm^3	103	1	62.428	0.036 127
lb/ft^3	16.019	0.016 019	1	5.787×10^{-4}
lb/in^3	27 680	27.680	1728	1

Force

SI units: N, kN Metric unit: kg_f
Imperial units: pdl (poundal), lb_f , UK ton_f

	N	kg_f	kN	pdl	lb_f	UK ton_f
N	1	0.1020	10^{-3}	7.233	0.2248	1.004×10^{-4}
kg_f	9.807	1	9.807×10^{-3}	70.93	2.2046	9.842×10^{-4}
kN	10^3	102.0	1	7233	224.8	0.1004
pdl	0.1383	0.0141	1.383×10^{-4}	1	0.0311	1.388×10^{-5}
lb_f	4.448	0.4536	4.448×10^{-3}	32.174	1	4.464×10^{-4}
UK ton_f	9964	1016	9.964	72070	2240	1

Additional unit: 1 dyne = 10^{-5} N = 7.233×10^{-5} pdl

Pressure

SI unit: pascal (1 Pa = 1 N/m^2) Metric units: mbar, bar, N/m^2
Imperial units: lb_f/in^2 (psi), in Hg, atmosphere (atm)

	mbar	bar	Pascal	lb_f/in^2	in Hg	atm
mbar	1	10^{-3}	100	0.014 50	0.029 53	9.869×10^{-4}
bar	10^3	1	10^5	14.50	29.53	0.9869
Pascal	0.01	10^{-5}	1	1.450×10^{-4}	2.953×10^{-4}	9.869×10^{-6}
lbf/in^2	68.95	0.068 95	6895	1	2.036	0.068 05
in Hg	33.86	0.033 86	3386	0.4912	1	0.033 42
atm	1013	1.013	1.013×10^5	14.70	29.92	1

Additional conversion factors:
1 N/m^2 = 1 Pascal (Pa)
1 lb_f/ft^2 = 47.88 Pa
1 kg_f/cm^2 = 9.807×10^4 Pa
1 torr = 1 mm Hg = 133.3 Pascal
1 in water = 0.07356 in Hg = 249.1 Pa
1 dyn/cm^2 = 0.1 Pa
1 kN/m^2 = 1000 Pa

Velocity

SI units: mm/s, m/s Metric unit: km/h
Imperial units: ft/s, mile/h

	mm/s	m/s	km/h	ft/s	mile/h
mm/s	1	10^{-3}	3.6×10^{-3}	3.281×10^{-3}	2.237×10^{-3}
m/s	10^3	1	3.6	3.2808	2.2369
km/h	277.78	0.277 78	1	0.911 34	0.621 37
ft/s	304.8	0.3048	1.0973	1	0.681 82
mile/h	447.04	0.447 04	1.6093	1.4667	1

Acceleration

SI unit: m/s^2 Metric unit: cm/s^2
Imperial unit: ft/s^2 Other unit: g

	m/s^2	cm/s^2	ft/s^2	g
m/s^2	1	100	3.2808	0.102
cm/s^2	0.01	1	0.032 808	0.001 02
ft/s^2	0.3048	30.48	1	0.0311
g	9.81	981	32.2	1

NB These values are based on an approximate value for the acceleration due to gravity (g) of 9.81 m/s^2. However, the exact value varies at different places around the earth and so correction should be made for the local value.

Mass Flow Rate

SI unit: g/s Metric units: kg/h, tonne/d
Imperial units: lb/s, lb/h, UK ton/d

	g/s	kg/h	tonne/d	lb/s	lb/h	UK ton/d
g/s	1	3.6	0.086 40	2.205×10^{-3}	7.937	0.085 03
kg/h	0.2778	1	0.024 00	6.124×10^{-4}	2.205	0.023 62
tonne/d	11.57	41.67	1	0.025 51	91.86	0.9842
lb/s	453.6	1633	39.19	1	3600	38.57
lb/h	0.1260	0.4536	0.010 89	2.778×10^{-4}	1	0.010 71
UK ton/d	11.76	42.34	1.016	0.025 93	93.33	1

Volume Flow Rate

SI unit: m^3/s Metric units: l/h, ml/s
Imperial units: gal/h, ft^3/s, ft^3/h

	l/h	ml/s	m^3/s	UK gal/h	ft^3/s	ft^3/h
l/h	1	0.2778	2.778×10^{-7}	0.2200	9.810×10^{-6}	0.035 32
ml/s	3.6	1	10^{-6}	0.7919	3.532×10^{-5}	0.1271
m^3/s	3.6×10^6	10^6	1	7.919×10^5	35.31	1.271×10^5
UK gal/h	4.546	1.263	1.263×10^{-6}	1	4.460×10^{-5}	0.1606
ft^3/s	1.019×10^5	2.832×10^4	0.028 32	2.242×10^4	1	3600
ft^3/h	28.32	7.867	7.867×10^{-6}	6.228	2.778×10^{-4}	1

Volume

SI units: mm^3, m^3 Metric units: ml, l
Imperial units: in^3, ft^3

	mm^3	ml	l	m^3	in^3	ft^3
mm^3	1	10^{-3}	10^{-6}	10^{-9}	6.102×10^{-5}	–
ml	10^3	1	10^{-3}	10^{-6}	0.061 024	3.531×10^{-5}
l	10^6	10^3	1	10^{-3}	61.024	0.035 315
m^3	10^9	10^6	10^3	1	61 024	35.315
in^3	16 387	16.387	0.016 387	1.639×10^{-5}	1	5.787×10^{-4}
ft^3	–	2.832×10^4	28.32	0.028 32	1728	1

Additional unit: 1 UK gallon = 4.546 litres (l)
1 US gallon = 3.785 litres (l)

Torque (Moment of Force)

SI Unit: N·m Metric unit: kg_f·m
Imperial units: pdl·ft, lb_f·ft

	N·m	kg_f·m	pdl·ft	lb_f·ft
N·m	1	0.1020	23.73	0.7376
kg_f·m	9.807	1	232.7	7.233
pdl·ft	0.042 14	4.297×10^{-3}	1	0.031 08
lb_f·ft	1.356	0.1383	32.17	1

Energy, Work, Heat

SI unit: J Metric units: kgf·m, kW·h
Imperial units: ft·lbf, cal, Btu

	J	kg_f·m	kW·h	ft·lb_f	cal	Btu
J	1	0.1020	2.778×10^{-7}	0.7376	0.2388	9.478×10^{-4}
kg_f·m	9.807	1	2.724×10^{-6}	7.233	2.342	9.294×10^{-3}
kW·h	3.600×10^{6}	367.1×10^{3}	1	2.655×10^{6}	859.8×10^{3}	3412
ft·lbf	1.356	0.1383	3.766×10^{-7}	1	0.3238	1.285×10^{-3}
cal	4.187	0.4270	1.163×10^{-6}	3.088	1	3.968×10^{-3}
Btu	1055	107.6	2.931×10^{-4}	778.2	252.0	1

Additional conversion factors:
1 therm = 10^5 Btu = 1.055×10^8 J
1 hp.h = 0.7457 kW·h = 2.684×10^6 J
1 erg = 10^{-7} J
1 thermie = 4.186×10^6 J
1 ft·pdl = 0.04214 J

Power

SI units: W, kW Imperial units: HP, ft·lb_f/s

	W	kW	HP	ft·lb_f/s
W	1	10^{-3}	1.341×10^{-3}	0.7376
kW	10^3	1	1.341	737.6
HP	745.7	0.7457	1	550
ft·lb_f/s	1.356	1.356×10^{-3}	1.818×10^{-3}	1

Additional conversion factor:
1 BTU/h = 0.2931 W

Dynamic Viscosity

SI unit: $N{\cdot}s/m^2$
Metric units: cP (centipoise),
P (poise)
[1 P = 100 g/m·s]

Imperial unit: $lb_m/ft{\cdot}h$

	$lb_m/ft{\cdot}h$	P	cP	Ns/m^2
$lb_m/ft{\cdot}h$	1	4.133×10^{-3}	0.4134	4.134×10^{-4}
P	241.9	1	100	0.1
cP	2.419	0.01	1	10^{-3}
Ns/m^2	2419	10	10^3	1

Additional unit: 1 Pascal second = 1 $N{\cdot}s/m^2$

Kinematic Viscosity

SI unit: m^2/s
Imperial unit: ft^2/s

Metric units: cSt (centistokes),
St (Stokes)

	ft^2/s	m^2/s	cSt	St
ft^2/s	1	0.0929	9.29×10^4	929
m^2/s	10.76	1	10^6	10^4
cSt	1.076×10^{-5}	10^{-6}	1	0.01
St	1.076×10^{-3}	10^{-4}	100	1

APPENDIX 3

Dynamic Characteristics of Instruments

The dynamic characteristics of a measuring instrument describe its behaviour between the time a measured quantity changes value and the time when the instrument output response attains a steady value. It should be noted that, as with static characteristics, any values for dynamic characteristics quoted in instrument data sheets only apply when the instrument is used under specified environmental conditions. Outside these calibration conditions, some variation in the dynamic parameters can be expected.

The dynamic characteristics can be formally defined mathematically as follows: In any linear, time-invariant measuring system, the following general relation can be written between input and output for time $t > 0$:

$$a_n \frac{d_n q_o}{dt_n} + a_{n-1} \frac{d_{n-1} q_o}{dt_{n-1}} + \ldots + a_1 \frac{dq_o}{dt} + a_0 q_o =$$

$$b_m \frac{d_m q_i}{dt_m} + b_{m-1} \frac{d_{m-1} q_i}{dt_{m-1}} + \ldots + b_1 \frac{dq_i}{dt} + b_0 q_i \quad \text{(A3.1)}$$

where q_i is the measured quantity, q_o is the output reading and $a_0 \ldots a_n$, $b_0 \ldots b_m$ are constants. If we limit consideration to that of step changes in the measured quantity only, then equation (A3.1) reduces to

$$a_n \frac{d_n q_o}{dt_n} + a_{n-1} \frac{d_{n-1} q_o}{dt_{n-1}} + \ldots + a_1 \frac{dq_o}{dt} + a_0 q_o = b_0 q_i \quad \text{(A3.2)}$$

Further simplification can be made by taking certain special cases of equation (A3.2) which collectively apply to nearly all measurement systems.

(a) Zero-order Instrument

If all the coefficients $a_1 \ldots a_n$ other than a_0 in equation (A3.2) are assumed zero, then:

$$a_0 q_o = b_0 q_i \quad \text{or} \quad q_o = \frac{b_0 q_i}{a_0} = K q_i \tag{A3.3}$$

where K is a constant known as the instrument sensitivity as defined earlier. Any instrument which behaves according to equation (A3.3) is said to be of zero-order type. A potentiometer, which measures motion, is a good example of such an instrument, where the output voltage changes instantaneously as the slider is displaced along the potentiometer track.

(b) First-order Instrument

If all the coefficients $a_2 \ldots a_n$ except for a_0 and a_1 are assumed zero in equation (A3.2) then:

$$a_1 \frac{dq_o}{dt} + a_0 q_o = b_0 q_i \tag{A3.4}$$

Any instrument which behaves according to equation (A3.4) is known as a first-order instrument. If d/dt is replaced by the D operator ($D \equiv d/dt$) in equation (A3.4), we get

$$a_1 D q_o + a_0 q_o = b_0 q_i$$

and rearranging:

$$q_o = \frac{(b_0/a_0) q_i}{(1 + \{a_1/a_0\} D)} \tag{A3.5}$$

Defining $K = b_0/a_0$ as the static sensitivity and $\tau = a_1/a_0$ as the time constant of the system, equation (A3.5) becomes:

$$q_o = \frac{K q_i}{1 + \tau D} \tag{A3.6}$$

If equation (A3.6) is solved analytically, the output quantity q_o in response to a step change in q_i varies with time in the manner shown in Figure 4.10. The time constant τ of the step response is the time taken for the output quantity q_o to reach 63 percent of its final value.

The thermocouple (see Chapter 13) is a good example of a first-order instrument. It is well known that, if a thermocouple at room temperature is plunged into boiling water, the output e.m.f. does not rise instantaneously to a level indicating 100 °C but instead approaches a reading indicating

100 °C in a manner similar to that shown in Figure 4.10. A large number of other instruments also belong to this first-order class.

Example A3.1
A balloon is equipped with temperature- and altitude-measuring instruments and has radio equipment which can transmit the output readings of these instruments back to ground. The balloon is initially anchored to the ground with the instrument output readings in steady state. The altitude-measuring instrument is approximately zero order and the temperature transducer first order with a time constant of 15 s. The temperature on the ground, T_0, is 10 °C and the temperature T_x at an altitude of x metres is given by the relation $T_x = T_0 - 0.01x$

(a) If the balloon is released at time zero, and thereafter rises upwards at a velocity of 5 m/s, draw a table showing the temperature and altitude measurements reported at intervals of 10 s over the first 50 s of travel. Show also in the table the error in each temperature reading.

(b) What temperature does the balloon report at an altitude of 5000 m?

Solution
This question requires the analytic solution of a first-order differential equation and it is assumed that the reader has covered the principles of this previously in a mathematics course. If the reader is not familiar with this type of mathematical working, he or she may need assistance to work through this solution. The solution uses the D operator and proceeds as follows. (NB a solution can also be obtained by using the Laplace transform and standard Laplace transform tables, which some readers may prefer).

Let the temperature reported by the balloon at some general time t be T_r. Then T_x is related to T_r by the relation

$$T_r = \frac{T_x}{1+\tau D} = \frac{T_0 - 0.01x}{1+\tau D} = \frac{10 - 0.01x}{1+15D}$$

It is given that $x = 5t$, thus

$$T_r = \frac{10 - 0.05t}{1+15D}$$

The transient or complementary function part of the solution ($T_x = 0$) is given by

$$T_{r_{cf}} = Ce^{-t/15}$$

where C is a constant derived from the initial conditions.

The particular integral part of the solution is given by:

$$T_{r_{pi}} = 10 - 0.05(t - 15)$$

Thus the whole solution is given by:

$$T_r = T_{r_{cf}} + T_{r_{pi}} = C e^{-t/15} + 10 - 0.05(t - 15)$$

Applying initial conditions:
At $t = 0$, $T_r = 10$
i.e. $10 = C e^{-0} + 10 - 0.05(-15)$
Thus $C = -0.75$ and the solution can be written as:

$$T_r = 10 - 0.75\, e^{-t/15} - 0.05(t - 15)$$

(a) Using the above expression to calculate T_r for various values of t, the following table can be constructed:

Time (s)	Altitude (m)	Temp. reading (°C)	Temp. error (°C)
0	0	10	0
10	50	9.86	0.36
20	100	9.55	0.55
30	150	9.15	0.65
40	200	8.70	0.70
50	250	8.22	0.72

(b) At 5000 m, $t = 1000$ s.
Calculating T_r from the above expression:

$$T_r = 10 - 0.75\, e^{-1000/15} - 0.05(1000 - 15)$$

The exponential term approximates to zero and so T_r can be written as:

$$T_r \approx 10 - 0.05(985) = -39.25\ °C$$

This result might have been inferred from the table above where it can be seen that the error is converging towards a value of 0.75. For large values of t, the transducer reading lags the true temperature value by a period of time equal to the time constant of 15 s. In this time, the balloon travels a distance of 75 m and the temperature falls by 0.75 °C. Thus for large values of t, the output reading is always 0.75 °C less than it should be.

(c) Second-order Instrument

If all coefficients $a_3 \ldots a_n$ other than a_0, a_1 and a_2 in equation (A3.2) are assumed zero, then we get

$$a_2 \frac{d^2 q_o}{dt^2} + a_1 \frac{dq_o}{dt} + a_0 q_o = b_0 q_i \qquad \text{(A3.7)}$$

Applying the D operator again:

$$a_2 D^2 q_o + a_1 D q_o + a_0 q_o = b_0 q_i$$

and rearranging:

$$q_o = \frac{b_0 q_i}{a_0 + a_1 D + a_2 D^2} \tag{A3.8}$$

It is convenient to re-express the variables a_0, a_1, a_2 and b_0 in equation (A3.8) in terms of three parameters K (static sensitivity), ω (undamped natural frequency) and ε (damping ratio)

$$\text{where} \quad K = b_0/a_0; \quad \omega = a_0/a_2; \quad \varepsilon = \frac{a_1}{2a_0 a_2}$$

Re-expressing equation (A3.8) in terms of K, ω and ε we get:

$$\frac{q_o}{q_i} = \frac{K}{D^2/\omega^2 + 2\varepsilon D/\omega + 1} \tag{A3.9}$$

This is the standard equation for a second-order system and any instrument whose response can be described by it is known as a second-order instrument. If equation (A3.9) is solved analytically for a step input signal, the shape of the response obtained depends on the value of the damping ratio parameter ε. The output step responses of a second-order instrument for various values of ε were shown in Figure 4.11 and discussed in detail in Section 4.3 of Chapter 4.

APPENDIX 4

Concepts of Probability

A4.1 SIMPLE PROBABILITY

The probability of occurrence of an event can be quantified on a scale of 0 to 1 or 0 percent to 100 percent. If a coin is tossed, then the probability of it coming down heads is 0.5 or 50 percent. If a six-sided die is thrown once, the probability of throwing a six is 1 in 6, i.e. 16.67 percent. In these simple examples, all possible outcomes are unbiassed, i.e. they are all equally likely to happen. The probability of a particular outcome A occurring when there are N equally possible outcomes can be expressed mathematically as:

$$P(A) = \frac{1}{N} \tag{A4.1}$$

Going a little further, suppose that we throw two dice together. What is the probability that the total of the two numbers obtained is nine? The total number of possible combinations is 36. Four of these combinations (3-6, 4-5, 5-4, 6-3,) total nine. Thus, the probability of throwing a total of nine with two dice is 4 in 36 (1 in 9) or 11.1 percent. The probability that such an outcome B occurs when there are n ways in which B can occur out of a total of N possible outcomes can be expressed mathematically as:

$$P(B) = \frac{n}{N} \tag{A4.2}$$

Expression (A4.2) is not in fact the best way of expressing probabilities because it implies that if we throw the dice nine times we will get a total score of nine once. Simple experimentation will disprove this. We might not get a total of nine at all or we might get it more than once out of the nine throws. In fact, throwing the dice nine times, we are quite likely to get a total of nine on 0, 1 or 2 of the throws, which is 0 percent, 11.1 percent or 22.2 percent of the throws. If we increase the sample size by say throwing the dice 90 times, we would be extremely unlucky to not score a nine at all and typically would score a nine on either 9, 10 or eleven of the throws, which is 10.0 percent, 11.1 percent or 12.2 percent. This shows that as we increase the sample size (number of throws), the actual number of times that we score a total of nine moves closer to 11.1 percent. However, only if we were to throw the dice an infinite number of times would we be certain that the number of times when we scored a nine would equal 11.1 percent of the total. It is thus better mathematically to define probability as a number that the result tends towards as the sample size tends towards infinity:

$$P(B) = \lim_{n \to \infty} \frac{n}{N} \tag{A4.3}$$

Many errors are also random in nature, but not in quite the same way as the examples above. If a digital temperature-measuring instrument with a measurement resolution of 0.1 °C were used to measure a constant temperature of 100 °C a large number of times, the temperature readings might range from 99.0 °C to 101.0 °C. Such variations in the output readings are due to random changes in the measurement system parameters. However, we should expect these random changes to affect the output reading by a large amount much less often than by a small amount (experimentation would confirm this). Thus small errors in the measurement will occur much more frequently than large errors, and therefore the distribution of the measurements between 99 °C and 101 °C will not be uniform. The greatest concentration in output readings will occur within a narrow band close to 100 °C, and the density of readings will get progressively thinner moving towards the extremity temperature readings of 99 °C and 101 °C. There will, for example, be many more readings of 99.9 °C than 99.1 °C.

A4.2 PROBABILITY OF THE COMPLEMENT ($P(\overline{A})$)

This is the probability of an event A not occurring and is given by

$$P(\overline{A}) = 1 - P(A) \tag{A4.4}$$

where $P(A)$ is the probability of event A occurring, expressed as a number in the range of 0.0 to 1.0 or as a percentage in the range of 0 percent to 100 percent.

If there is a 5 percent chance that a light bulb will fail in the first 500 hours of operation ($P(A) = 0.05$), then from equation (A4.4), $P(\overline{A}) = 0.95$, i.e. there is a 95 percent probability that the light bulb will not fail in the first 500 hours of operation.

A4.3 JOINT PROBABILITY ($P(AB)$)

The joint probability is the probability of two independent events A and B both occurring together and is given by:

$$P(AB) = P(A) \cdot P(B) \tag{A4.5}$$

where $P(A)$ and $P(B)$ are the probabilities respectively of the events A and B. Thus the joint probability is the product of the individual probabilities. Expression (A4.5) is known as the product rule or series rule and can be extended to cover any number of independent events, i.e.

$$P(AB \ldots N) = P(A) \cdot P(B) \ldots P(N) \tag{A4.6}$$

A4.4 SINGLE EVENT PROBABILITY ($P(A + B)$)

This is the probability of any one of two independent events A and B (but not both) occurring, and is given by:

$$P(A + B) = P(A) + P(B) - P(AB) \tag{A4.7}$$

If A and B are mutually exclusive, i.e. they cannot both occur together, then this expression simplifies to:

$$P(A + B) = P(A) + P(B) \tag{A4.8}$$

A4.5 CONDITIONAL PROBABILITY ($P(A \mid B)$)

This is the probability of event A occurring, given that event B has occurred: it is expressed by the notation $P(A|B)$. The conditional probability of an event A can be evaluated from:

$$P(A|B) = \frac{P(AB)}{P(B)} \tag{A4.9}$$

In general, $P(A|B) < P(A)$. If:

$$P(A|B) = P(A|\overline{B}) = P(A),$$

this shows that $P(A)$ is unaffected by whether or not B occurs, i.e. A and B are independent events.

APPENDIX 5

Curve Fitting by Regression Techniques

A5.1 INTRODUCTION TO REGRESSION TECHNIQUES

Regression techniques consist of finding a mathematical relationship between measurements of two variables y and x, such that the value of one variable y can be predicted from a measurement of the other variable x. It is assumed for this that random errors only affect the y values and that the values of x are exact. Regression procedures are simplest if a straight-line relationship exists between the variables, which can then be estimated by linear least squares regression.

In many cases, inspection of the raw data points plotted on a graph shows that a straight-line relationship between the points does not exist. However, knowledge of physical laws governing the data can often suggest a suitable alternative form of relationship between the two sets of variable measurements.

In some cases, the measured variables can be transformed so that a linear relationship is obtained. For example, suppose that two variables y and x are related according to $y = a \cdot x^c$. A linear relationship from this can be derived as

$$\log(y) = \log(a) + c \cdot \log(x)$$

Thus if a graph is constructed of $\log(y)$ plotted against $\log(x)$, the parameters of a straight-line relationship can be estimated by linear least squares regression.

Where a linear relationship does not exist, physical laws often suggest a suitable relationship where one variable y is related to another variable x by a power series of the form:

$$y = a_0 + a_1 \cdot x + a_2 \cdot x^2 + \ldots + a_p \cdot x^p$$

Estimation of the parameters $a_0 \ldots a_p$ is very difficult if p has a large value. Fortunately, a relationship where p only has a small value can be fitted to most data sets. Quadratic least squares regression is used to estimate parameters where p has a value of 2, and for larger values of p, polynomial least squares regression is used for parameter estimation.

Where the appropriate form of relationship between variables in measurement data sets is not obvious either from visual inspection or from consideration of physical laws, a method which is effectively a trial and error one has to be applied. This consists of estimating the parameters of successively higher-order relationships between y and x until a curve is found which fits the data sufficiently closely. What level of closeness is acceptable is considered in the later section (A5.5) on confidence tests.

A5.2 LINEAR LEAST SQUARES REGRESSION

If a linear relationship between y and x exists for a set of n measurements $y_1 \ldots y_n$, $x_1 \ldots x_n$, then this relationship can be expressed as $y = a + b{\cdot}x$, where the coefficients a and b are constants. The purpose of least squares regression is to select the optimum values for a and b such the line gives the best fit to the measurement data.

The deviation of each point (x_i, y_i) from the line can be expressed as d_i, where $d_i = y_i - (a + b{\cdot}x_i)$.

The best-fit line is obtained when the sum of the squared deviations, S, is a minimum, i.e. when

$$S = \sum_{i=1}^{n}[d_i^2] = \sum_{i=1}^{n}[y_i - a - b{\cdot}x_i]^2 \text{ is a minimum}$$

The minimum can be found by setting the partial derivatives $\delta S/\delta a$ and $\delta S/\delta b$ to zero and solving the resulting two simultaneous (normal) equations

$$\delta S/\delta a = \sum 2[y_i - a - b{\cdot}x_i][-1] = 0 \qquad \text{(A5.1)}$$

$$\delta S/\delta b = \sum 2[y_i - a - b{\cdot}x_i][-x_i] = 0 \qquad \text{(A5.2)}$$

The values of a and b at the minimum point can be represented by $\hat{a}$ and $\hat{b}$. These parameters $\hat{a}$ and $\hat{b}$, which are known as the least squares estimates of a and b, can be calculated as follows:

From equation (A5.1)

$$\sum y_i = \sum \hat{a} + \hat{b}\sum x_i = n\hat{a} + \hat{b}\sum x_i, \text{ and thus: } \hat{a} = \frac{\sum y_i - \hat{b}\sum x_i}{n} \tag{A5.3}$$

From equation (A5.2):

$$\sum (y_i x_i) = \hat{a}\sum x_i + \hat{b}\sum (x_i^2) \tag{A5.4}$$

Now substitute $\hat{a}$ in equation (A5.4), using equation (A5.3)

$$\sum (y_i x_i) = \frac{\left(\sum y_i - \hat{b}\sum x_i\right)}{n}\sum x_i + \hat{b}\sum (x_i^2)$$

Collecting terms in $\hat{b}$:

$$\hat{b}\left[\sum (x_i^2) - \frac{\left(\sum x_i\right)^2}{n}\right] = \sum (x_i y_i) - \frac{\sum x_i \sum y_i}{n}$$

Rearranging:

$$\hat{b}\left[\sum (x_i^2) - n\left\{\left(\sum x_i/n\right)\right\}^2\right] = \sum (x_i y_i) - n\sum (x_i/n)\sum (y_i/n)$$

This can be expressed as:

$$\hat{b}\left[\sum (x_i^2) - nx_m^2\right] = \sum (x_i y_i) - nx_m y_m,$$

using x_m and y_m to represent the mean values of x and y. Thus:

$$\hat{b} = \frac{\sum (x_i y_i) - nx_m y_m}{\sum (x_i^2) - nx_m^2} \tag{A5.5}$$

and from equation (A5.3):

$$\hat{a} = y_m - \hat{b}x_m \tag{A5.6}$$

Example A5.1
In an experiment to determine the variation of the specific heat of a substance with temperature, the following table of results was obtained:

Temperature (°C)	40	45	50	55	60	65	70	75	80	85	90	95	100
Specific heat (kJ\kg °C)	1.38	1.43	1.46	1.49	1.56	1.57	1.59	1.64	1.69	1.72	1.78	1.83	1.85

Fit a straight line to this data set using least squares regression and estimate the specific heat at a temperature of 72 °C.

Solution
Let y represent the specific heat and x represent the temperature. Then a suitable straight line is given by $y = a + b \times x$. We can now proceed to calculate estimates for the coefficients a and b using equations (A5.5) and (A5.6) above.

The first step is to calculate the mean values of x and y. These are found to be $x_m = 70$ and $y_m = 1.6146$. Next, we need to tabulate $(x_i y_i)$ and (x_i^2) for each pair of data values:

i	x_i	y_i	$x_i y_i$	x_i^2
1	40	1.38	55.20	1600
2	45	1.43	64.35	2025
3	50	1.46	73.00	2500
⋮				
13	100	1.85	185.00	10 000

Now calculate the quantities required in equation (A5.5) from this table:

$$n = 13; \quad \sum x_i y_i = 1504.90; \quad \sum (x_i^2) = 68\,250$$

$$n x_m y_m = 1469.286; \quad n x_m^2 = 63\,700$$

Thus, substituting the values in (A5.5):

$$\hat{b} = \frac{1504.90 - 1469.286}{68\,250 - 63\,700} = 0.007\,827\,2$$

Hence, using (A5.6), $\hat{a} = 1.6146 - 0.007\,827\,2 \times 70 = 1.066\,696$ Thus, $y = 1.066\,696 + 0.007\,827\,2x$

Using this equation, at a temperature (x) of 72 °C, the specific heat (y) is 1.63 kJ\kg·°C.

Note that in this solution, we have only specified the answer to an accuracy of three figures, which is the same accuracy as the measurements. Any greater number of figures in the answer would be meaningless.

Least squares regression is often appropriate for situations where a straight-line relationship is not immediately obvious, for example where $y \propto x^2$ or $y \propto \exp(x)$.

A5.3 QUADRATIC LEAST SQUARES REGRESSION

Quadratic least squares regression is used to estimate the parameters of a relationship $y = a + b \cdot x + c \cdot x^2$ between two sets of measurements $y_1 \ldots y_n$ and $x_1 \ldots x_n$. The deviation of each point (x_i, y_i) from the line can be expressed as d_i, where $d_i = y_i - (a + b \cdot x_i + c \cdot x_i)^2$. The best-fit line is obtained when the sum of the squared deviations, S, is a minimum,

$$\text{i.e. when } \underset{i=1}{\overset{n}{S}} = \sum[d_i^2] = \sum_{i=1}^{n}[y_i - a - b \cdot x_i - c \cdot x_i^2]^2 \text{ is a minimum.}$$

The minimum can be found by setting the partial derivatives $\delta S/\delta a$, $\delta S/\delta b$ and $\delta S/\delta c$ to zero and solving the resulting simultaneous equations, as for the linear least squares regression case above. Standard computer programs to estimate the parameters a, b and c by numerical methods are widely available and therefore a detailed solution is not presented here.

A5.4 POLYNOMIAL LEAST SQUARES REGRESSION

Polynomial least squares regression is used to estimate the parameters of the pth order relationship $y = a_0 + a_1 \cdot x + a_2 \cdot x^2 + \ldots + a_p \cdot x^p$ between two sets of measurements $y_1 \ldots y_n$ and $x_1 \ldots x_n$. The deviation of each point (x_i, y_i) from the line can be expressed as d_i, where $d_i = y_i - (a_0 + a_1 \cdot x_i + a_2 \cdot x_i^2 + \ldots + a_p \cdot x_i^p)$. The best fit line is obtained when the sum of the squared deviations, S, is a minimum. i.e. when

$$S = \sum_{i=1}^{n}[d_i^2] \text{ is a minimum}$$

The minimum can be found as before by setting the p partial derivatives $\delta S/\delta a_0 \ldots \delta S/\delta a_p$ to zero and solving the resulting simultaneous equations. Again, as for the quadratic least squares regression case, standard computer programs to estimate the parameters $a_0 \ldots a_p$ by numerical methods are widely available and therefore a detailed solution is not presented here.

A5.5 CONFIDENCE TESTS IN CURVE FITTING

Having applied least squares regression to estimate the parameters of a chosen relationship, some form of follow-up procedure is clearly required to assess how well the estimated relationship fits the data points. One fundamental requirement in curve fitting is that the maximum deviation d_i of any data point (y_i, x_i) from the fitted curve is less than the calculated maximum measurement error level. For some data sets, it is impossible to find a relationship between the data points that satisfies this requirement. This normally occurs when both variables in a measurement data set are subject to random variation, such as where the two sets of data values are measurements of human height and weight. Correlation analysis is applied in such cases to determine the degree of association between the variables.

Assuming that a curve can be fitted to the data points in a measurement set without violating the fundamental requirement that only one, not both, of the measured variables is subject to random errors, a further simple curve-fitting confidence test is to calculate the sum of squared deviations S for the chosen y/x relationship and compare it with the value of S calculated for the next-higher-order regression line which could be fitted to the data. Thus if a straight-line relationship is chosen, the value of S calculated should be of a similar magnitude to that obtained by fitting a quadratic relationship. If the value of S were substantially lower for a quadratic relationship, this would indicate that a quadratic relationship was a better fit to the data than a straight-line one and further tests would be needed to examine whether a cubic or higher-order relationship was a better fit still.

Other more sophisticated confidence tests exist such as the F-ratio test are covered in more advanced texts such as [1].

REFERENCES

1. Chatfield, C., *Statistics for Technology*, Chapman and Hall, London (1983).

APPENDIX 6

Typical Structure of a Quality Manual

—

CONFIDENTIAL

MORRIS FABRICATIONS LIMITED

QUALITY ASSURANCE MANUAL

MEASUREMENT AND CALIBRATION SYSTEMS

THIS MANUAL IS A CONFIDENTIAL DOCUMENT

IT MUST BE KEPT SECURE AND MUST NOT BE DISCLOSED TO NON-COMPANY PERSONNEL EXCEPT TO BONA-FIDE CUSTOMERS AS NECESSARY

COPY NUMBER : 4

AUTHORIZED HOLDER : Works Manager - West Cross Site

Revision number: 07 Date: 08 Sep 1996

—

CONFIDENTIAL

CONTENTS

Section number

Revision number: 07 Date: 08 Sep 1996

CONFIDENTIAL

REVISION NUMBERS OF MANUAL PAGES

Page number	Revision number	Date
front page	07	08 Sep 1996
contents page	07	08 Sep 1996
1.01	01	02 Jul 1986
1.02	03	17 Nov 1986
2.01	02	09 Sep 1986
2.02	06	08 Sep 1989
2.03	04	23 Feb 1988
3.01	05	31 Mar 1989
3.02	04	23 Feb 1988
4.01	01	02 Jul 1986

CONFIDENTIAL

AUTHORIZED HOLDERS OF MANUAL

Copy number	Job title of holder
1	Quality Assurance Director
2	Works Manager - Ringwood site
3	Quality Assurance manager - Ringwood site
4	Works Manager - West Cross site
5	Quality Assurance manager - West Cross site
6	Works Manager - Fiveoaks site
7	Quality Assurance manager - Fiveoaks site

NO FURTHER COPIES OF THIS MANUAL SHALL BE MADE UNLESS THESE ARE AUTHORIZED BY THE QUALITY ASSURANCE DIRECTOR AND DETAILS OF SUCH FURTHER COPIES AND THEIR AUTHORIZED HOLDERS ARE ENTERED ON THIS PAGE.

Page number : 1.01 Revision number: 01 Date: 02 Jul 1986

CONFIDENTIAL

AMENDMENT PROCEDURE

Any suggestions for amendments to this manual, and the quality assurance procedures described therein, should be addressed to the Quality Assurance Director. These will be considered, and any positive response will be in the form of an official revision of the manual. **NO ALTERATION WHATSOEVER MUST BE MADE TO THIS MANUAL EXCEPT FOR SUCH OFFICIAL REVISIONS MADE IN ACCORDANCE WITH THE FOLLOWING PROCEDURE.**

Procedure for amendment of manual:

1. Amendments as necessary will be issued by the Quality Assurance Director.
2. Each amendment package issued will consist of a set of pages, together with instructions about how these are to be integrated into the existing manual. These instructions will typically require some totally new pages to be added and some existing pages to be replaced with newer versions.
3. The amendment package will include a page which gives the correct revision levels of all pages in the manual. After updating the document, the manual holder should check all pages carefully to ensure that they are all of the correct revision level. After use, this page should be filed at the front of the manual.
4. All sheets removed from the manual in accordance with these instructions must be destroyed.
5. When the amendments to the manual have been completed, the 'Amendment Receipt Slip' provided with the amendment package should be signed and dated and returned to the Quality Assurance Director.

Page number : 1.02 Revision number: 03 Date: 17 Nov 1986

CONFIDENTIAL

MEASUREMENT SYSTEMS SUBJECT TO CONTROL

(a) Ringwood site

Measurement requirements

This should include details of all measurements on the site which must be made to satisfy quality control and assurance requirements. For each measurement to be made, the environmental conditions existing, the measurement-limit required and the type of instrument therefore specified must be described.

Instruments used

Here, a description of each type of instrument used should be given. This should include instructions about the proper way of using each instrument and give the necessary information about any environmental control or other special precautions to be taken. A method of marking on the instrument itself such instructions about proper use and special precautions required should be prescribed if practical.

The system of serial numbers used to uniquely identify each instrument must also be defined.

The name of the person responsible for each instrument must be stated.

Calibration

The instruments used on site should be set down in tabular form here, using the following columns:

Serial number	Type	Required calibration frequency	Person responsible

Training courses

Initial training and refresher courses to be attended by personnel using instruments are specified here.

CONFIDENTIAL

(b) West Cross site

Information as above

CONFIDENTIAL

(c) Fiveoaks site

Information as above

CONFIDENTIAL

CALIBRATION PROCEDURES

Standard instruments

Here, a list of the standard instrument to be used for calibrating each type of process instrument in the system should be given, in a table with the following headings:

Type of process instrument	Type of standard instrument

Where calibration is not to be carried out internally in the company, the name of the sub-contractor nominated to perform the calibration must be specified instead of defining the standard instrument to be used.

Handling of standard instruments and traceability

For each type of standard instrument to be used, the following information must be given here:

1. How it must be stored and handled
2. The environmental conditions under which it must be used
3. The traceability of the instrument back to National Reference Standards, in the form of the list of instruments in the intermediate calibration chain. The whereabouts of the documentary evidence of this traceability, in the form of calibration certificates etc., must be given.

Page number: 3.01 Revision number: 05 Date: 31 Mar 1989

CONFIDENTIAL

Recording of calibration results

In this section, the format in which calibration results are to be recorded, (i) in the site instrument manual (see section 9.7 and figure 9.1), and (ii) on the instruments themselves (where appropriate), should be given.

The document should specify, as a minimum, the following information for each instrument:

serial number
name of person responsible for calibration
required calibration frequency
date of last calibration
calibration results

Procedure following calibration

The procedure to be followed if an instrument is found to be outside calibration limits is set out here. A foolproof procedure for marking such instruments and preventing their further use until they have been corrected must be defined.

Training courses

Details of training courses and/or refresher courses to be attended by personnel involved in calibration duties should be set down here

CONFIDENTIAL

CALIBRATION PROCEDURE REVIEWS

Here, the procedure used to review the continued effectiveness of the calibration system in operation is described. The information on which this review is based will consist of the incidence of customer complaints about quality plus any data available from final product inspection stations prior to delivery to customers. If any breakdowns in quality control are attributed to poor measurements, the instrument calibration system in operation must be carefully examined to determine what must be changed in order to prevent future deterioration in the quality of measurements.

The frequency at which such reviews are to take place must be defined here also.

Finally in this section, the results of each review of the calibration system must be written down. Any action taken to modify the calibration system, in order to correct for deficiencies in measurements, must also be documented.

INDEX